72 **Springer Series in Solid-State Sciences**

Edited by Peter Fulde

Springer Series in Solid-State Sciences

Editors: M. Cardona P. Fulde K. von Klitzing H.-J. Queisser

Volumes 1–39 are listed on the back inside cover

40 **Semiconductor Physics** An Introduction
3rd Edition By K. Seeger

41 **The LMTO Method**
Muffin-Tin Orbitals and Electronic Structure
By H. L. Skriver

42 **Crystal Optics with Spatial Dispersion, and Excitons**
By V. M. Agranovich and V. L. Ginzburg

43 **Resonant Nonlinear Interactions of Light with Matter**
By V. S. Butylkin, A. E. Kaplan, Yu. G. Khronopulo, and E. I. Yakubovich

44 **Elastic Media with Microstructure II**
Three-Dimensional Models
By I. A. Kunin

45 **Electronic Properties of Doped Semiconductors**
By B. I. Shklovskii and A. L. Efros

46 **Topological Disorder in Condensed Matter**
Editors: F. Yonezawa and T. Ninomiya

47 **Statics and Dynamics of Nonlinear Systems**
Editors: G. Benedek, H. Bilz, and R. Zeyher

48 **Magnetic Phase Transitions**
Editors: M. Ausloos and R. J. Elliott

49 **Organic Molecular Aggregates,** Electronic Excitation and Interaction Processes
Editors: P. Reineker, H. Haken, and H. C. Wolf

50 **Multiple Diffraction of X-Rays in Crystals**
By Shih-Lin Chang

51 **Phonon Scattering in Condensed Matter**
Editors: W. Eisenmenger, K. Laßmann, and S. Döttinger

52 **Superconductivity in Magnetic and Exotic Materials**
Editors: T. Matsubara and A. Kotani

53 **Two-Dimensional Systems, Heterostructures, and Superlattices**
Editors: G. Bauer, F. Kuchar, and H. Heinrich

54 **Magnetic Excitations and Fluctuations**
Editors: S. Lovesey, U. Balucani, F. Borsa, and V. Tognetti

55 **The Theory of Magnetism II**
Thermodynamics and Statistical Mechanics
By D. C. Mattis

56 **Spin Fluctuations in Itinerant Electron Magnetism** By T. Moriya

57 **Polycrystalline Semiconductors,**
Physical Properties and Applications
Editor: G. Harbeke

58 **The Recursion Method and Its Applications**
Editors: D. Pettifor and D. Weaire

59 **Dynamical Processes and Ordering on Solid Surfaces**
Editors: A. Yoshimori and M. Tsukada

60 **Excitonic Processes in Solids**
By M. Ueta, H. Kanzaki, K. Kobayashi, Y. Toyozawa, and E. Hanamura

61 **Localization, Interaction, and Transport Phenomena**
Editors: B. Kramer, G. Bergmann, and Y. Bruynseraede

62 **Theory of Heavy Fermions and Valence Fluctuations**
Editors: T. Kasuya and T. Saso

63 **Electronic Properties of Polymers and Related Compounds**
Editors: H. Kuzmany, M. Mehring, and S. Roth

64 **Symmetries in Physics** Group Theory Applied to Physical Problems
By W. Ludwig and C. Falter

65 **Phonons: Theory and Experiments II**
Experiments and Interpretation of Experimental Results By P. Brüesch

66 **Phonons: Theory and Experiments III**
Phenomena Related to Phonons
By P. Brüesch

67 **Two-Dimensional Systems: Physics and New Devices**
Editors: G. Bauer, F. Kuchar, and H. Heinrich

68 **Phonon Scattering in Condensed Matter V**
Editors: A. C. Anderson and J. P. Wolfe

69 **Nonlinearity in Condensed Matter**
Editors: A. R. Bishop, D. K. Campbell, P. Kumar and S. E. Trullinger

70 **From Hamiltonians to Phase Diagrams**
The Electronic and Statistical-Mechanical Theory of sp-Bonded Metals and Alloys
By J. Hafner

71 **High Magnetic Fields in Semiconductor Physics**
Editor: G. Landwehr

72 **One-Dimensional Conductors**
By S. Kagoshima, H. Nagasawa, and T. Sambongi

73 **Quantum Solid-State Physics**
Editors: S. V. Vonsovsky and M. I. Katsnelson

74 **Quantum Monte Carlo Methods** in Equilibrium and Nonequilibrium Systems
Editor: M. Suzuki

75 **Pseudo Potential for the Electronic Structure of Semiconductors**
By M. L. Cohen and J. R. Chelikowsky

76 **Electronic Properties of Conjugated Polymers**
Editors: H. Kuzmany, M. Mehring, and S. Roth

S. Kagoshima H. Nagasawa
T. Sambongi

One-Dimensional Conductors

With 161 Figures

Springer-Verlag Berlin Heidelberg New York
London Paris Tokyo

Professor Seiichi Kagoshima

Department of Pure and Applied Sciences, University of Tokyo
Komaba 3-8-1, Meguro, Tokyo 153, Japan

Professor Hiroshi Nagasawa

Institute of Physics, Tsukuba University
Tennodai, Tsukuba, Ibaraki 305, Japan

Professor Takashi Sambongi

Department of Physics, Hokkaido University
Kita-ku, Sapporo, Hokkaido 060, Japan

Title of the original Japanese edition: *Ichijigen denki dendōtai*

ISBN-13:978-3-642-83181-2 e-ISBN-13:978-3-642-83179-9
DOI: 10.1007/978-3-642-83179-9

Library of Congress Cataloging-in-Publication Data. Kagoshima, Seiichi, 1945- One-dimensional conductors. (Springer series in solid-state sciences ; 72) Translation of: Ichijigen denki dendōtai. Bibliography: p. Includes index. 1. One-dimensional conductors. I. Nagasawa, Hiroshi, 1936-. II. Sambongi, Takashi, 1937-. III. Title. IV. Series. QC176.8.E4K3513 1987 530.4'1 87-26315

Softcover reprint of the hardcover 1st edition 1988

2153/3150-543210

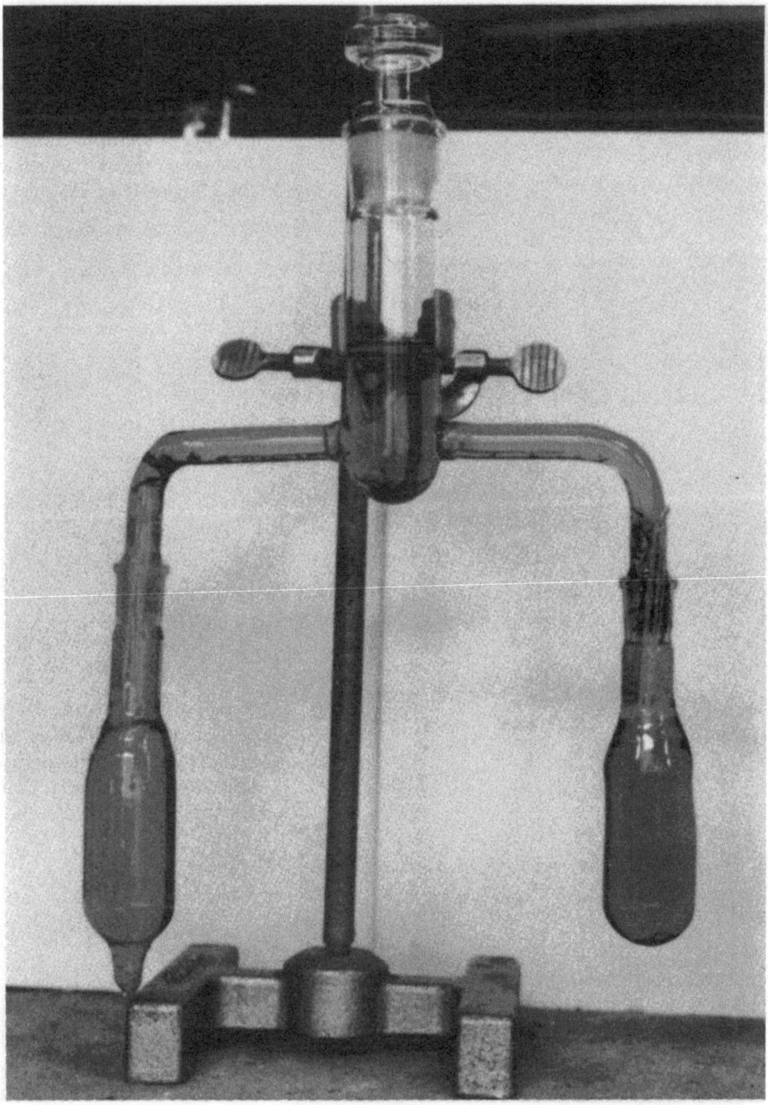

Single-crystal growth of the organic one-dimensional conductor TTF-TCNQ by the diffusion method. Single crystals grown are found in the bridge-tube. See the text Sect. 3.1. (Photograph donated by Dr. Hiroyuki Anzai of Electrotechnical Laboratory)

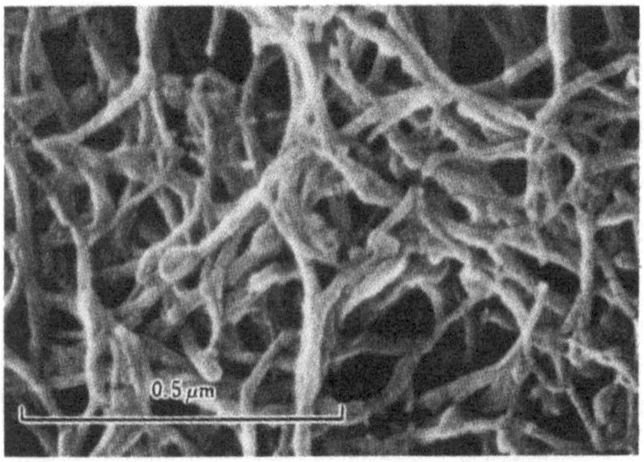

Microscopic structure of a film of the linear chain polymer $(CH)_x$ observed by an electron microscope. See the text Sect. 6.1. (Photograph donated by Professor Hideki Shirakawa of the University of Tsukuba)

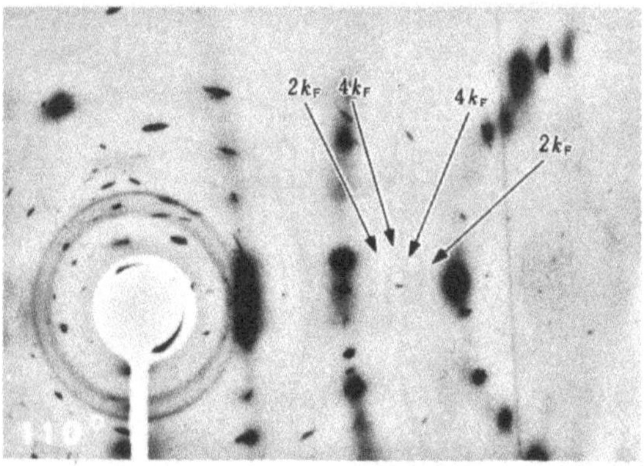

X-ray diffuse scattering pattern of TTF-TCNQ at 110 K. See the text Sect. 3.4. (Photograph donated by Dr. Robert Comès of Université Paris-Sud)

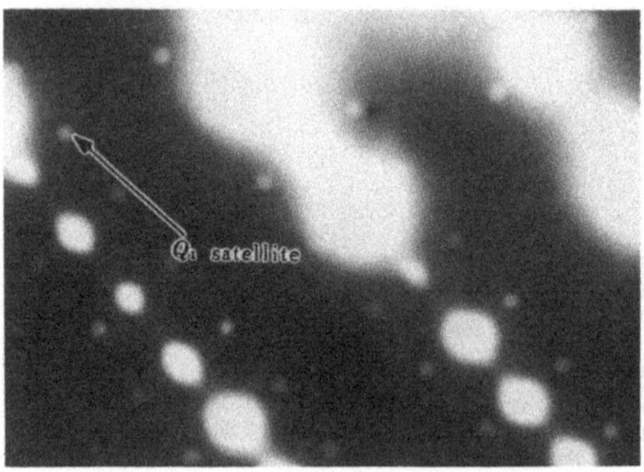

Electron diffraction pattern of $NbSe_3$ below 140 K. See the text Sect. 4.2. (Photograph donated by Dr. Kitomi Tsutsumi of Iwate University)

Preface

This volume deals with physical properties of electrically one-dimensional conductors. It includes both a description of basic concepts and a review of recent progress in research. One-dimensional conductors are those materials in which an electric current flows easily in one specific crystal direction while the resistivity is very high in transverse directions. It was about 1973 when much attention began to be focussed on them and investigations started in earnest. The research was stimulated by the successful growth of crystals of the organic conductor TTF-TCNQ and of the inorganic conductor KCP. New concepts, characteristic of one dimension, were established in the investigations of their properties. Many new one-dimensional conductors were also found and synthesized.

This field of research is attractive because of the discovery of new materials, phenomena and concepts which have only recently found a place in the framework of traditional solid-state physics and materials science. The relation of this topic to the wider field of solid-state sciences is therefore still uncertain. This situation is clearly reflected in the wide distribution of the fields of specialization of researchers. Due to this, and also to the rapid progress of research, no introductory book has been available which covers most of the important fields of research on one-dimensional conductors.

The purposes of this volume are, first, to serve as an introduction to this field for graduate and undergraduate students of physics and chemistry, and, second, to provide a general survey of activities in this area for researchers in various fields of solid-state physics and chemistry.

This volume is organized as follows: The opening chapter gives an overall profile of one-dimensional conductors, and Chap. 2 deals with fundamental properties of a one-dimensional electron system with the emphasis on theoretical aspects. Some mathematical equations appear in this chapter but most of them remain at the level of those found in conventional textbooks on solid state physics. This chapter has the form of an exercise in applying three-dimensional solid state physics to one-dimensional systems. Subsequent chapters present experimental aspects of research into real one-dimensional conductors. The authors have tried to explain, as much as possible, principles of measurements for students unfamiliar with practical experimental research. Stress is laid upon typical one-dimensional conductors, TTF-TCNQ, MX_3, and KCP. Chapters 6 and 7 give a brief introduction to the properties of linear-chain polymers and a linear chain mercury compound which has a character slightly different from the others.

In the various sections of this volume, we have aimed to develop our discussions following a unified way of presentation. Because of this, the presentation of the investigations may not always be completely faithful to the original. Moreover, because of space limitations, we have rather boldly omitted some of the problems that have less direct bearing on the main course of the discussions. We also had to forego a description of many interesting results of investigations. We trust that the original authors of those papers and the readers of this volume will understand this necessity. The Japanese edition on which this book is based was written at the suggestion of the editors of the Shokabo Series in Materials Science, Professor Taira Suzuki, Professor Sôsin Chikazumi, and Professor Sadao Nakajima. On rewriting the volume for the Springer Series in Solid-State Sciences, we revised the content and updated each chapter in the light of recent developments in this field. We were keenly aware of the need for an introductory book in this field and would like to express our sincere thanks to those above and to Dr. H. Lotsch and Dr. A.M. Lahee of Springer-Verlag for giving us this opportunity and for their encouragement. We acknowledge also the following researchers who kindly donated us glossy prints of their beautiful photographs shown in the frontispiece: Dr. Hiroyuki Anzai of the Electrotechnical Laboratory, Professor Hideki Shirakawa of the University of Tsukuba, Dr. Robert Comès of Université Paris-Sud and Dr. Kitomi Tsutsumi of Iwate University.

September 1987 *The authors*

Contents

1. What Are One-Dimensional Conductors?

"One-dimensional conductors" are materials whose electrical conductivity is high only along one direction. In this volume we shall focus our attention on materials that show a metallic conductivity. In general, the presence of an anisotropy in the electrical conductivity of a material is not a sufficient condition for it to be termed a one-dimensional (1d) conductor. The 1d conductors described in this volume are those in which a band model for electrons is valid along the 1d axis and, along other directions, some other conduction model, for example a hopping conduction model, is applied; that is to say we shall be considering those conductors which have a "one-dimensional conduction band". More precisely, 1d conductors are materials which satisfy the following conditions: $l_{\|} \gtrsim d_{\|}$ and $l_{\perp} \ll d_{\perp}$, where $l_{\|}$ and $d_{\|}$ are the electron mean free path and the interatomic distance along the 1d axis, respectively, and $l_{\perp}$ and $d_{\perp}$ are the corresponding quantities perpendicular to the 1d axis.

As a naive picture of 1d conductors we might consider, for example, the following description: Atoms are aligned along the 1d axis with an interatomic distance which is less than twice the van der Waals radius; the overlap of electron wave functions is then so large that a conduction band is formed. Some real materials are composed of these 1d atomic chains. If in the direction perpendicular to the atomic chain the interchain distance is larger than twice the van der Waals radius and thus the overlap of wave functions small, the conduction band model is not applicable. Another possible circumstance is the following: Even if the interatomic distances are nearly the same in both directions, the extent of the overlap of the electron wave functions may show strong directional dependence. This can arise when the wave functions themselves show anisotropy as for example do d-electrons in atoms or π-electrons in molecules.

A main motivation for investigating 1d conductors is to establish which of their properties are characteristic of "one-dimension". In ordinary text books for solid state physics the description of the properties of metals and semiconductors may begin with a 1d conduction band for the sake of simplicity; the 1d model is used in these cases in order to simplify the mathematics. However, it turns out that we are still a long way from a full understanding of the properties of electronic systems in one dimension.

Experimental studies of 1d electronic systems only started at the beginning of 1970's, i.e. at the time when it first became possible to synthesize

1d conductors. During the course of these investigations many properties characteristic of 1d electronic systems have been revealed. An important example is the "Peierls transition", which is a metal-insulator transition typical of 1d conductors. It has been demonstrated that the essence of this transition is the fluctuation and condensation of a new collective state of electrons called a "charge density wave (CDW)" which tends to occur in low-dimensional electronic systems. The CDW is found itself to be capable of carrying current and, under ideal conditions, to give rise to an electrical conduction without resistance. In addition, studies of the dynamics of CDWs have demonstrated the existence of another type of current carrier called a "phase soliton".

One-dimensional conductors are interesting not only from a view point of fundamental research. It is predicted that they are promising materials for the realization of our ideal of "high-temperature superconductivity". It is believed that in order to have critical temperatures much higher than at present some electronic excitations, instead of phonons, must take part in the mechanism of superconductivity. A model compound for this idea has been proposed, namely linear chains of platinum atoms surrounded by ligands like dye molecules. This material, if it can really be synthesized, is predicted to have a critical temperature of the order of 3000 K. It is not assured that this prediction is correct, nor that the material will ever be obtained, but nonetheless, the possibility is interesting and worth persuing.

Investigations of 1d conductors began with the successful growth of single crystals of the organic charge transfer salt called TTF-TCNQ, which is composed of carbon, nitrogen, sulphur, and hydrogen atoms. The first synthesis of TTF-TCNQ was achieved in 1970 and single crystals were made in 1973. Measurements in the same year of the electrical conductivity showed that with decreasing temperature the conductivity increases much more rapidly than expected in normal metals. Furthermore, TTF-TCNQ undergoes a metal-insulator transition at about 53 K. The maximum conductivity is of the order of 10^5–$10^6\ \Omega^{-1}\,m^{-1}$, which is much higher than values observed in conventional organic conductors. The strong temperature dependence is also a novel feature. TTF-TCNQ has attracted the attention of many research workers due probably to the early claim that a high temperature superconductivity may be realized in this material. It is said that in meetings of American Physical Society of those days, so many questions and comments were presented that the chairman could hardly regulate the sessions. The possibility of high temperature superconductivity was eventually discarded following energetic studies by many researchers worldwide. However, step by step the properties of the Peierls transition and CDW have been clarified. Besides the study of TTF-TCNQ, other compounds such as $K_2Pt(CN)_4Br_{0.3} \cdot 3.2H_2O$, and MX_3, which have 1d chains of platinum atoms and transition-metal atoms, respectively, were made and studied in detail. Furthermore, linear chain polymers $(SN)_x$ and $(CH)_x$, which correspond closely to a naive model of 1d conductors, were synthesized. In the case of $(CH)_x$ especially, some researchers are considering possible applications in electronic engineering.

In studies of 1d conductors it must be emphasized that a close collaboration is taking place between chemistry and physics, especially in experimental studies. It is unnecessary to remark that the synthesis of samples plays a curcial role in materials science, a fact which is truly proved in the field of 1d conductors. Such rapid progress in research would have been impossible without the major contribution of chemists to the materials synthesis. At the same time solid state physicists, who want to start from a microscopic point of view and to have single crystals, will have influenced the chemists who themselves usually try to synthesize new and useful compounds starting from a molecular point of view.

One could say, to a first approximation, that investigations of the Peierls transition and the CDW have been finished. Since 1980, in particular, the main effort has been focused on studies of a newly discovered superconductivity in the organic 1d conductor $(TMTSF)_2PF_6$ and its family. The aim of these studies will be to obtain a deeper understanding of the superconductivity in 1d conductors and to clarify the interrelations among the possible electronic states in 1d: CDW, spin-density waves (SDW), superconductivity and the ordinary metallic state. As far as applications are concerned, the research will be centred on the synthesis of possible high temperature superconductors and, as a more realistic problem, on the design and synthesis of new linear-chain polymeric materials.

The scope of this volume is the following. In Chap. 2 we learn basic properties of a 1d electronic system form a theoretical point of view. Discussion will be kept to the minimum necessary to understand the succeeding chapters. If the level of Chap. 2 is not high enough for some readers, they can refer to the articles listed in the References. Succeeding chapters provide a description of the properties of each 1d conductor from an experimental point of view.

This volume makes use for the most part of the SI system of units. However, the energy will be usually expressed in eV and, in case of optical measurements, in cm^{-1}. In those cases the values in Joules (J) are shown in curly brackets. Also the kbar unit may be used in expressing the pressure. When these expressions appear frequently on a page, only eV, cm^{-1}, and kbar will be used. Values and equations related to electromagnetism are written with the definitions shown in the Appendix. Also given there is a transformation table for units of energy etc. The figures shown in this volume are often somewhat modified versions of the original ones in order to avoid possible confusion for beginners. For example, data points in original figures are usually omitted and only smooth curves are shown.

2. Fundamental Properties of Electronic Systems in One-Dimension

We shall make here a general survey of the basic properties of conduction electrons responsible for one-dimensional (1d) electrical conduction. Section 2.1 describes characteristic properties of a 1d conduction band by making differences clear from conventional 3d bands. It is also shown how the metal-insulator transition known as a Peierls transition takes place. In Sect. 2.2 we will learn of the origins and behaviour of the charge-density wave. This is a collective state of electrons typical of low-dimensional electronic systems and it plays an important role in the Peierls transition. We will use the mean field approximation in these sections. In Sect. 2.3 we shall investigate qualitatively the important effects of fluctuations in low-dimensional systems. Fluctuations and 3d effects in real 1d conductors will also be examined here. In the last section, 2.4, we will consider how the discussion presented in the preceding sections must be modified when we take into account the Coulomb interactions between conduction electrons.

Mathematical formulae are limited to the level of elementary text books for solid state physics and explanations are intended to give an intuitive picture of the physics.

2.1 The One-Dimensional Conduction Band and the Peierls Instability

2.1.1 Conduction Electrons in a One-Dimensional Band

We shall begin with the simplest possible model to describe the properties of materials showing a 1d metallic conductivity. The free-electron model is used and electronic wave functions are expressed by plane waves. Furthermore the conduction electrons are constrained to move in one dimension only. These approximations are used until Sect. 2.3. In Sect. 2.4 possible interactions among electrons are taken into account and the tight-binding model, instead of the free-electron model, is employed. See [2.1] for textbooks which discuss these methods.

The state of an electron moving in one dimension is represented by a plane wave $\phi(r) = \exp(\mathrm{i}kr)$, where k is the wave vector. The kinetic energy of the electron is given by

$$E_k = \frac{\hbar^2}{2m}k^2 \quad , \tag{2.1}$$

where $2\pi\hbar$ and m are Planck's constant and the electron mass, respectively. A periodic boundary condition gives discrete values for k,

$$k = \pm l\frac{2\pi}{L} \quad , \quad l = 0, 1, 2, \dots \quad . \tag{2.2}$$

Here, L is the length of the system used in the periodic boundary condition. The normalized wave function becomes $\phi(r) = (l/\sqrt{L})\exp(ikr)$. Each state specified by a particular k value in (2.2) can admit, according to Pauli principle, one electron with up-spin and another with down-spin. At absolute zero N electrons in the length L occupy states up to the Fermi energy E_F with wave vectors between $k = -k_F$ and k_F as shown in Fig. 2.1, and these electrons form a conduction band. The Fermi wave vector k_F and the Fermi energy E_F are determined by the number density of electrons, $n = N/L$.

At a finite temperature $T\neq 0$, the probability of occupation of a state with an energy E is given by the Fermi distribution function

$$f(E) = \frac{1}{\exp(E - \zeta)\beta + 1} \quad , \tag{2.3}$$

where $\beta = 1/k_B T$ and ζ is the chemical potential. When we introduce the density of states of electrons, $D(E)$, the chemical potential is determined by the condition

$$N = \int_0^\infty f(E)D(E)dE \quad . \tag{2.4}$$

At $T = 0$ this equation is rewritten as follows using the relations $f(E) = 1$ for $E<\zeta$ and $f(E) = 0$ for $E>\zeta$,

$$N = \int_0^\zeta D(E)dE = \int_0^{E_F} D(E)dE \quad , \quad T = 0 \quad . \tag{2.5}$$

The chemical potential at $T = 0$ is the Fermi energy.

The density of states for the 1d band is given as

$$\begin{aligned} D(E)dE &= \mathcal{D}(k)dk = \mathcal{D}(k)\frac{dk}{dE}dE \\ &= \frac{2\sqrt{m}L}{\sqrt{2}\pi\hbar}\frac{1}{\sqrt{E}}dE \quad . \end{aligned} \tag{2.6}$$

The factor 2 arises from the spin degeneracy. Substituting this equation into (2.4), we obtain the following expression for the Fermi energy E_F:

$$E_F = \frac{\pi^2\hbar^2N^2}{8mL^2} = \frac{\pi^2\hbar^2}{8m}n^2 \quad . \tag{2.7}$$

When we define the Fermi wave number by the equation $E_F = \hbar^2k_F^2/2m$, k_F is given as

$$k_F = \tfrac{1}{2}\pi n \quad . \tag{2.8}$$

On introducing a periodic potential due to the lattice, the k-axis of Fig. 2.1 is divided into sections of length of $2\pi/a$. The section between $-\pi/a$ and π/a is called the first Brillouin zone. The periodic potential, regarded as a perturbation, gives rise to a splitting of the band at the edges, $k = \pm\pi/a$, of the Brillouin zone. The width of the band gap is given by $E_g = 2|V|$, where V is the magnitude of the potential. If k_F of the electron system coincides with π/a, the system does not show metallic behaviour but is instead an insulator. Since, in this case, the relation $k_F = \pi/a$ gives an electron density $n = 2/a$, one may understand this situation from a 'classical' point of view: In each period a two electrons with antiparallel spins are present (as in a full atomic s-shell). This provides a naive picture for insulators. It is also possible to regard this system as an intrinsic semiconductor.

In the following we shall assume that $n<2/a$ and, accordingly, the system has metallic properties. Furthermore the periodic potential of the lattice will be neglected until it becomes essential in Sect. 2.2.

We note below some points which are important in the study of 1d systems. One should be careful in using results derived in usual textbooks for solid state physics because many of them apply only to three dimensions.

(i) The density of states: $D(E) \sim E^{-1/2}$ in 1d. In 2- and 3d $D(E) =$ constant and $\sim E^{1/2}$, respectively. Therefore for example, some quantities having the factor of dD/dE change their signs in going from 3d to 1d.

(ii) The Fermi wave number: In 1d, $k_F \sim n$ whereas in 2- and 3d $k_F \sim n^{1/2}$ and $\sim n^{1/3}$, respectively.

(iii) Electronic transitions: When an electron changes its state by interacting with a phonon, only one transition process is allowed in 1d – from k_F to $-k_F$ (or vice versa), because the energy of the phonon is negligibly small compared to electron's energy. The phonon taking part in the interaction has a wave number close to $2k_F$. Other transitions are forbidden as far as first order processes are concerned; such processes could not satisfy both energy and wave number conservation. In the case of 2- and 3d, all phonons can interact with electrons unless their wave numbers are larger than $2k_F$. This fact is closely related to the more common occurrence of CDWs in 1d systems. The same fact has a bearing on some of the experimental data which will be described in later chapters, for example, an unusually narrow linewidth of electron paramagnetic resonance.

2.1.2 The Peierls Instability and the Peierls Transition

a) The Polarization Function

A characteristic property of electronic systems in 1d is their instability for a potential having the wave number $2k_F$. The amplitude of electron-density waves increases divergently when the wave number approaches $2k_F$. This instability gives rise to a collective state of electrons called a charge-density wave (CDW) and causes a metal-insulator transition called a Peierls transition, which is characteristic of 1d conductors.

We shall examine the response of a free-electron system to a weak perturbing potential with the wave number Q and the frequency ω [2.2]. The potential V_Q oscillating spatially with the wave number Q will cause a spatial modulation ϱ_Q of the electron density. If V_Q is sufficiently small, ϱ_Q will be proportional to V_Q. We introduce a coefficient of proportionality $\chi(Q)$, which is called a polarization function or a density response function [2.3]. We shall derive a formula for $\chi(Q)$. The wave function of a free electron with the wave number k and the energy E_k is given by $\phi_k = (1/\sqrt{V})\exp(\mathrm{i}kr + \mathrm{i}E_k t/\hbar)$, where V is the volume of the system and $E_k = \hbar^2 k^2/2m$.[1]

The Hamiltonian for the electron in the absence of any perturbation is given by

$$H_0 = \frac{p^2}{2m} \quad . \tag{2.9}$$

The perturbation potential

$$V(r,t) = V_Q \mathrm{e}^{\mathrm{i}Qr} \mathrm{e}^{\mathrm{i}\omega t} \mathrm{e}^{\alpha t} \quad , \tag{2.10}$$

completes the total Hamiltonian $H_0 + V(r,t)$ where α ensures that the perturbation is zero for $t \to -\infty$. A wave function ψ_k for a new eigenstate in the perturbed potential is expanded as

$$\psi_k = \phi_k + a_{k+Q}(t)\phi_{k+Q} \quad . \tag{2.11}$$

The coefficient a_{k+Q} ($\ll 1$) is given by

$$a_{k+Q}(t) = \frac{V_Q \mathrm{e}^{\mathrm{i}\omega t} \mathrm{e}^{\alpha t}}{E_k - E_{k+Q} + \hbar\omega - \mathrm{i}\hbar\alpha} \quad . \tag{2.12}$$

The electron density can be expressed as

---

[1] In the case of a pure 1d system, the length L should be used instead of the volume V. However, we must employ here the mathematical expressions appropriate for electrons in real 3d materials. An example is the expression of the Coulomb potential acting on the electrons due to the lattice.

$$\varrho(r) = (1/N)\sum_k |\psi_k|^2 \quad , \tag{2.13}$$

where the factor $1/N$ is introduced in order to give the density per one electron in an N-electron system. Substituting ψ_k into $\varrho(r)$ and neglecting higher order terms, we obtain the following,

$$\varrho(r,t) \cong \frac{1}{NV}\sum_k (1 + a_{k+Q}(t)\mathrm{e}^{\mathrm{i}Qr} + a^*_{k+Q}(t)\mathrm{e}^{-\mathrm{i}Qr}) \quad . \tag{2.14}$$

The first term on the right hand side gives an average density of electrons, which is uniform in space. Omitting this term we rewrite the remaining spatially varying part as $\delta\varrho(r,t)$. Equation (2.14) combined with (2.12) gives the following.

$$\delta\varrho(r,t) = \frac{1}{NV}\sum_k \left(\frac{V_Q \mathrm{e}^{\mathrm{i}Qr}\mathrm{e}^{\mathrm{i}\omega t}\mathrm{e}^{\alpha t}}{E_k - E_{k+Q} + \hbar\omega - \mathrm{i}\hbar\alpha} + \mathrm{c.c.}\right) \quad . \tag{2.15}$$

When we consider a perturbation which is real, we obtain the following expression by adding $V^*_Q\mathrm{e}^{-\mathrm{i}Qr}\mathrm{e}^{-\mathrm{i}\omega t}\mathrm{e}^{\alpha t}$ to Eq. (2.10).

$$\delta\varrho(r,t) = \frac{1}{NV}\sum_k \Bigg\{\left(\frac{1}{E_k - E_{k+Q} + \hbar\omega - \mathrm{i}\hbar\alpha} + \frac{1}{E_k - E_{k-Q} + \hbar\omega - \mathrm{i}\hbar\alpha}\right) \times \mathrm{e}^{\mathrm{i}Qr}\mathrm{e}^{\mathrm{i}\omega t}\mathrm{e}^{\alpha t} + \mathrm{c.c.}\Bigg\} \quad . \tag{2.16}$$

Now the Fermi distribution of electrons must be taken into account. As it is evident in (2.11), an electron in the state k requires empty states at $k \pm Q$ in order to form $\delta\varrho(r)$. Thus the terms in the parenthesis of (2.16) should be rewritten as

$$(\quad) = \frac{f_k(1 - f_{k+Q})}{E_k - E_{k+Q} + \hbar\omega - \mathrm{i}\hbar\alpha} + \frac{f_k(1 - f_{k-Q})}{E_k - E_{k-Q} + \hbar\omega - \mathrm{i}\hbar\alpha} \quad . \tag{2.17}$$

We then rewrite the second term on the right hand side by changing the suffices, $k - Q \rightarrow k$ and $k \rightarrow k + Q$. This is possible because these terms are summed up in (2.16) with respect to k. Finally we obtain

$$\delta\varrho(r,t) = -\frac{1}{NV}V_Q\sum_k \frac{(f_{k+Q} - f_k)\mathrm{e}^{\mathrm{i}Qr}\mathrm{e}^{\mathrm{i}\omega t}\mathrm{e}^{\alpha t}}{E_k - E_{k+Q} + \hbar\omega - \mathrm{i}\hbar\alpha} + \mathrm{c.c.} \tag{2.18}$$

The parameter α is of course artificial and can be made negligibly small. Hence, for a static potential, (2.18) in the limit of $\omega \rightarrow 0$ and $\alpha \rightarrow 0$ is appropriate for the problem. A relation between the Fourier component V_Q of $V(r)$ and ϱ_Q of $\delta\varrho(r)$, where $\delta\varrho(r) = (1/V)\sum_Q \varrho_Q \exp(\mathrm{i}Qr)$, gives the expression for $\chi(Q)$;

$$\chi(Q) \equiv \frac{1}{N}\sum_k \frac{f_{k+Q} - f_k}{E_k - E_{k+Q}} \quad , \tag{2.19}$$

where ϱ_Q and V_Q are related to one another by

$$\varrho(Q) = -V_Q\chi(Q) \quad . \tag{2.20}$$

We shall now examine the properties of $\chi(Q)$, considerig first the case of $T = 0$. Each term in (2.19) has a non-vanishing value only when either $E_k < E_\mathrm{F}$ and $E_{k+Q} > E_\mathrm{F}$ or $E_k > E_\mathrm{F}$ and $E_{k+Q} < E_\mathrm{F}$. In the former case $f_k = 1$ and $f_{k+Q} = 0$, and in the latter $f_k = 0$ and $f_{k+Q} = 1$. A diagram such as Fig. 2.1 will make it easy to perform the calculation. Replacing summation with integration, we obtain

$$\chi(Q) = \frac{2m}{\pi n\hbar^2 Q}\ln\left|\frac{Q+2k_\mathrm{F}}{Q-2k_\mathrm{F}}\right| \quad . \tag{2.21}$$

Derivations of the equivalent expressions for 2- and 3d are left to the reader. The Q-dependence of $\chi(Q)$ is shown qualitatively for each dimensionality in Fig. 2.2 [2.4]. In 1d $\chi(Q)$ diverges at $Q = 2k_\mathrm{F}$ but in 2- and 3d it has only a jump and a divergence respectively in its derivatives at $2k_\mathrm{F}$.

The origin of the singularity in $\chi(Q)$ at $2k_\mathrm{F}$ can be understood as follows. In 1d each term of (2.19) can diverge only when $Q \cong \pm 2k_\mathrm{F}$, $k \cong \mp k_\mathrm{F}$ and $Q \cong 0$, $k \cong \pm k_\mathrm{F}$. Other cases do not contribute to the sum even when

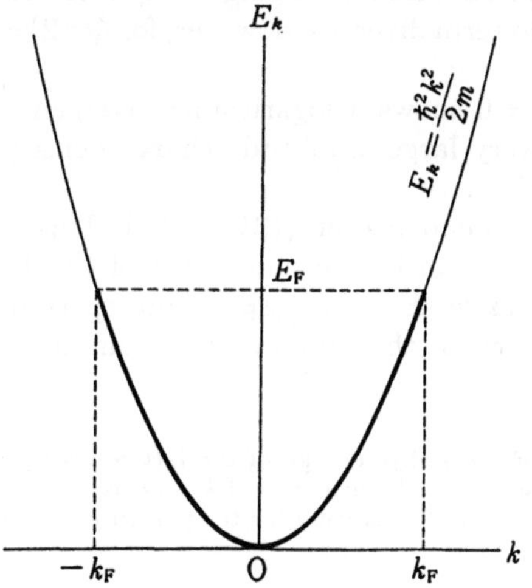

Fig. 2.1. Conduction band for free electrons in one dimension

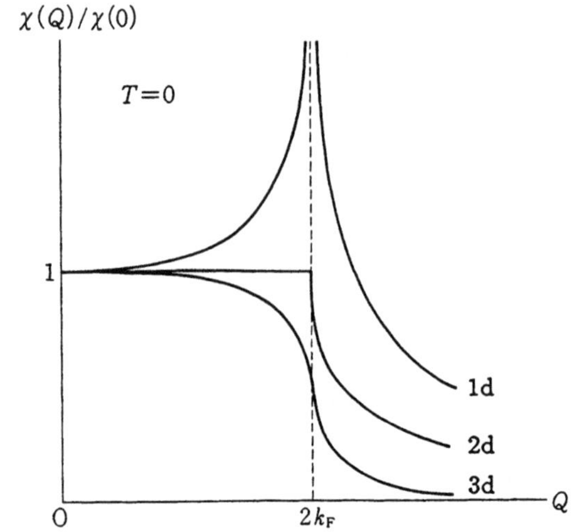

Fig. 2.2. Polarization function $\chi(Q)$ at absolute zero in each dimension

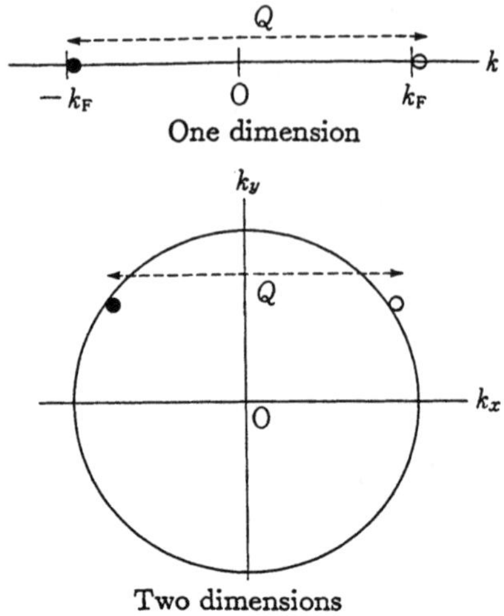

Fig. 2.3. Fermi surface in one- and two-dimensions

$E_k \cong E_{k+Q}$ because then $f_{k+Q} - f_k$ in the numerator becomes exactly equal to zero. The apparent singularity of each term at $Q \cong 0$ gives a finite value after integration but the divergence at $Q = 2k_\mathrm{F}$ remains.

The situation in 2- and 3d is not different from this with respect to the singularity of each term at $|k| \cong k_\mathrm{F}$.

As can be seen from the drawing of 1d and 2d Fermi surfaces shown in Fig. 2.3, the number of diverging terms for a given Q is of the order of $1/k_\mathrm{F}$ in 1d but only of $1/k_\mathrm{F}^2$ in 2d. Parallel conclusions can be reached for the 3d case. Thus we can qualitatively understand why the singularity at $Q = 2k_\mathrm{F}$ weakens with increasing dimensionality. In 2- and 3d the singularity is found only in the derivatives of $\chi(Q)$ at $2k_\mathrm{F}$. No term diverges, however, for $Q > 2k_\mathrm{F}$ in any dimensionality.

We have learned that $\chi(Q)$ at $T = 0$ shows a logarithmic divergence at $Q = 2k_\mathrm{F}$ in 1d. Thus we expect a very large amplitude charge density wave, ϱ_{2k_F}, for a very weak potential V_{2k_F}.

We turn now to the temperature dependence of $\chi(2k_\mathrm{F})$ [2.5]. This is dominated by the Fermi-distribution function in the numerator of (2.19). We evaluate (2.19) only in the energy range $|E - E_\mathrm{F}| \le \varepsilon_\mathrm{B}$, where ε_B is an appropriate energy width ($\varepsilon_\mathrm{B} \ll E_\mathrm{F}$),[2] because the summation is dominated

[2] The energy ε_B is usually of the order of the Debye energy of the lattice because the potential $V(r,t)$, which mixes the electronic state k with that of $k + Q$, is exerted by lattice vibrations (see next section). Therefore (2.24) is valid for temperatures much lower than the Debye temperature.

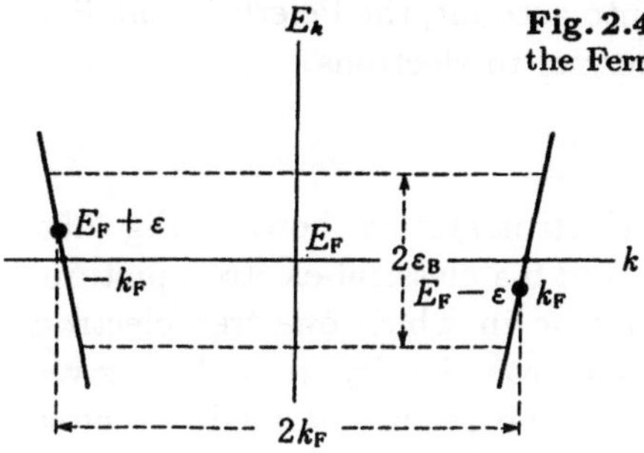

Fig. 2.4. Approximate relation between E_k and k near the Fermi level

mostly by terms with E_k and E_{k+Q} close to E_F. The dispersion relation is approximated by a straight line in the vicinity of E_F as shown in Fig. 2.4. If we introduced ε such that $E_k = E_F + \varepsilon(\varepsilon<\varepsilon_B)$, then E_{k+2k_F} is given as $E_{k+2k_F} = E_F - \varepsilon$. The dominator (of 2.19) becomes $E_k - E_{k+2k_F} = 2\varepsilon$, and the numerator is written as,

$$\frac{1}{\exp(-\varepsilon\beta)+1} - \frac{1}{\exp(\varepsilon\beta)+1} = \tanh\frac{\varepsilon\beta}{2} \quad . \tag{2.22}$$

Replacement of the summation with integration in Eq. (2.19) gives

$$\chi(2k_F) = \frac{D(E_F)}{2N} \int_0^{\varepsilon_B\beta/2} \frac{\tanh x}{x} dx \tag{2.23}$$

where $D(E_F)$ denotes the density of states at the Fermi level. Using the approximation[2] $\beta\varepsilon_B \gg 1$, we obtain the following,

$$\chi(2k_F) = \frac{D(E_F)}{2N} \ln\frac{1.14\varepsilon_B}{k_BT} \quad . \tag{2.24}$$

Here the arbitrary parameter ε_B does not affect the essential point of the following discussion.

The temperature dependence of $\chi(2k_F)$ is approximated as $\chi(2k_F) \sim \ln(1/k_BT)$. This approximation tells us that $\chi(2k_F)$ diverges logarithmically with decreasing temperature. When $k_BT \gtrsim E_F$, $\chi(2k_F)$ is expected to decrease with increasing temperature, because the numerator of Eq. (2.19) must decrease. However, in this case $\chi(2k_F)$ cannot be expressed in a simple form.

The instability in 1d electronic systems for $Q \to 2k_F$ and $T \to 0$ which gives rise to the density wave ϱ_{2k_F}, is called a Peierls instability. In the free-electron system without interactions, which we have been discussing until now, there is of course no potential present to drive the Peierls' instability.

However when a phonon system is taken into account, the Peierls instability can really occur since phonons give a potential to electrons.

b) The Peierls Transition

A lattice system is introduced into the 1d electron system shown in Fig. 2.5a and b. Figure 2.5a depicts a band-picture and b a classical-electron picture. For the sake of simplicity we consider a case in which one free electron is present per atom (or molecule). The electron density, $n = 1/a$, gives $k_F = (1/4)(2\pi/a) = (1/4)a^*$, where a^* is the reciprocal lattice vector, namely the size of the first Brillouin zone. The following discussion applies to any electron density $n<2/a$ since this is always associated with metallic behaviour.

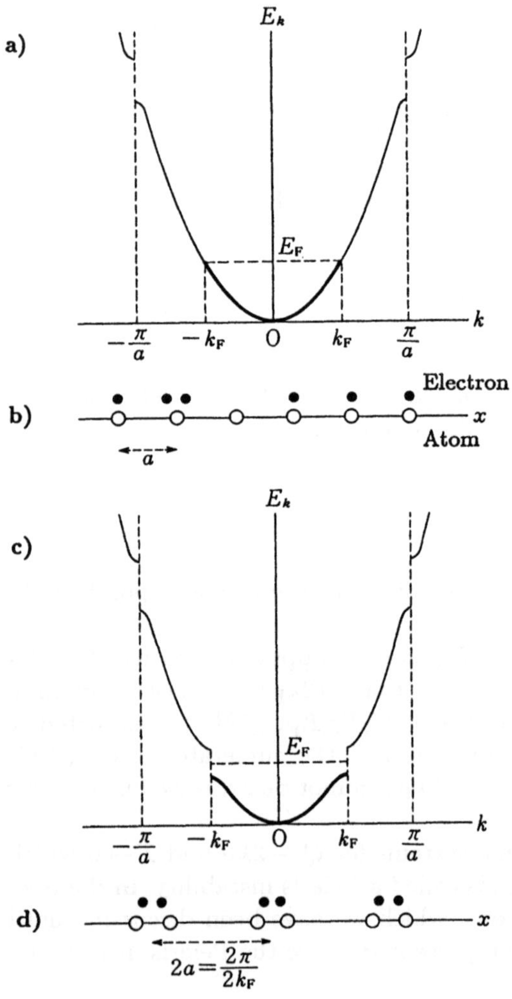

Fig. 2.5a–d. The Peierls transition in a one-dimensional electron system. **(a)** Band model for the metallic state. **(b)** Localized-electron model. **(c)** Band model for the insulating state caused by the Peierls transition. **(d)** Localized-electron model

First, we shall consider the case of static lattice distortions. A periodic lattice distortion with the wave number Q is described as,

$$u = u_Q \cos(Qx) \tag{2.25}$$

where u is the displacement of the ions from their equilibrium position. This lattice distortion produces a potential V which acts on the electron system:

$$V \equiv V_Q \cos(Qx) = g u_Q \cos(Qx) \quad , \tag{2.26}$$

where g denotes the magnitude of electron-lattice interactions. The electron system forms a density wave $\varrho_Q = -V_Q \chi(Q)$ corresponding to the potential V. The electron system is expected to lower its energy by forming the density wave. On the other hand, the periodic lattice distortion of (2.25) will increase the strain energy by an amount of δU :

$$\delta U = (1/2) c u_Q^2 \quad , \tag{2.27}$$

where c denotes the elastic stiffness. If the decrease of the electron energy outweighs the increase of the lattice energy δU, the total system will stabilize itself by forming the lattice distortion u and the electron-density wave, ϱ_Q, spontaneously. This phenomenon really occurs for $Q = 2k_\mathrm{F}$ in 1d.

We shall turn now to the problem of the balance in energy. A potential with the wave number Q creates a band gap at $k = Q/2$ in the conduction band of electrons as described in Sect. 2.1.1. A perturbation calculation gives the width of the band gap as $E_\mathrm{g} = 2\Delta = 2|V_Q| = 2g|u_Q|$. The electron energy bands lying below and above $k = Q/2$ are bent downwards and upwards respectively. For the case $Q = 2k_\mathrm{F}$ and at $T = 0$ electrons occupy all states below the gap and no others. The decrease of electron energy near the gap will result in the decrease in total energy of electrons in the band. This decrease is easily calculated. The perturbed energy is given by

$$E_k^\pm = \tfrac{1}{2}(E_k^0 + E_{k+Q}^0) \pm \tfrac{1}{2}\sqrt{(E_k^0 - E_{k+Q}^0)^2 + 4|V_Q|^2} \quad , \tag{2.28}$$

where E_k^0 and E_{k+Q}^0 are the unperturbed energies and the sign $\pm$ denotes the upper and the lower bands, respectively. The change in the electron energy due to the perturbation is calculated as

$$\delta K = \sum_k E_k^\pm f(E_k^\pm) - \sum_k E_k^0 f(E_k^0) \quad , \tag{2.29}$$

where the equation is generalized to finite temperatures. The approximation $V_Q \ll E_\mathrm{F}$ gives

$$\begin{aligned}\delta K &\cong -|V_Q|^2 \sum_k \frac{f(E^0_{k+Q}) - f(E^0_k)}{E^0_k - E^0_{k+Q}} \\ &= -|V_Q|^2 N\chi(Q) \quad , \end{aligned} \tag{2.30}$$

which is valid up to the order of V_Q^2.

The change in the total energy of electron and lattice systems is therefore

$$\begin{aligned}\delta K + \delta U &= -|V_Q|^2 N\chi(Q) + \frac{c}{2}u_Q^2 \\ &= -\Delta^2 \left\{ N\chi(Q) - \frac{c}{2g^2} \right\} \quad . \end{aligned} \tag{2.31}$$

The polarization function $\chi(Q)$ diverges logarithmically for $Q \to 2k_F$ and $T \to 0$. On the other hand the elastic stiffness c must remain finite. Therefore $\delta K + \delta U$ becomes negative below a certain temperature T_P and the Peierls instability takes place whenever g is non-zero.[3]

Thus the instability gives rise to a band gap at the Fermi level and the system turns from a conductor into an insulator. The periodic lattice distortion with the wave number $2k_F$ is called a Peierls distortion.

In a classical picture of localized electrons, the system after the transition contains two electrons per wavelength of the lattice distortion. Each period of the wave is regarded as a separate site for electrons. The lowest energy level of each site accepts two electrons which are the maximum allowed by the Pauli principle. Thus the system becomes insulating.

The metal-insulator transition arising from the Peierls instability is called a Peierls transition.[4]

From (2.31) we can derive an expression for the transition temperature T_P, by setting $\delta K + \delta U = 0$: First we obtain a relation $\chi(2k_F) = (c/2)Ng^2$. This, combined with (2.24), gives the following:

$$k_B T_P^{MF} = 1.14\varepsilon_B \exp(-1/\lambda) \quad , \quad \lambda \equiv g^2 D(E_F)/c \quad , \tag{2.32}$$

where λ is the dimensionless parameter of electron-lattice interactions. The superscript MF of T_P indicates that the result is derived on the basis of a mean-field approximation. The effects of fluctuation which are neglected in the mean-field approximation, will be considered in Sect. 2.3.

---

[3] A more exact treatment for finite temperatures requires a discussion in terms of a free energy formulation.

[4] R.E. Peierls [2.6] introduced this concept in order to explain electronic states in bismuth etc.

2.1.3 Electron-Phonon Coupling and Kohn Anomalies

In the last section we have seen that the static lattice distortion and the charge density wave with wave number $Q = 2k_F$ appear below the Peierls transition temperature T_P^{MF}. Above T_P^{MF}, instead of the static distortion, it is found that the lattice vibration (phonon) with $Q = 2k_F$ increases its amplitude and decreases its frequency with decreasing temperature toward T_P^{MF}. This is the phenomenon of phonon softening. The softening of $2k_F$ phonons due to interactions with an electron system is called a Kohn anomaly [2.7]. For a 1d electron system the frequency of soft phonons becomes zero just at T_P^{MF} giving rise to the static lattice distortion with $Q = 2k_F$. The Kohn anomaly can be regarded as a precursor to the Peierls transition.

a) Electron-Phonon Interactions

We shall investigate electron-phonon interactions in order to see how the Kohn anomaly occurs. Lattice vibrations create an electric polarization of the lattice due to the displacements of the positive ions. This polarization produces an electric field which acts on electrons. Thus the electrons and the lattice interact with each other through this polarization field. A Hamiltonian describing the interaction is usually expressed within the formalism of second quantization.

First we regard the lattice as an elastic continuum of positive charges; this is the so-called jellium model. A lattice displacement, namely a displacement of the elastic body of positive charges, is described as

$$u(x,t) = \frac{1}{\sqrt{V}} \sum_Q \eta_Q(t) \exp(\mathrm{i}Qx) \quad . \tag{2.33}$$

This displacement causes an electric polarization $P = Z|e|n_i u$, where n_i is the number density of ions. The density of positive charges $\delta\varrho_i$ is then given by $\delta\varrho_i = -\mathrm{div}P$. Poisson's equation, $-\nabla^2\phi = \delta\varrho_i/\varepsilon_0$, determines the potential $\phi(x)$ due to this charge density. Here we use only the dielectric constant of vacuum since the Coulomb interaction among electrons is neglected; we are using the simple approximation that the potential due to the positive ions is not screened by electrons. The potential $\phi(x)$ is then given as

$$\phi(x) = -\frac{\mathrm{i}}{\varepsilon_0\sqrt{V}} \sum_Q \frac{Z|e|n_i\eta_Q e^{\mathrm{i}Qx}}{Q} \quad . \tag{2.34}$$

Writing the electron wave function as

$$\psi(x) = \frac{1}{\sqrt{V}} \sum_k a_k \exp(\mathrm{i}kx) \quad , \tag{2.35}$$

we obtain the potential energy,

$$H' = \int e\phi(x)\psi(x)^*\psi(x)d^3r \quad . \tag{2.36}$$

This is the interaction energy between electrons and phonons. H' can be evaluated to give

$$H' = \frac{\mathrm{i}Ze^2 n_\mathrm{i}}{\varepsilon_0\sqrt{V}} \sum_Q \sum_k \frac{a_k^\dagger a_{k-Q}\eta_Q}{Q} \quad . \tag{2.37}$$

On the basis of the second quantization the parametrs $a_k^\dagger$ and a_k will be regarded respectively, as the creation and the annihilation operators for electrons having the wave number k. In terms of the creation and the annihilation operators of phonons, $b_Q^\dagger$ and b_Q, η_Q is written as

$$\eta_Q = \sqrt{\frac{\hbar}{2Mn_\mathrm{i}\omega_Q}}(b_Q + b_{-Q}^\dagger) \quad . \tag{2.38}$$

The electron-phonon interaction Hamiltonian is given by

$$H' = \frac{1}{\sqrt{V}} \sum_Q \sum_k g(-Q) a_{k+Q}^\dagger a_k (b_Q + b_{-Q}^\dagger) \quad , \tag{2.39}$$

with

$$g(Q) = \mathrm{i}\sqrt{\frac{\hbar}{2Mn_\mathrm{i}\omega_Q}} \frac{Ze^2 n_\mathrm{i}}{\varepsilon_0 Q}$$

where $g(Q)$ is the parameter of electron-phonon interaction [see (2.26)].

Combining this with Hamiltonians for the bare electrons and phonons, the total Hamiltonian becomes

$$\begin{aligned} H = {} & \sum_k E_k a_k^\dagger a_k + \sum_Q \hbar\omega_Q b_Q^\dagger b_Q \\ & + \frac{1}{\sqrt{V}} \sum_{k,Q} g(Q) a_{k+Q}^\dagger a_k (b_Q + b_{-Q}^\dagger) \quad . \end{aligned} \tag{2.40}$$

This is known as the Fröhlich Hamiltonian for non-interacting electrons and the jellium model of the lattice [2.8]. The Kohn anomaly is derived from this Hamiltonian.

b) The Kohn Anomaly

The frequency of phonons having the wave number Q (wavelength $2\pi/Q$) is determined by the magnitude of the restoring force for the corresponding lattice distortion. The origin of the restoring force is the Coulomb interaction between ions. The lattice distortion associated with a phonon of wave number Q, provides the electron system with a potential V_Q, which gives rise to a density wave ϱ_Q of electrons. ϱ_Q is written as $\varrho_Q = -\chi(Q)V_Q$ [see (2.20)]. The density wave of electrons must reduce interionic forces by screening the electric force. Thus the restoring force decreases and the phonon frequency becomes smaller than it would be in the absence of electron-phonon interactions.

We calculate the phonon frequency in the presence of electron-phonon interactions by using the Fröhlich Hamiltonian, (2.40). The equation of motion for the phonon operator η_Q is

$$-\hbar^2\ddot{\eta}_Q = [[\eta_Q, H], H] \quad . \tag{2.41}$$

After some calculations we obtain from this

$$\ddot{\eta}_Q = -\omega_Q^2\eta_Q - g_{-Q}\sqrt{\frac{2\omega_Q}{Mn_\mathrm{i}V\hbar}}\sum_k a_{k-Q}^\dagger a_k \quad . \tag{2.42}$$

If we have N electrons in a volume V, the electron density can be written as $|\psi(r)|^2 = (N/V)\sum_Q \varrho_Q \exp(\mathrm{i}Qr)$. In the second quantization ϱ_Q, regarded as the number operator, becomes $\varrho_Q = (1/N)\sum_k a_{k-Q}^\dagger a_k$. Thus Eq. (2.42) is rewritten as

$$\ddot{\eta}_Q = -\omega_Q^2\eta_Q - g_{-Q}\sqrt{\frac{2\omega_Q}{Mn_\mathrm{i}\hbar V}}N\varrho_Q \quad . \tag{2.43}$$

The potential V_Q will be given by

$$V_Q = g_Q\eta_Q\sqrt{\frac{2Mn_\mathrm{i}\omega_Q}{\hbar V}} \tag{2.44}$$

after writing H' (2.39) in the form $H' = N\sum_Q \varrho_Q V_{-Q}$. Thus the equation of motion (2.43) becomes

$$\ddot{\eta}_Q = -\omega_Q^2\eta_Q + \frac{2|g_Q|^2\omega_Q N}{\hbar V}\chi(Q)\eta_Q \quad . \tag{2.45}$$

The new phonon frequency Ω_Q renormalized by the electron-phonon interaction is

$$\Omega_Q^2 = \omega_Q^2 - \frac{2|g_Q|^2 \omega_Q N}{\hbar V} \chi(Q) \leq \omega_Q^2 \quad , \tag{2.46}$$

where ω_Q is the frequency of bare phonons without the interaction.

For the case of a 1d electron system, $\chi(Q)$ diverges logarithmically as $Q \to 2k_\mathrm{F}$ and $T \to 0$ as can be seen from (2.21) and (2.24); thus the frequency ϕ_{2k_F} of phonons with wave number $Q = 2k_\mathrm{F}$ is particularly small compared to that for other wave numbers. Furthermore the frequency approaches zero with decreasing temperature. In 2- and 3d $\chi(Q)$ does not diverge but decreases rapidly for $Q > 2k_\mathrm{F}$ as shown in Fig. 2.2. This fact is reflected in a weak anomaly in the phonon frequencies at $Q = 2k_\mathrm{F}$. Figure 2.6 depicts the form of the Kohn anomly in each dimension [2.7]. The Kohn anomly in 1d is sometimes called the giant Kohn anomaly.

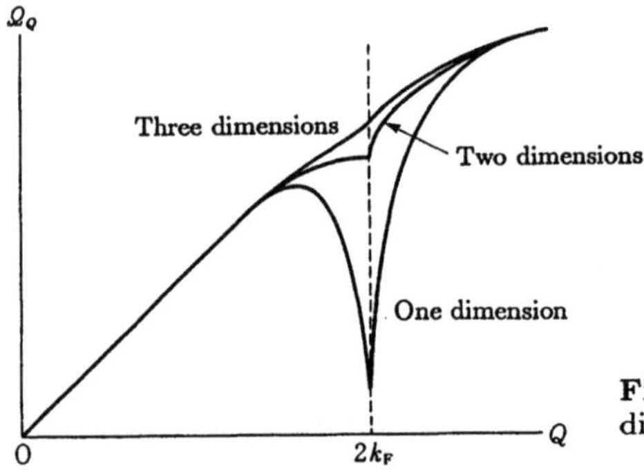

Fig. 2.6. Kohn anomaly of phonons in each dimension

The temperature at which in 1d Ω_{2k_F} becomes zero, is the Peierls transition temperature T_P^MF. Setting $\Omega_{2k_\mathrm{F}} = 0$ in (2.46) and using (2.24) for $\chi(2k_\mathrm{F})$, we obtain

$$k_\mathrm{B} T_\mathrm{P}^\mathrm{MF} = 1.14 \varepsilon_\mathrm{B} \exp(-1/\lambda) \quad , \quad \text{with}$$

$$\lambda = \frac{|g_{2k_\mathrm{F}}|^2 D(E_\mathrm{F})}{\hbar \omega_{2k_\mathrm{F}} V} \quad . \tag{2.47}$$

The parameter λ must be the same as λ in (2.32). Comparing this equation with Eq. (2.32), we see that the elastic stiffness c in (2.32) therefore corresponds to ω_{2k_F}. It is natural that the elastic stiffness corresponds to the phonone frequency.

From (2.47) for T_P^MF, we obtain the temperature dependence of $\Omega_{2k_\mathrm{F}}^2$ as

$$\Omega_{2k_\mathrm{F}}^2 = \lambda \omega_{2k_\mathrm{F}}^2 \ln(T/T_\mathrm{P}^\mathrm{MF}) \quad . \tag{2.48}$$

c) The Order Parameter of the Peierls Transition

Two quantities which describe the change of state that occurs in the Peierls transition are the frequency of soft phonons, Ω_{2k_F}, and the width of the bandgap, $E_g = 2\Delta$. We can employ Δ as an order parameter of the Peierls transition because $\Delta = 0$ for $T \geq T_P^{MF}$ and $\Delta > 0$ for $T < T_P^{MF}$. We shall examine the temperature dependence of Δ below T_P^{MF}.

For $T \leq T_P^{MF}$, the restoring force for the lattice distortion with $Q = 2k_F$ becomes zero because of screening due to electron-density waves. A static lattice distortion appears. The balance of forces is represented on the right hand side of (2.43) or (2.45). Using (2.45) and noting the non-vanishing expectation value $\langle \eta_{2k_F} \rangle \neq 0$ for $T < T_P^{MF}$, we obtain an equation for the force balance as follows,

$$\omega_{2k_F}^2 \langle \eta_{2k_F} \rangle = \frac{2|g_{2k_F}|^2 \omega_{2k_F} N}{\hbar V} \chi(2k_F; \langle \eta_{2k_F} \rangle) \langle \eta_{2k_F} \rangle \quad . \tag{2.49}$$

Here, $\chi(2k_F; \langle \eta_{2k_F} \rangle)$ is the polarization function in the presence of the gap Δ due to $\langle \eta_{2k_F} \rangle \neq 0$. This should be calculated with use of (2.28) which describes the electronic energy in the presence of the band gap.

We calculate the polarization function in this case. The energy is measured from $E = E_F$ and the calculation is done within an energy range $E_F - \varepsilon_B < E < E_F + \varepsilon_B$ (i.e. $k \sim k_F$) as was the case in (2.23). The dispersion relation near E_F is approximated by a linear relation between E_k^0 and k,

$$E_k^0 = E_F + \varepsilon_k^0 \quad , \quad E_{k+Q}^0 = E_F - \varepsilon_k^0 \quad . \tag{2.50}$$

Equation (2.28) is written as

$$E_k^{\pm} = \sqrt{(\varepsilon_k^0)^2 + |V_Q|^2} = \pm\sqrt{(\varepsilon_k^0)^2 + \Delta^2} \quad . \tag{2.51}$$

Following the same procedure as in (2.22) and (2.23), we obtain the following:

$$\chi(2k_F; \langle \eta_{2k_F} \rangle) = \frac{D(E_F)}{2N} \int_0^{\varepsilon_B} \tanh\left(\frac{E_k^+}{2k_B T}\right) \frac{d\varepsilon_k^0}{E_k^+} \quad . \tag{2.52}$$

Thus (2.49) can be rewritten using (2.52) and λ of (2.47) as

$$\frac{1}{\lambda} = \int_0^{\varepsilon_B} \tanh\left(\frac{E_k^+}{2k_B T}\right) \frac{d\varepsilon_k^0}{E_k^+} \quad . \tag{2.53}$$

This equation gives a relation between T and Δ contained in E_k^+. The form of this equation is the same as the form of that which is used in the derivation of the superconductivity gap. We shall therefore make use of the result for the superconductivity. The gap is given at $T = 0$ as [2.9]

$$\Delta(0) = 2\varepsilon_{\mathrm{B}} \exp(-1/\lambda) = 1.76 k_{\mathrm{B}} T_{\mathrm{P}}^{\mathrm{MF}} \quad . \tag{2.54}$$

This is obtained when $\Delta(0) \ll \varepsilon_{\mathrm{B}}$, where ε_{B} is of the order of the Debye energy. The condition is usually satisfied in real systems. Figure 2.7 shows the temperature dependence of Δ. Also shown is the temperature dependence of the frequency $\Omega_{2k_{\mathrm{F}}}$ of soft phonons above $T_{\mathrm{P}}^{\mathrm{MF}}$. The gap Δ is zero above $T_{\mathrm{P}}^{\mathrm{MF}}$ and increases continuously with decreasing temperature below $T_{\mathrm{P}}^{\mathrm{MF}}$. This suggests that the Peierls transition, like the superconducting transition, is a second-order phase transition whose order parameter is the gap Δ. Of course we could also use $\langle u_{2k_{\mathrm{F}}} \rangle$ the amplitude of the lattice distortion, or $\langle \varrho_{2k_{\mathrm{F}}} \rangle$, (which is proportional to $\langle u_{2k_{\mathrm{F}}} \rangle$), as an order parameter.

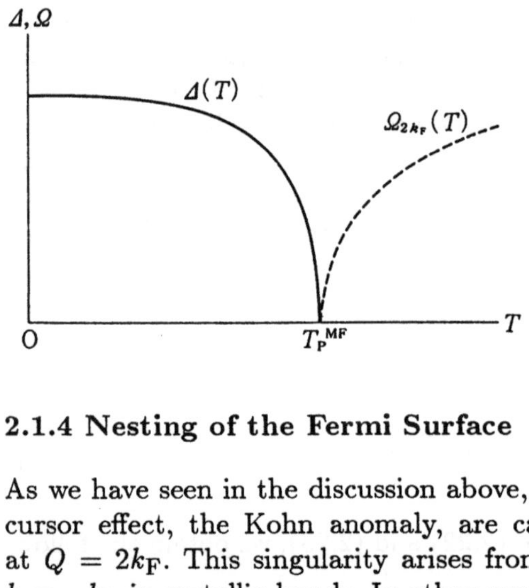

Fig. 2.7. Temperature dependence of the order parameter for the Peierls transition and of the frequency $\Omega_{2k_{\mathrm{F}}}$ of soft phonons

2.1.4 Nesting of the Fermi Surface

As we have seen in the discussion above, the Peierls transition and its precursor effect, the Kohn anomaly, are caused by the singularity of $\chi(Q)$ at $Q = 2k_{\mathrm{F}}$. This singularity arises from the fact that $E_k = E_{k+2k_{\mathrm{F}}}$ for $k = -k_{\mathrm{F}}$ in metallic bands. In other words, the origin of the singularity is the presence of the touching area of the Fermi surface for a displacement of the original Fermi surface with the magnitude $Q = 2k_{\mathrm{F}}$ as shown in Fig. 2.8a.

In order to extend our discussion to 2- and 3d, we regard the Fermi surface in 1d as a flat plane perpendicular to the sheet as shown in Fig. 2.8b. In 1d, wave numbers perpendicular to the 1d axis are meaningless. However, this fact allows us to introduce any wave number perpendicular to this axis because it has nothing to do with the description of the 1d electronic states. In other words, we may say that the uncertainty in electron wave number

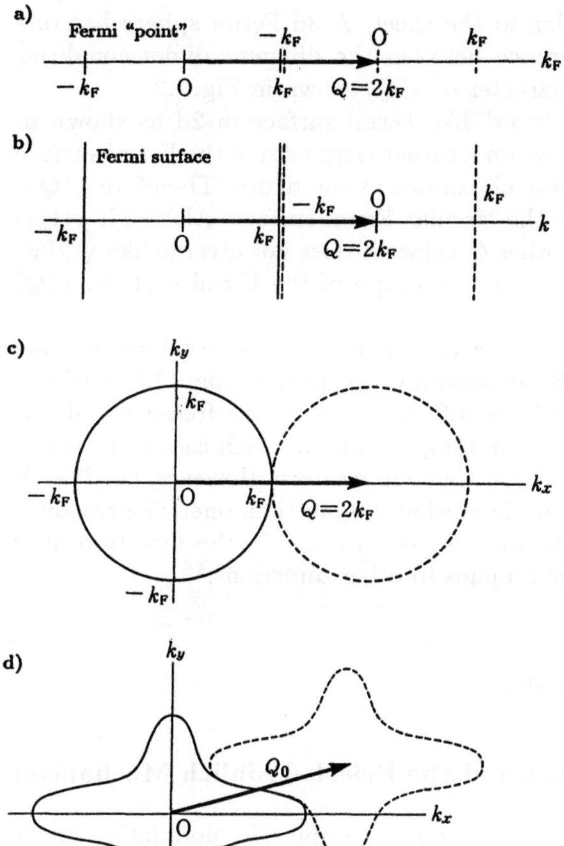

Fig. 2.8a–d. Nesting of the Fermi surface. (a,b) One dimension. (c,d) Examples in two-dimension

perpendicular to the 1d axis is infinite since the positional uncertainty of 1d electrons is zero in those directions.

The pictorial representation in Fig. 2.8b clarifies the discussion given in the first paragraph of this section. The Fermi surface at $k_{\|} = k_F$ coincides perfectly with the translated surface at $k_{\|} = -k_F$ for the translation with $Q = 2k_F$. Such an overlap of planes by translation is called a nesting. The nesting is perfect in 1d.

What about the nesting in 2- and 3d? Figure 2.8c shows a circular Fermi surface (exactly speaking, a Fermi line) in 2d. By means of parallel arguments to those used in transforming Fig. 2.8a into b, we can regard 2d Fermi line as being elongated into a circular cylinder perpendicular to the sheet and having the cross-section shown in Fig. 2.8c. It is evident that the nesting (i.e. area of contact) with respect to $Q = 2k_F$ occurs only on a straight line perpendicular to the sheet. A 3d Fermi sphere has one nesting point only. This difference between the different dimensionalities corresponds to the different character of $\chi(Q)$ shown in Fig. 2.2.

Now we shall consider a "star"-like Fermi surface in 2d as shown in Fig. 2.8d. The nesting now occurs on a rather large area of the Fermi surface for the translational wave vector Q_0 shown in the figure. Therefore $\chi(Q_0)$ will be larger than $\chi(2k_F)$ for the circular Fermi surface. Although $\chi(Q_0)$ becomes larger than $\chi(Q)$ for other Q values it does not diverge like $\chi(2k_F)$ in 1d. Similarly in 3d, dependig on the shape of the Fermi surface, $\chi(Q)$ may increase at certain Q values.

When $\chi(Q)$ in 2- or 3d increases at a certain Q due to a large nesting area for this wave vector, the Kohn anomaly as expected from (2.40), will be pronounced. Furthermore if $\chi(Q)$ is sufficiently large, the frequency of soft phonons may go to zero at a certain temperature in which case a static lattice distortion u_Q will appear below this temperature. However, the Peierls transition does not take place in dimensions higher than one. The reason is that the gap-opening at the Fermi level occurs only in the direction of Q and the metallic Fermi surface remains in other directions.[5]

2.2 Charge Density Waves

2.2.1 Charge Density Waves and the Peierls-Fröhlich Mechanism

As we have learned in 2.1.2 and 2.1.3, a large amplitude modulation of the charge density together with a lattice distortion appear with the wave number $2k_F$ below the Peierls transition temperature T_P. These two waves must be regarded as components of a mixed wave and not as two independent

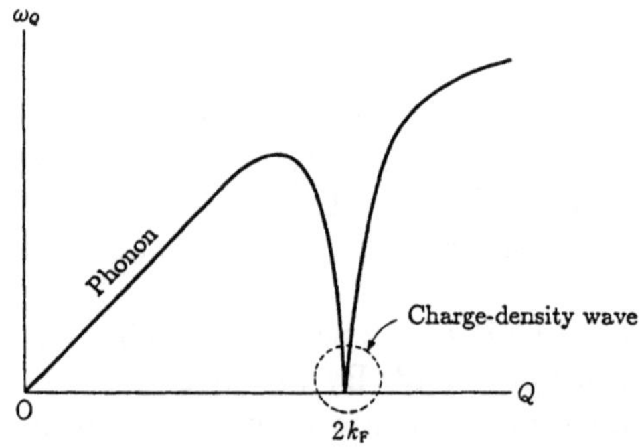

Fig. 2.9. Relation between the phonon softening and the charge density wave

[5] If the shape of the Fermi surface is complicated, the resultant Fermi surface may appear different from the original one because of the folding of the Fermi surface into the new Brillouin zone.

waves since they are strongly coupled to one another. We will call the combined wave a charge-density wave (CDW). Figure 2.9 shows the dispersion relation for phonons having a Kohn anomaly. The CDW corresponds to the state of soft phonons having a frequency $\omega = 0$ at the wave number $2k_F$.

An electrical conductor in which single-particle electronic processes are responsible for conduction becomes semiconducting when the Peierls transition takes place and the CDW appears. As each site is occupied by two electrons, one electron must hop into a higher energy level if it is to move to the neighbouring site; thus current flow necessitates an excitation energy. However, when all electrons move as a whole accompanied by the lattice distortion, this activation energy will not be necessary.[6] This fact corresponds, in a band model, to a translational motion of the whole electron system accompanied by the band gap which is defined in the coordinate system fixed to the electrons. Electron energies observed in the laboratory frame become those of Fig. 2.10 when the center of mass of the electron system moves with the velocity v_s and has a momentum $\hbar q = mv_s$ with respect to the laboratory frame. If the increment in the kinetic energy of the electron at the Fermi level is less than the gap, i.e. $\hbar k_F v_s < \Delta$, namely $v_s < \Delta / \hbar k_F$, electrons are not scattered and the electrical resistance vanishes.

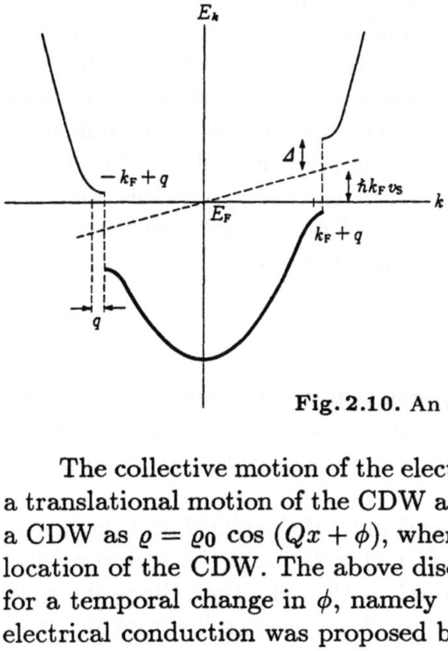

Fig. 2.10. An electron system moving with the band gap

The collective motion of the electrons described here can be regarded as a translational motion of the CDW as shown in Fig. 2.11. When we describe a CDW as $\varrho = \varrho_0 \cos(Qx + \phi)$, where $Q = 2k_F$, the phase ϕ represents the location of the CDW. The above discussion suggests a vanishing resistance for a temporal change in ϕ, namely the motion of ϕ. Such a mechanism of electrical conduction was proposed by Fröhlich as an explanation of super-

[6] The effective mass of the current carrier becomes heavier because it is accompanied by the lattice distortion.

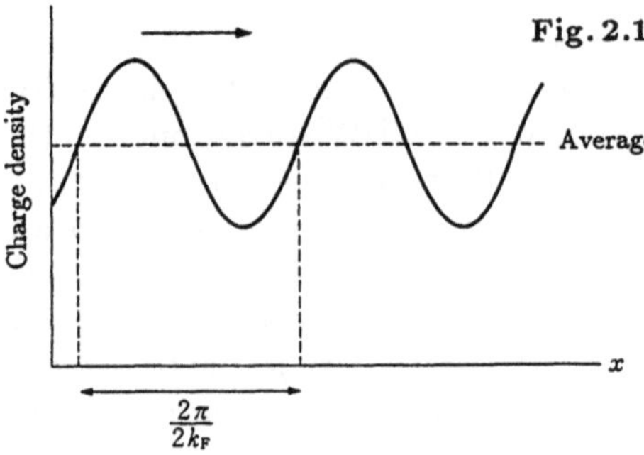

Fig. 2.11. Sliding motion of a charge density wave

conductivity [2.8]. Fröhlich's model was once discarded in the face of the success of the so-called BCS theory [2.10]. However, Fröhlich's model has been revived in studies of 1d conductors and the conduction mechanism due to the translational motion of CDW is sometimes called the Peierls-Fröhlich mechanism. The resistanceless motion of the CDW arises from a translational invariance of the Fröhlich Hamiltonian, described in 2.1.3, with respect to an arbitrary magnitude of translation. The translational invariance of the Fröhlich Hamiltonian is a natural result of the jellium model employed there. Intuitively speaking, the potential energy of a CDW is constant provided it is located in a homogeneous jellium of positive charges; the resistance for the translational motion of this CDW must be zero.

The translational invariance is satisfied however not only in the case of the jellium model. If the lattice parameter a and the wave length λ of the CDW make the ratio λ/a irrational, the situation becomes equivalent to the jellium model. We can understand this point as follows: The CDW is represented by $\varrho = \varrho_0 \cos(Qx + \phi)$, where $Q = 2k_F$, and the potential $V(x)$ is a periodic function with the period of a. The potential energy of the electrons forming the CDW may thus be written as

$$\int \varrho V(x) dx \quad . \tag{2.55}$$

This quantity is zero unless $V(x)$ has a component expressed by $\cos(Qx)$. In the present case it does not because λ/a is irrational. Thus the potential energy is independent of the value of ϕ which denotes the location of the CDW.

In this situation we say that the wavelength of the CDW and the lattice constant, or equivalently the wave number $Q = 2k_F$ and the reciprocal lattice vector $a^* \equiv 2\pi/a$, are *incommensurable* with each other. Since the wave number $2k_F$ is written as $2k_F = n\pi$ (2.8), $2k_F$ becomes incommensurable when n is irrational and the CDW can slide without any resistance.

The above definition of the incommensurability states that the ratio of $2k_F$ to a^* must be strictly irrational; this is a mathematical definition. In physics, however, we can regard those ratios as incommensurable which are rational but cannot be expressed by simple fractions such as 1, $\frac{1}{2}$, $\frac{2}{3}$, etc. The reason is the following: If the ratio is given by a complicated rational number, higher order terms in the potential energy such as

$$\int \varrho^m V(x)dx \quad , \quad m \gg 1 \tag{2.56}$$

may give a finite result which is a function of ϕ, but nonetheless such energies due to higher-order interactions must be small. When these energies are negligibly small compared to the kinetic energy of electrons, phonon energies, thermal energies etc., the problem can be regarded as equivalent to the incommensurable case.

The translational invariance is broken in the following cases: (i) a commensurable case in the physical meaning, (ii) when impurities in the lattice interact with the CDW, (iii) when the CDWs on each of the 1d chains, whose assembly forms real materials, interact with each other. In case (i), for example when $\lambda/a = 1$ as shown in Fig. 2.12a, the potential $V(x)$ has a component cos (Qx); the potential energy after integration is proportional to cos ϕ. The CDW is locked in a position of $\phi = \pi$ because here the potential energy is a minimum. In case (ii), the CDW will be "locked" or "pinned" so that a peak of the wave coincides with the impurity as shown in Fig. 2.12b when, for example, an attractive interaction exists between the impurity and the electrons. Case (iii) becomes equivalent to case (i) with

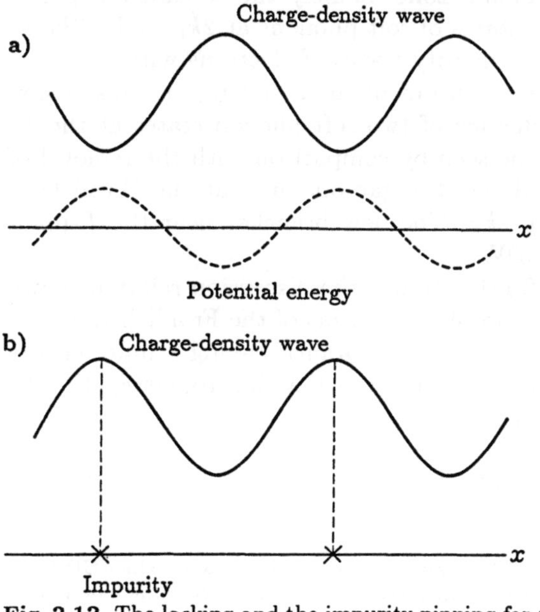

Fig. 2.12. The locking and the impurity pinning for the sliding motion of charge density waves

$\lambda/a = 1$ when we consider a single CDW and regard the surrounding CDWs as potential sources for this.[7]

The suppression of the sliding motion of CDWs by the mechanisms mentioned above is called a pinning or a locking of the CDW. Usually case (ii) is named an impurity pinning. The case (iii) is called a 3d locking although it is also a kind of commensurability locking as in (i).

Charge-density waves can carry an electric current without resistence unless pinning or locking mechanisms are present. These mechanisms inhibit the electrical conduction by CDW under a weak applied electric field. The dielectric constant, however, is still expected to become large since the location of the CDW can easily be displaced from the equilibrium position by the applied field although the displacement remains finite. Furthermore the conductivity for an *ac* field will be finite since the CDW can oscillate. We have now learned the concepts important in the dynamics of CDWs. More detailed investigations are presented in the next section.

2.2.2 Dynamics of Charge Density Waves

a) Phasons

The CDW corresponds to a soft phonon with $\omega = 0$ at the wave number $2k_\mathrm{F}$, Fig. 2.9. The size of the first Brillouin zone becomes $2k_\mathrm{F}$ below the Peierls transition temperature T_P, because of the presence of the static lattice distortion having the wave number $2k_\mathrm{F}$. Therefore the points at $\pm 2k_\mathrm{F}$ become new centres of the Brillouin zone, namely the so-called Γ-point. This fact makes degenerate two states of soft phonons at $2k_\mathrm{F}$ and $-2k_\mathrm{F}$ in the original zone. We expect that density waves of electrons with the wave number $2k_\mathrm{F}$ will produce two branches of phonons having new dispersion relations by removing the degeneracy of two soft phonon states at the Γ-point. A naive explanation can be seen by comparison with the removal of degeneracy in the electron band (i.e. the gap-opening) at the Fermi level by the lattice distortion having $2k_\mathrm{F}$. The new branches near the Γ-point describe the excitation of the CDW.

Lee, Rice, and *Anderson* [2.11] obtained the dispersion relation shown in Fig. 2.13 by investigating this problem in terms of the Fröhlich Hamiltonian for the jellium model. They named one branch having a finite energy at the Γ-point the A_+ mode and the other branch with zero energy, the A_- mode. Their dispersion relations are:

$$A_+ : \quad \omega_+^2 = \lambda\omega_Q^2 + \frac{4}{3}\frac{m}{m^*}v_\mathrm{F}^2 q^2 \quad ,$$

[7] In fact the whole collection of CDWs can slide along the 1d axis. When CDWs of positive charges and of negative ones coexist, however, the whole set cannot be moved by an applied electric field.

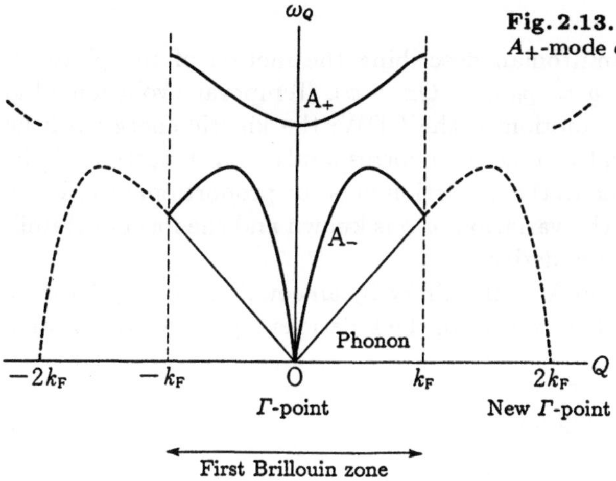

Fig. 2.13. A_--mode (phason mode) and A_+-mode of charge density waves

$$A_- : \quad \omega_-^2 = \frac{m}{m^*} v_F^2 |q|^2 \quad , \tag{2.57}$$

where λ, ω_Q and v_F are respectively the dimensionless electron-phonon interaction parameter, the bare-phonon frequency before softening and the Fermi velocity of conduction electrons before the gap-opening. The effective mass m^* of electrons in the CDW is given by

$$m^*/m = 1 + \frac{4\Delta^2}{\lambda \hbar^2 \omega_Q^2} \quad , \tag{2.58}$$

where 2Δ denotes the width of the Peierls gap. The relation $m^* > m$ means that the CDW involves also a lattice distortion.

Lee, Rice, and Anderson examined what kinds of excitation of the CDW corresponds to the A_+ and the A_- modes [2.11]. They found that, when the CDW is represented as $\varrho = \varrho_0 \cos(Qx + \phi)$, where $Q = 2k_F$, the A_+ mode corresponds to a spatial modulation in the amplitude ϱ_0 of the CDW, with a small wave number $q \ll Q$, and the A_- mode to that in the phase ϕ. The excitation in the phase ϕ is called a phason or a phase phonon. The excitation energy of the phason at $q = 0$ is zero. This corresponds to a change of ϕ homogeneous in space, namely the sliding motion of the CDW without resistance. From the foregoing discussion we know that the sliding motion without resistance is inhibited by the presence of a pinning or locking mechanism. In those cases, therefore, the energy of the A_- mode at $q = 0$, Fig. 2.13, becomes finite. A gap will appear at the Γ-point in the phason excitation spectrum. It is the motion of the phase that is interesting in relation to the electrical conduction.

b) *Motion of the Phase*

We shall formulate a Hamiltonian describing the motion of the phase ϕ. A CDW is described as $\varrho = \varrho_0 \cos(Qx + \phi)$. Temporal evolution of ϕ corresponds to the sliding motion of the CDW; the kinetic energy will be proportional to $\dot{\phi}^2$. Spatial change in ϕ corresponds to a distortion of the wave. An elastic energy due to the distortion must be proportional to $(\nabla\phi)^2$. Thus the total energy for the variation of ϕ is known and the form of Hamiltonian for the motion of ϕ is derived.

If we define the position X of the CDW by an equation $\cos(QX + \phi) = 1$, we obtain $X = -\phi/Q$. The density of the kinetic energy T for the motion of the CDW is given by

$$T = \tfrac{1}{2} nm^* \dot{X}^2 = \tfrac{1}{2} nm^* (\dot{\phi}/Q)^2 \quad , \tag{2.59}$$

where n is the number density of electrons. By introducing a phason velocity v of the A_- mode and using the relations $\omega_- = (m/m^*)^{1/2} v_F q = vq$, $v' = v^2/v_F$, and $2k_F = \pi n$, we obtain

$$T = \frac{\hbar}{4\pi v'} \dot{\phi}^2 \quad . \tag{2.60}$$

Next we derive an expression for the elastic energy U due to the spatial change of ϕ with the wave number q. It is to be noted that the A_- mode, whose excitation has a dispersion relation like acoustic phonons, corresponds to the excitation of elastic waves in ϕ; ϕ being the displacement. An equation of motion $M\ddot{\phi} = c\nabla^2\phi$ for elastic waves gives $v = \sqrt{c/M}$. Since the kinetic energy, $(M/2)\dot{\phi}^2$, corresponds to T derived above, M must correspond to $\hbar/2\pi v'$ giving rise to a relation $c = \hbar v^2/2\pi v'$. Substituting this c into the elastic energy, $(c/2)(\nabla\phi)^2$, we obtain

$$U = \frac{\hbar v^2}{4\pi v'} (\nabla\phi)^2 \quad . \tag{2.61}$$

This is the potential energy for the motion of ϕ. Thus a Lagrangian is obtained as

$$\mathcal{L} = \frac{\hbar}{4\pi v'} \int \{\dot{\phi}^2 - v^2(\nabla\phi)^2\} dx \quad . \tag{2.62}$$

Deriving a momentum conjugate p from $p = \partial\mathcal{L}/\partial\dot{\phi}$, we obtain the following Hamiltonian,

$$H = \frac{\pi v'}{\hbar} \int \left\{ p^2 + \frac{\hbar^2}{4\pi^2} \left(\frac{v}{v'}\right)^2 (\nabla\phi)^2 \right\} dx \quad . \tag{2.63}$$

This is called the phase Hamiltonian [2.12]. From this Hamiltonian we obtain the equation of motion for ϕ,

$$\ddot{\phi} - v^2\nabla^2\phi = 0 \quad . \tag{2.64}$$

Solutions are generally represented by a linear combination of arbitrary functions $\Phi(x - vt)$ and $\Psi(x + vt)$. However we can assume here a solution of the form $\phi = \phi_0 \exp(\mathrm{i}\omega t - \mathrm{i}qx)$ because we are interested in elastic waves whose displacements are those of ϕ. We obtain dispersion relation $\omega = vq$. It is natural that this result gives the dispersion relation for the A_- mode.

c) Pinning

We shall now introduce the mechanisms for pinning or locking into the phase Hamiltonian. A potential energy due to these mechanisms will be written as $\varrho_0 V(\phi)$. Adding this to the Hamiltonian, we obtain the following equation of motion:

$$\ddot{\phi} - v^2\nabla^2\phi + 2\pi v\varrho_0\frac{\partial V}{\partial \phi} = 0 \quad . \tag{2.65}$$

The potential $V(\phi)$ is a periodic function with the period of 2π. We consider the case in which the variation of ϕ is small compared to 2π; $V(\phi)$ can then be approximated as $V(\phi) \sim (V''/2)\phi^2$. By assuming a solution of the form $\phi = \phi_0 \exp(\mathrm{i}\omega t - \mathrm{i}qx)$, we obtain

$$\omega = (\omega_\mathrm{F}^2 + v^2q^2)^{1/2} \quad , \tag{2.66}$$

where $\omega_\mathrm{F}^2 \equiv 2\pi v\varrho_0 V''$. This dispersion relation implies a presence of the energy gap at $q = 0$ with the width $\hbar\omega_\mathrm{F}$. The excitation energy ω_F is necessary to excite a sliding motion of the CDW. This is the pinning or locking energy. Explicit forms of the potential $V(\phi)$ are now given for the three cases.

Impurity pinning: We assume the range of an impurity potential to be short. The potential $V(x)$ is written as $V(x) = \sum_i V_0\Delta(x - x_i)$ where x_i denotes the position of the i-th impurity. The potential energy is then

$$\begin{aligned} H' &= \sum_i \int dx\,\varrho(x)V(x) \\ &= \varrho_0 V_0 \sum_i \cos\left[Qx_i + \phi(x_i)\right] \quad . \end{aligned} \tag{2.67}$$

Some suitable approximations must be made to obtain linear solutions because of the nonlinearity of the total Hamiltonian. Solutions are obtained in two extremal cases; namely a very strong pinning and a very weak pinning [2.13].

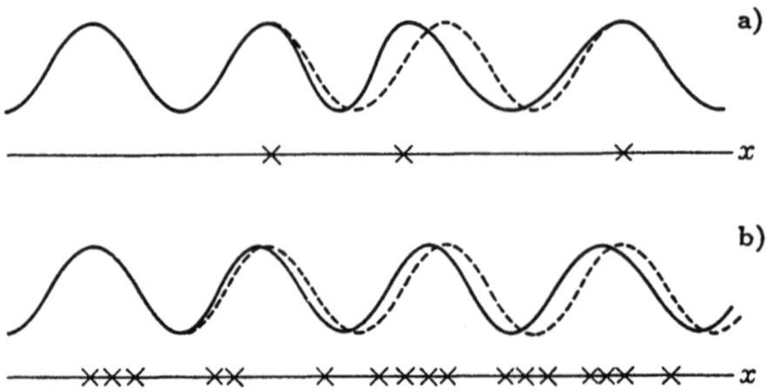

Fig. 2.14a,b. Impurity pinning. **(a)** Strong pinning. **(b)** Weak pinning

The strong pinning occurs when the impurity potential is strong enough or when the impurity density is sufficiently low. In this case, the phase ϕ is locked in $\phi(x_i) = \pi - Qx_i$ at each impurity site so that the potential energy H' becomes minimum. Accordingly the CDW will be significantly distorted as shown in Fig. 2.14a. The weak pinning corresponds to a weak impurity potential or a high density of impurities. If the phase is locked in $\phi(x_i) = \pi - Qx_i$, a loss of the elastic energy due to the distortion of CDW exceeds the gain of the potential energy. Thus we expect as shown in Fig. 2.14b that the phase $\phi(x)$ varies moderately for many impurities in order to realize the configuration having the minimum energy. In both the case of strong pinning and of weak pinning, we determine first the phase $\phi_0(x)$ for the most stable state and then derive equations of motion for displacements from $\phi_0(x)$.

Commensurability locking: When the ratio Q/a^*, where Q is the wave number of the CDW and a^* the reciprocal lattice vector, is equal to $1/N$, we say that the system has a commensurability of the Nth order. In this case a translation of the CDW by a distance a does not change the situation, as shown in Fig. 2.15; the potential energy is invariant for the change of X by a (i.e. the change of ϕ by aQ), where $X(= -\phi/Q)$ denotes the position of CDW as defined in part b) of this section. Since $aQ = 2\pi Q/a^* = 2\pi/N$, the potential energy will be given as

$$H' = \varrho_0 V_0\{1 - \cos(N\phi)\} \quad . \tag{2.68}$$

Expansion of this nonlinear term around an equilibrium value $\phi = 0$ gives the result of (2.66), so long as the approximation $\phi \ll 2\pi$ is employed.

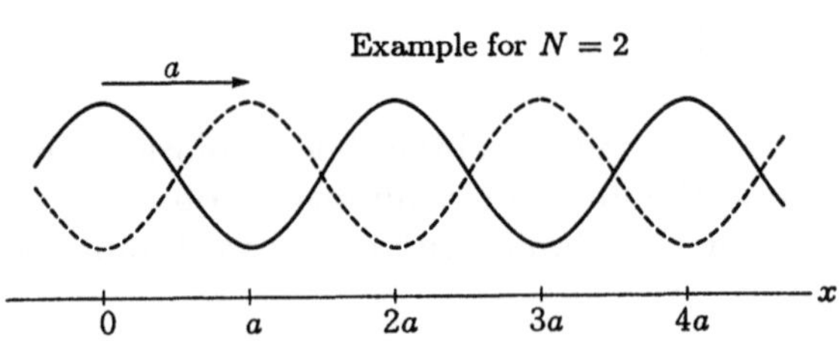

Fig. 2.15. Commensurability locking and the translational symmetry of a charge density wave

Three-dimensional locking: This case is equivalent to that of first-order commensurability locking.

d) Electrical Conduction Due to the Sliding Motion of Charge Density Waves

A spatially homogeneous change in ϕ of a CDW contributes to the electrical conduction. The equation of motion turns into a derivative equation in time t only, because $\nabla\phi = 0$ in this case. A general expression is given as

$$\ddot{\phi} + \Gamma\dot{\phi} + \omega_{\mathrm{F}}^2\phi = \frac{Qe}{m^*}E \quad . \tag{2.69}$$

The second term in the left side, which was not taken into account in the previous discussion, denotes the damping which arises for example from a damping of phonons participating in the formation of CDW. This term is necessary for the discussion of finite temperatures. The third term denotes the restoring force for a small change of ϕ around its equilibrium value. The right hand side arises from an external force due to an applied electric field.

Writing the electric field as $E = E_0 \exp(-\mathrm{i}\omega t)$, we obtain the solution for (2.69). Using the expression for the current density, $j = ne\dot{\phi}/Q$, we obtain the complex conductivity $\sigma_{\mathrm{F}}(\omega)$ as [2.14]

$$\sigma_{\mathrm{F}}(\omega) = \frac{ne^2\omega}{\mathrm{i}m^*(\omega_{\mathrm{F}}^2 - \omega^2 - \mathrm{i}\Gamma\omega)} \quad . \tag{2.70}$$

The complex dielectric constant, $\varepsilon_{\mathrm{F}}(\omega) = \mathrm{i}\sigma_{\mathrm{F}}(\omega)/\varepsilon_0\omega$, is given by

$$\varepsilon_{\mathrm{F}}(\omega) = \frac{\omega_{\mathrm{p}}^2(m/m^*)}{\omega_{\mathrm{F}}^2 - \omega^2 - \mathrm{i}\Gamma\omega} \quad , \tag{2.71}$$

where $\omega_{\mathrm{p}} = (ne^2/\varepsilon_0 m)^{1/2}$ denotes the plasma frequency. The experimentally observed conductivity $\mathrm{Re}\{\sigma_{\mathrm{F}}(\omega)\}$, is given by

$$\mathrm{Re}\{\sigma_{\mathrm{F}}(\omega)\} = \frac{ne^2}{m^*\omega_{\mathrm{F}}} \frac{(\omega/\omega_{\mathrm{F}})^2(\Gamma/\omega_{\mathrm{F}})}{\{1-(\omega/\omega_{\mathrm{F}})^2\}^2 + (\omega/\omega_{\mathrm{F}})^2(\Gamma/\omega_{\mathrm{F}})^2} \quad . \tag{2.72}$$

Figure 2.16 shows the ω dependence of $\mathrm{Re}\{\sigma_{\mathrm{F}}(\omega)\}$. The conductivity at $\omega = 0$ (i.e. the *dc* conductivity) is always zero because of the presence of pinning or locking mechanisms, i.e. $\omega_{\mathrm{F}} \neq 0$.

If we assume the strength of the pinning to be zero, i.e. $\omega_{\mathrm{F}} \to 0$, $\mathrm{Re}\{\sigma(0)\}$ becomes proportional to $1/\Gamma$ and the conductivity at $T = 0$ diverges since Γ

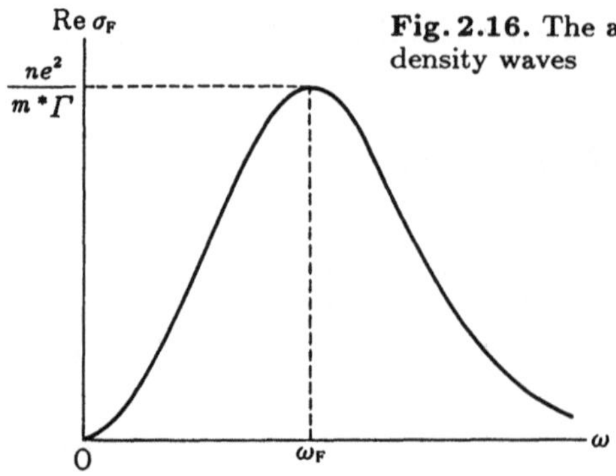

Fig. 2.16. The ac electrical conductivity for pinned charge density waves

goes to zero.[8] In real materials, in which ω_F never vanishes, the conductivity has its maximum at $\omega = \omega_F$ where the CDW resonates under restoring forces due to pinning or locking mechanisms.

The abscissa of Fig. 2.16 can be regarded as an energy axis. Therefore the abscissa is also expected to represent $T - T_P$ and so the point $\omega = 0$ will correspond to the Peierls transition temperature T_P. This naive consideration makes it possible to predict the temperature dependence of the electrical conductivity due to fluctuations of CDWs above T_c, which has not been considered until now [2.16]. We shall investigate this problem again in Chapter 3 in relation to experimental results.

The dielectric constant at $\omega = 0$ becomes

$$\varepsilon_F(0) = \left(\frac{\omega_P}{\omega_F}\right)^2 \frac{m}{m^*} \quad . \tag{2.73}$$

Although m/m^* may be much less than 1, $\varepsilon_F(0)$ is expected to have a very large value when ω_F is small enough, because ω_F is much less than ω_P. This point will be considered in Chap. 3.

It should be emphasized that the equation of motion for ϕ has been derived in this chapter using a crude simplification of pinning or locking potentials. A more exact treatment of the problem has given the result: Re $\{\sigma(\omega)\} = 0$ for $\omega < \omega_F$ in the case of strong pinning [2.13].

e) Phase Solitons and Electrical Conductivity

On the basis of a linear approximation we have solved the nonlinear equation of motion for ϕ. Sometimes, however, we can solve the nonlinear equation exactly; we obtain a solution which represents solitary waves, or solitons.

[8] Strictly speaking $\sigma(\omega)$ becomes $-ne^2/im^*\omega$ when $\omega_F = 0$ and $\Gamma = 0$. This corresponds to the London equation for superconductivity [2.15]. It means that the *dc* resistivity becomes zero.

The solitary wave is a spatially localized wave of distortions, and has the property that it can propagate in a medium without changing its shape. This property arises from the nonlinearity of the medium.

The importance of the nonlinearity is intuitively understood as follows: We consider a wave packet with an arbitrary shape. This wave packet is composed of a linear combination of many waves having different wave numbers. Each component wave is of the form $A(q)\exp(\mathrm{i}\omega t - iqx)$ which is a solution of a linear wave equation. The frequency of a wave having q is not necessarily proportional to q because the medium is usually dispersive. The phase velocity, ω/q, of each wave component, $A(q)\exp(\mathrm{i}\omega t - \mathrm{i}qx)$, is dependent on q. Therefore in general, the wave packet will change its shape with time. The nonlinearity plays an important role in restoring this change.

The solitary wave was found for the first time in water waves and, nowadays, is known to appear also in plasma waves and in lattice waves in some kinds of crystal. The displacement in these solitary waves is described by that of the media. For the solitary wave discussed in this section, however, the displacement is given by the phase ϕ of the CDW. It is called a phase soliton.

When the locking potential is approximated by $\varrho_0 V_0(1 - \cos\phi)$, the equation of motion for ϕ is

$$\ddot{\phi} - v^2\nabla^2\phi + \omega_\mathrm{F}^2 \sin\phi = 0 \quad , \tag{2.74}$$

where we set $\omega_\mathrm{F} \equiv 2\pi v \varrho_0 V_0$. The third term on the left hand side makes the equation nonlinear.

Equations of this form are known as sine-Gordon equations. We obtain a solution $\phi(x,t) = s(\xi)$ by making the Galilei transformation $x \to \xi = x - ut$ and $t \to t' = t$. An integration with respect to ξ, making use of relations $\partial/\partial x = d/d\xi$ and $\partial/\partial t = -u(d/d\xi)$, gives

$$\frac{1}{2}(v^2 - u^2)\left(\frac{ds}{d\xi}\right)^2 = -\omega_\mathrm{F}^2 \cos(s) + c_1 \quad . \tag{2.75}$$

When c_1 is set to $c_1 \equiv \omega_\mathrm{F}^2$, the following is obtained,

$$\frac{ds}{d\xi} = \pm\frac{4\omega_\mathrm{F}}{\sqrt{v^2 - u^2}} \sin(s/4)\cos(s/4) \quad . \tag{2.76}$$

Integration by separation of variables gives the solution,

$$\tan\left(\frac{s}{4}\right) = \exp\left(\pm\frac{\omega_\mathrm{F}}{\sqrt{v^2 - u^2}}\xi + c_2\right) \quad . \tag{2.77}$$

The final solution of (2.74) is obtained as

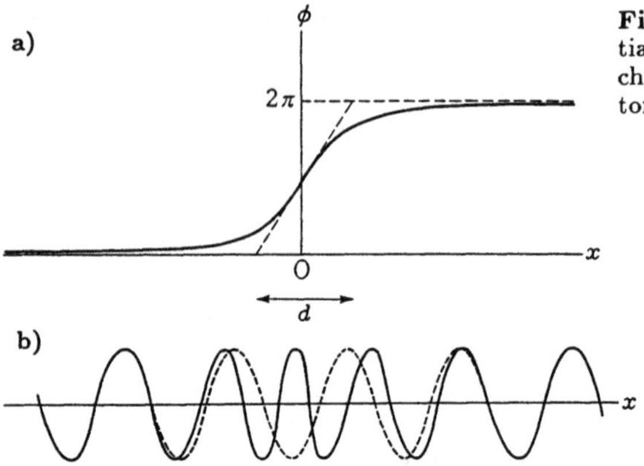

Fig. 2.17a,b. Phase soliton. **(a)** Spatial change in the phase ϕ. **(b)** A charge density wave with a local distortion in the phase

$$\phi(x,t) = s(\xi) = 4\tan^{-1}\left[\exp\left\{\pm\frac{\omega_F}{\sqrt{v^2-u^2}}(x-ut)+c_2\right\}\right] \quad . \tag{2.78}$$

Figure 2.17a shows the shape of $\phi(x,0)$ for $c_2 = 0$ and $t = 0$. The phase ϕ changes by 2π in a region having the size of $d \equiv v/\omega_F$. This part propagtes as the solitary wave. The shape of ϕ shown in Fig. 2.17a is invariant for the evolution of both x and t which satisfy the relation $x - ut =$ constant, namely for the propagation with the velocity u. This is characteristic of solitary waves. Figure 2.17b depicts the CDW corresponding to this variation of ϕ in space.

A certain amount of energy will be necessary to create such a deformation in the CDW. The energy of the soliton is found by substituting the above solution into the Hamiltonian [2.17] giving

$$E_\pm = \frac{Mv^2}{(1-u^2/v^2)^{1/2}} \qquad \text{with} \quad M = 8nm^*/Q^2d \quad . \tag{2.79}$$

This expression is reminiscent of that for the kinetic energy of relativistic particles having the rest mass M. We may say that the phase soliton looks like a relativistic particle. It is to be noted, however, that the maximum velocity is not the light velocity but that of the phason, v.

A phase soliton also carries some electrical charge. For a CDW represented by $\varrho = \varrho_0 \cos(Qx + \phi)$, the wave vector Q is given by $Q = 2k_F = n\pi$ except in regions within a soliton. Since ϕ varies spatially within a soliton, the CDW is expressed as $\varrho = \varrho_0 \cos(Qx + \nabla\phi \cdot x + \phi_0)$ on the assumption of a slow variation of ϕ in space. $Q + \nabla\phi$ corresponds to the local value of $2k_F$ inside the soliton. Thus the local density of charge $n' = n + \delta n$ is given by

$$\text{local } 2k_F = Q + \nabla\phi = (n + \delta n)\pi \quad . \tag{2.80}$$

Finally we obtain

$$\delta n = \nabla\phi/\pi = \nabla\phi n/Q \quad . \tag{2.81}$$

A phase soliton with $\nabla\phi>0$ has an excess charge $\delta n>0$ and that with $\nabla\phi<0$ has $\delta n<0$. The total charge of a soliton is given by integration of δn over the length d of the soliton,

$$\int dn = \int(\nabla\phi/\pi)dx = \int_0^{2\pi}(d\phi/\pi) = \pm 2 \quad . \tag{2.82}$$

Thus each soliton has a charge equivalent to that of two electrons or two holes.

The requirement of charge conservation in a CDW leads to the conclusion that the number of phase solitons, n_ϕ^+, having $\nabla\phi>0$ must be equal to that, n_ϕ^-, having $\nabla\phi<0$,

$$n_\phi^+ = n_\phi^- \quad . \tag{2.83}$$

Phase solitons with $\nabla\phi>0$ and $\nabla\phi<0$ are called solitons and anti-solitons, or ϕ particles and anti-ϕ particles, respectively.

A creation energy of Mv^2 is needed to produce a phase soliton. Once it is created, however, it can propagate with an arbitrary velocity less than v. Therefore when the strength of an applied electric field is increased beyond a certain value corresponding to the creation energy, pairs of ϕ and anti-ϕ particles will be produced. Thermal energies can also create such pairs.

The ϕ particles and anti-ϕ particles so produced will propagate in opposite directions to each other since they are oppositely charged. A certain amount of net electrical current is carried by them.

The electrical conduction by phase solitons will be investigated again in Chaps. 3 and 4 in connection with experimental data.

2.3 Fluctuations and Three-Dimensionality

The discussion in the previous sections was based on the mean-field approximation, which neglects the fluctuations which accompany phase transitions. The observed behaviour is however generally affected by fluctuations near the transition temperature. The temperature range in which fluctuations play an important role is narrow in the case of 3d systems; usually $|T-T_c|/T_c \lesssim 0.1$, where T_c is the transition temperature. In low-dimensional systems, however, we cannot neglect fluctuations even in qualitative discussions since they become large in a wide range of temperature.

2.3.1 Fluctuations in One Dimension

It is well-known that purely 1d systems dominated by short-range interactions have no long-range order at finite temperatures. That is to say, no phase is defined in such systems, and, therefore, no phase transition can occur at finite temperatures [2.18].

We consider this problem using a simple model [2.19]. Figure 2.18 shows a chain of Ising spins in 1d. The spins have a state of either $s = 1/2$ or $s = -1/2$. For nearest-neighbour interactions the total energy of the system is

$$E = -J \sum_{i=1}^{N-1} s_i s_{i+1} \quad , \tag{2.84}$$

where J denotes the energy of the exchange interaction between neighbouring spins. When $J>0$, the ferromagnetic state shown in Fig. 2.18a is stabilized at $T = 0$, and when $J<0$, the antiferromagnetic state is stable. The former case, $J>0$, will be discussed in the following since the essential points are the same for both cases.

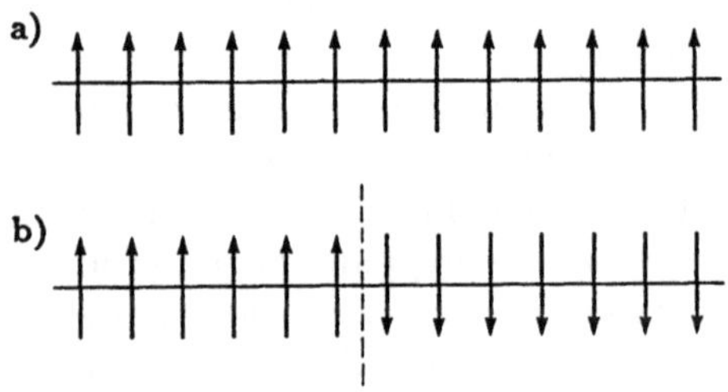

Fig. 2.18
Ising spin chains in one dimension

We assume that the spin chain is divided into two ferromagnetic domains with opposite spin direction as shown in Fig. 2.18b. The interaction energy between two spins across the domain boundary makes the total energy larger than that in the uniform ferromagnetic state by an amount of $2Js^2$. The stabilization of systems at finite temperatures must be discussed in terms of the free energy, $F = E - TS$, where S is the entropy. In the present system the equation $S = k_B \ln N$ gives the entropy arising from the degrees of freedom for the location of the domain boundary. Therefore the increment in the free energy from the uniform ferromagnetic state is written as

$$\Delta F = 2Js^2 - k_B T \ln N \quad . \tag{2.85}$$

Provided the size of the system represented by N is large enough, ΔF will always be negative because the first term is finite. Whenever $T>0$, the uniform ferromagnetic state is unstable, and the spin chain is divided into many domains. We obtain the average domain size, $\xi(T) = Na$, where a is the interspin distance by settig $\Delta F = 0$,

$$\xi(T) \cong a \exp(2Js^2/k_B T) \quad . \tag{2.86}$$

Spins within each domain have strong correlations with each other. However those separated by distances greater than the domain size, $\xi(T)$, have no correlation. The corerlation length $\xi(T)$, goes to infinity as $T \to 0$. Furthermore, the pattern of the many ferromagnetic domains in the system fluctuates in time. The relaxation time (i.e. the mean lifetime) of each pattern also goes to infinity as $T \to 0$. A state having $\xi(T)$ and the relaxation time infinitely long forms a ferromagnetic "phase". Although mathematical expressions for $\xi(T)$ are model dependent, the qualitative result derived here is a general one.

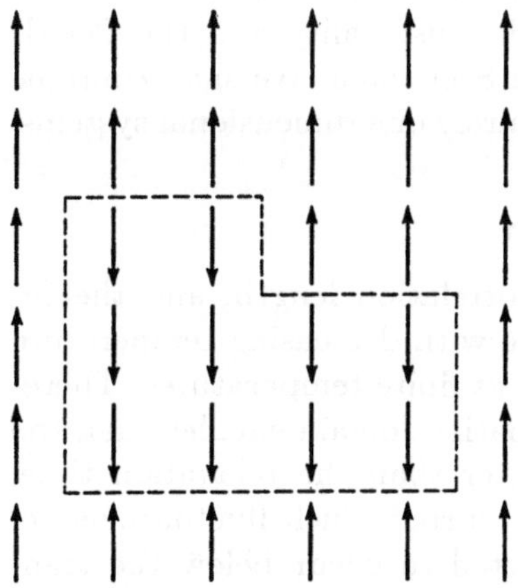

Fig. 2.19. Ising spin lattice in two-dimension

What happens in two-dimensions? Figure 2.19 shows a 2d lattice of Ising spins. We assume that those spins surrounded by the dashed line point in the opposite direction to that of remaining spins. The increment of the interaction energy is of the order of $2Js^2P$, where P is the number of perimeter sites. The entropy is given as $S = k_B \ln 3^P$, because each perimeter site has one emerging line that may connect to any of three nearest neighbours. The increment in the free energy is then

$$\begin{aligned} \Delta F &= 2Js^2P - k_B T \ln 3^P \\ &= P(2Js^2 - k_B T \ln 3) \quad . \end{aligned} \tag{2.87}$$

The uniform ferromagnetic state is stable below the temperature $T_c = 2Js^2/k_B \ln 3$ because $\Delta F > 0$ in this case. A phase transition takes place at $T = T_c$. The situation is found to be similar in 3d where again phase transition can occur at finite temperatures.

Thus we have learned that 1d systems having short-range interactions undergo no phase transition at finite temperatures. The correlation length and relaxation time of the order parameter only increase towards infinity with $T \rightarrow 0$. The amplitude of the order parameter is also known to increase with $T \rightarrow 0$.

The order parameter of the Peierls transition is the band gap, $E_g = 2\Delta$, or the amplitude of lattice distortions with $Q = 2k_F$, $\langle u_{2k_F} \rangle$, or the amplitude of waves of charge density, $\langle \varrho_{2k_F} \rangle$. The interaction dominating the phase transition is the electron-phonon interaction which is a short-range interaction. Therefore the Peierls transition cannot occur at finite temperatures because of the fluctuations characteristic of 1d systems. Real materials, however, are three dimensional; lattice systems remain 3d even when the conduction band in the lattice is purely 1d. In addition, possible Coulomb interactions among CDWs can be a source of the three dimensionality. Hence real conductors may in fact undergo the Peierls transition at finite temperatures. The relationship between three-dimensionality and the Peierls transition will be discussed in more detail in Sect. 2.3.3. We shall continue in the next section with the investigation of purely one-dimensional systems.

2.3.2 Fluctuations and the Pseudo-Gap

In purely 1d systems, the amplitude, the correlation length, and the relaxation time of the order parameter increase with decreasing temperature although no Peierls transition can take place at finite temperatures. Therefore when the system is observed within a spatial domain smaller than the correlation length and on a time scale shorter than the relaxation time, it will look as if the Peierls transition has occurred. Such fluctuations accompanying the Peierls transition are expected to occur below the transition temperature T_P^{MF} calculated by the mean field approximation. For $T < T_P^{MF}$, the band gap, $E_g = 2\Delta$, as an order parameter, fluctuates in time and space. We may say that the electronic state is fluctuating between the metallic and the insulating state. This fluctuation is reflected particularly in the electronic density of states near the Fermi level, where the band gap is expected to open.

The theory of Lee, Rice, and Anderson gives the density of states near the Fermi level as shown in Fig. 2.20 [2.20]. The density of states for $T \sim T_P^{MF}$ is essentially similar to that in the metallic state. On the other hand, that for $T \lesssim T_P^{MF}$ becomes considerably smaller in the energy range $|E - E_F| < \Delta$ thus showing a characteristic of insulators. Such a fluctuating gap is called a pseudo-gap. The system still keeps its metallic character for $T < T_P^{MF}$ because the density of states, $D(E)$, is finite in the pseudo-gap. However, the smallness of $D(E)$ in the pseudo-gap should be reflected in some of the physical quantities which depend upon $D(E)$.

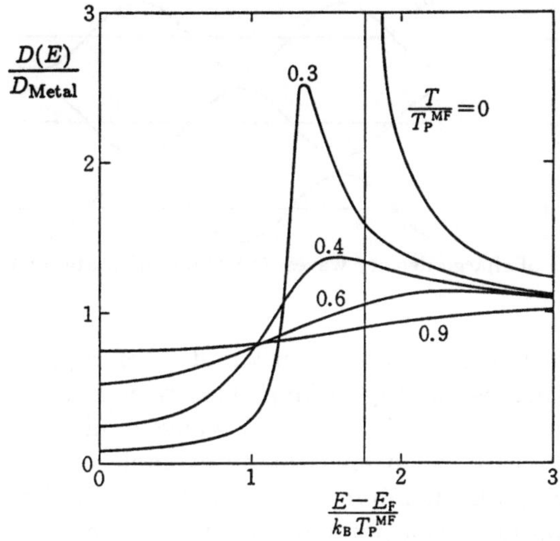

Fig. 2.20. Density of states in the pseudo-gap

2.3.3 Three-Dimensionality and the Peierls Transition

The presence of three-dimensionality in real systems makes it possible for a system to undergo a real Peierls transition at a certain temperature below which the three-dimensionality becomes important. The Peierls transition temperature in real systems is written as T_P. For $T_\mathrm{P}<T<T_\mathrm{P}^\mathrm{MF}$, the fluctuation of an order parameter increases with $T \to T_\mathrm{P}$; for example, the density of states decreases in the pseudo-gap. We can expect that the amplitude of the CDW is sufficiently large near T_P because, as shown in Fig. 2.7, the magnitude of the order parameter itself grows below T_P^MF which is much higher than T_P. What happens at T_P will be the growth of the correlation length and of the relaxation time of the CDW. In other words, when we describe the CDW as $\varrho = \varrho_0(x,t) \cos\{2k_\mathrm{F}x + \phi(x,t)\}$, the correlation function of the amplitude, $\langle\varrho_0(x,t)\varrho_0(0,0)\rangle$, already has a long correlation in time and space for $T_\mathrm{P} \lesssim T \leq T_\mathrm{P}^\mathrm{MF}$. On the other hand the phase correlation function, $\langle\phi(x,t)\phi(0,0)\rangle$, increases in space and time as the temperature approaches T_P.

Thus we may qualitatively describe the process of a real Peierls transition as a condensation of a phase fluctuation such as that shown in Fig. 2.21a into the form shown in Fig. 2.21b as T_P is approached from above [2.21].

How does the phase ϕ of the CDW condense below T_P? First, in the absence of pinning or locking mechanisms, ϕ can have an arbitrary value, i.e. the sliding motion of the CDW occurs without any resistance. One of the factors which gives a definite value of the phase ϕ is the impurity pinning. The phase ϕ takes a value $\phi(x)$ which minimizes the total energy of the system, i.e. the sum of the deformation energy of the CDW and the

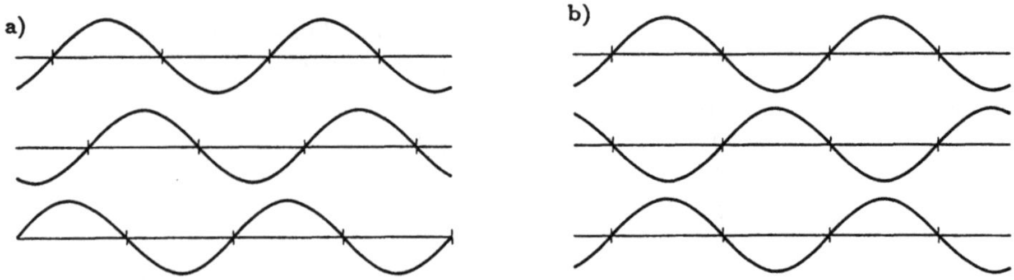

Fig. 2.21a,b. Freezing in of the phase of charge density waves. **(a)** "Molten" state. **(b)** An example of the frozen state

interaction energy with the impurities, for a random distribution of impurities. The second factor is the commensurability locking. This gives a series of possible values, $\phi = 2\pi n/N$ ($n = 0, \pm 1, \pm 2, \dots$) which minimizes the energy. The third possibility is the 3d locking. As an example of this we shall consider the simplest case, a Coulomb interaction between CDWs. The nearest-neighbour interaction favours a phase difference of π between neighbouring CDWs. Although the phase difference is not necessarily π in general, e.g. for long-range interactions, some definite value must be preferred. The absolute value of the phase ϕ, with respect to the lattice system, however, is not determined in contrast to the first two cases.

In real materials, the Peierls transition is driven at a finite temperature T_P by a 3d interaction of the kind described above, and the CDWs form an ordered 3d structure. It is to be noted, however, that the first mechanism, the impurity pinning, is always present to some extent and affects the correlation length of the array of ordered CDWs – both along the 1d axis and the perpendicular directions. When the impurity pinning is strong enough, the phase of the CDWs will be almost entirely determined by the distribution of impurities. Three-dimensional interactions can then only produce a poorly ordered 3d CDW structure with a rather short correlation length. If, however, the impurity potentials are weak compared to the 3d interactions the CDW may still possess a long-range order.

This problem has been investigated theoretically with the following result [2.22]: Three-dimensional long-range ordering is not realized in systems whose dimensionality is lower than three however weak the impurity potential is. Therefore real conductors whose dimensionality is close to one cannot have a perfect long-range order of CDWs at any temperature below the Peierls transition temperature.

Nevertheless the correlation length in many materials is often so long along some direction that experiments cannot detect its finiteness as we will see in later chapters. This suggests that the impurity effect (i.e. impurity density and magnitude of potentials) is not crucial in most real materials. Those materials clearly showing the impurity effect are $TTF_{1-x}TSeF_x$–TCNQ and $K_2Pt(CN)_4Br_{0.3} \cdot 3.2H_2O$ and these will be described later.

2.4 Roles of the Coulomb Interaction

We have neglected until now the Coulomb interactions between conduction electrons on each of the 1d chains. We investigate now the effects which might arise from the inclusion of these interactions. When the Coulomb interaction is absent, electrons can move freely along the 1d chain provided they satisfy the Pauli principle; they gain kinetic energy by forming a conduction band. A Coulomb repulsion increases the electronic energy when two electrons come close to each other. When the Coulomb interaction is not strong however, this increment in energy will be small compared to electrons' kinetic energy and the resultant electronic state will be little different from that in the absence of Coulomb repulsion. However, the situation changes entirely if the Coulomb energy is large.

We shall now try to make our model for the conduction band more realistic in order to take account of the Coulomb interaction among conduction electrons.

2.4.1 The Tight-Binding Model

In the previous sections we have employed the free-electron model to describe electronic states. However, we can also represent a conduction band in terms of wave functions for bound electrons, because the conduction band arises from an overlap of wave functions for electrons originally bound to individual atoms; this is the tight-binding mode [2.1]. According to the problem being tackled, the more suitable of these two models is chosen. The tight-binding model is usually used when realistic band structures are required or when the mean-free path of the conduction electrons is not long enough compared to the interatomic distance for a free-electron model to be used. In this section we consider the tight-binding model for two reasons; firstly, we want to compare the magnitude of the Coulomb interaction energy with that of the kinetic energy and secondly, the mean-free path of carriers in many 1d conductors is found to be only a few times as long as the interatomic distance.

We introduce the wave function $\phi_{\mathrm{a}}(x-l)$ for an electron on an isolated atom located at l. The wave function of an electron on a 1d chain of such atoms is given by

$$\psi_k(x) = \frac{1}{\sqrt{N_i}} \sum_l \exp(\mathrm{i}kl)\phi_{\mathrm{a}}(x-l) \quad , \tag{2.88}$$

where N_i is the total number of atoms in the chain. It is evident that $\psi_k(x)$ satisfies the Bloch condition. This wave function makes it possible to calculate the expectation value of the Hamiltonian in the crystal, $H = (-\hbar^2/2m)\nabla^2 + V(x)$. Introducing an interatomic distance, i.e. the lattice parameter a, we obtain the following,

$$\begin{aligned} E_k &= \int \psi_k^*(x) H \psi_k(x) dx \\ &= \sum_n \exp(-ikna) \int \phi_a^*(x-na) H \phi_a(x) dx \quad . \end{aligned} \tag{2.89}$$

If we assume that the extent of the wave functions $\phi_a(x)$ is so small that only the terms, $n = 0$ and $n = \pm 1$, are important in the above equation, we can easily obtain a solution,

$$\begin{aligned} E_k &= -\varepsilon_0 - 2t_{\|} \cos(ka) \quad , \\ \varepsilon_0 &= -\int \phi_a^*(x) H \phi_a(x) dx \quad , \\ t_{\|} &= -\int \phi_a^*(x-a) H \phi_a(x) dx \quad , \end{aligned} \tag{2.90}$$

where $t_{\|}$ is called a transfer integral. Figure 2.22 depicts the cosine band thus obtained. Its band width is $4t_{\|}$. The energy near $k = 0$ is approximated as

$$E_k \cong -\varepsilon_0 - 2t_{\|} + t_{\|} k^2 a^2 \quad , \quad (ka \ll 1) \quad . \tag{2.91}$$

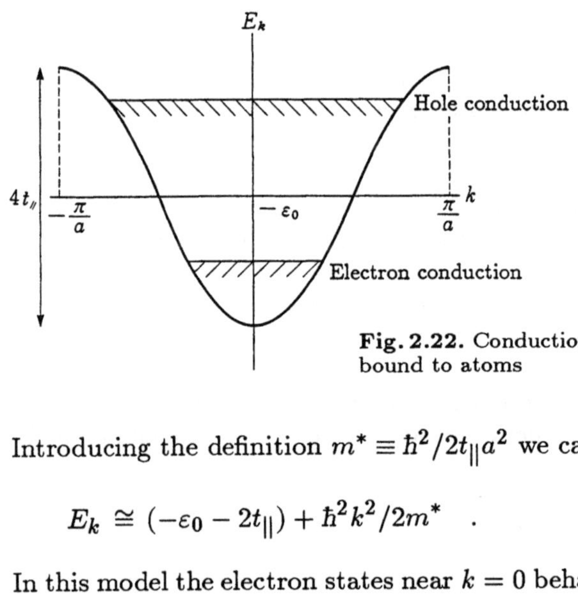

Fig. 2.22. Conduction band in the model of electrons bound to atoms

Introducing the definition $m^* \equiv \hbar^2/2t_{\|}a^2$ we can rewrite the above as

$$E_k \cong (-\varepsilon_0 - 2t_{\|}) + \hbar^2 k^2/2m^* \quad . \tag{2.92}$$

In this model the electron states near $k = 0$ behave like free electrons having the effective mass m^*. If the electrons are highly localized this makes $t_{\|}$ small and m^* large, as one would expect for less mobile particles.

The effective mass of general k is written as $1/m^* \equiv (d^2 E_k/dk^2)/\hbar^2$ giving rise, in this model, to a positive (or zero) mass, $m^* \geq 0$, for $|k| \leq \pi/2a$

and a negative mass, $m^*<0$, for $|k|>\pi/2a$. Therefore a tight-binding band which is more than half full gives rise to hole conduction and when less than half full to electron conduction.

2.4.2 Short-Range Coulomb Interactions

How are electronic states in the conduction band modified by the Coulomb interaction between electrons? First, we shall consider the so-called Hubbard model [2.23] which is suitable for a short-range interaction due to screening of the Coulomob potential. This model is based on the tight-binding model in the presence of a Coulomb repulsion U that acts only between two electrons on the same atom. The Coulomb interaction U must be taken into account when it is not negligible compared to the kinetic energy of the electrons. The kinetic energy to be compared with U is represented by the band width $4t_{\|}$. The reason is as follows. When we say that an electron is present on an atom, we have in mind a localized picture for the electron. The wave function for this localized electron is composed of a linear combination of the ψ_k of (2.88). Therefore the energy of the localized electron is represented approximately by the maximum energy, $4t_{\|}$, available in the tight-binding model.

When the Coulomb repulsion U is present, electrons become less mobile than for the case $U = 0$. This can be understood as follows. We shall assume that the electrons with a density of 1 electron/atom are present on a linear chain of atoms whose interatomic distance is a. The electrons will occupy half of the band as shown in Fig. 2.23. Then we add another electron to the system. This electron will increase its potential energy by an amount U whichever atom it enters, because each atom is already occupied by an electron. In the band model this electron will not occupy the state A shown in Fig. 2.23a, but another state which is higher in energy than A by the amount of U.

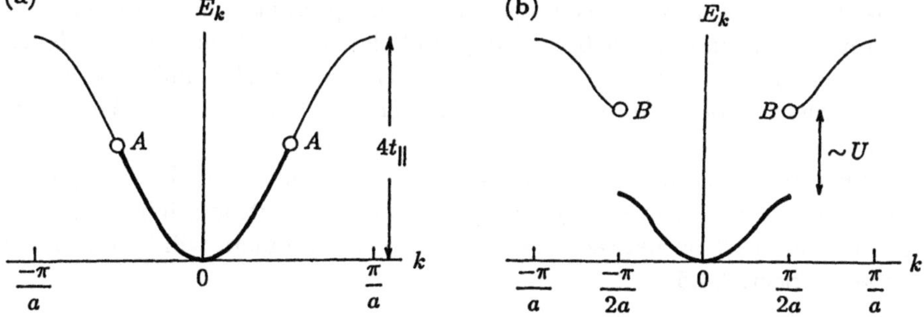

Fig. 2.23a,b. Structure of the tight-binding band with a width $4t_{\|}$. The thick and thin curves represent occupied and empty states, respectively. The electron density is 1 electron/atom. **(a)** $U = 0$. **(b)** $U \neq 0$

According to the results of theoretical calculations, the band is reconstructed as shown in Fig. 2.23b. The band splits at $k = \pm\pi/2a$ and the newly added electron enters the state B which is separated from lower states by a gap. This gap is known as a Hubbard gap and its magnitude is of the order of U. Comparison of Fig. 2.23b with part a suggests that the band width becomes narrower than the original width by the amount of $\sim U$.

When the electron density is less than 1 electron/atom, the lower band in Fig. 2.23b is partially occupied and, therefore, the system remains metallic. However, the electrons are less mobile than the case of $U = 0$ since the bandwidth is narrower. With increasing $U/4t_{\|}$, the bandwidth is expected to become narrower still giving rise eventually to nearly localized electron states.

The nearly localized electrons are expected to be distributed equidistantly on the linear chain of atoms. We shall consider the case of an electron density of 1 electron/3 atoms. The wave functions of each electrons will extend on average over three neighbouring atoms. However, a tail of the wave function may enter the 'territory' of the neighbouring electron resulting in a slight increase of the Coulomb energy. When the atomic chain suffers a periodic distortion with the period of $3a$ as shown in Fig. 2.24, the electron will be attracted to the triad of atoms. The electron wave function will become more confined within the triad giving rise to a decrease in the Coulomb energy between neighbouring electrons.

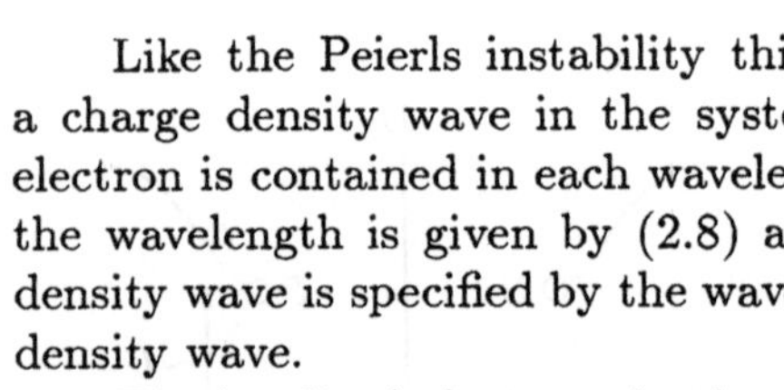

Fig. 2.24. Localized-electron picture for $4k_F$ charge density waves

Like the Peierls instability this kind of mechanism can bring about a charge density wave in the system. In the present case, however, one electron is contained in each wavelength of the charge density wave. Hence the wavelength is given by (2.8) as $\lambda = 1/n = 2\pi/4k_F$ and this charge density wave is specified by the wave number $4k_F$. It is named a $4k_F$ charge density wave.

The localized electrons in the $4k_F$ charge density wave have unpaired spins. The system may undergo a magnetic phase transition originating in an exchange interaction between the electrons. This problem will be discussed further in Sect. 2.4.5.

2.4.3 Long-Range Coulomb Interactions

What is expected to occur when the force range of the Coulomb repulsion between conduction electrons is not restricted to within each atom but reaches far beyond next neighbours? We denote the Coulomb repulsion between an electron on the ith atom and another on the jth atom by V_{ij}. The assumption $V_{ij}/4t_{\|}\gg1$ $(i-j<\infty)$ is again used in order to obtain qualitative results. If V_{ij} is a long-range interaction such as a bare Coulomb interaction, all electrons repel each other. In the case of the short-range repulsion, U, it was meaningful to use the band model because in this electrons can move freely unless two of them come onto the same atom. In the present case, however, V_{ij} is larger for two electrons distant from each other than the energy $4t_{\|}$ gained by a free motion. All electrons will become localized in such a way as to minimize the total Coulomb energy. The resultant configuration of electrons is easily obtained: they are located equidistantly from one another.

The ordered structure of electrons thus brought about is called a Wigner crystal [2.25] of electrons. The apparent structure is the same as that shown in Fig. 2.24; it is specified by the wave number $4k_F$. In the case of the short-range Coulomb repulsion discussed in Sect. 2.4.2, the lattice spontaneously distorts due to the Peierls instability. However, the present case is different from that. The lattice distortion here is driven by electrons which tend to distribute themselves equidistantly. Nevertheless we may say that an apparent $4k_F$ CDW appears [2.26].

If $4t_{\|}=0$, the system is of an insulating character. When $4t_{\|}\neq0$, however small it is compared to V_{ij}, electrons will more or less be able to move. This is particularly so when the electron density is low and hence the mean distance between neighbouring electrons is long, in which case the movement of an electron to its neighbouring site will cuase only a small increment in the Coulomb energy. A certain amount of electrical conduction is expected to occur due to excitations of electrons by, for example, thermal motion although the system may not show a metallic character [2.27].

2.4.4 Coulomb Interactions and Charge Density Waves

In the last two sections we have investigated the role of Coulomb interactions in the limit where the Coulomb energy is much larger than the band energy. The conclusion reached is that this can produce a $4k_F$ CDW. Such an extreme situation however, is not thought to be realized in any real 1d conductor. Furthermore in all real cases the force range of the Coulomb interaction is most probably intermediate. A problem to be investigated experimentally with real materials is the interrelation between the $2k_F$ CDW arising from the electron-phonon interaction and the $4k_F$ CDW due to the strong Coulomb repulsion.

As we will see in Chap. 3, any material in which both the $2k_F$ and the $4k_F$ CDW appear, has two kinds of one-dimensional chain. Therefore a possibility is that one of the two types of chain has a weak interaction among electrons giving rise to the $2k_F$ CDW while the other, having a strong Coulomb repulsion, brings about the $4k_F$ CDW.

Another possibility, however, must also be remembered, i.e. that the $2k_F$ and the $4k_F$ CDW may coexist on a single kind of 1d chain [2.28]. In such a case, one of three waves, a $2k_F$ CDW, a $4k_F$ CDW, and a $2k_F$ spin-density wave (SDW) to be discussed in Sect. 3.6.4 will be dominant depending on the relative magnitudes of the electron-phonon interaction, the Coulomb interaction between electrons, and the interchain interaction. However, waves other than the dominant type can coexist. Theories have shown that a $2k_F$ CDW can grow in spite of a presence of the electron-electron Coulomb repulsion, and that its growth is assisted by the interchain interaction [2.28].

2.4.5 The Spin-Peierls Transition

Properties characteristic of 1d conductors are the appearance of a $2k_F$ CDW and the occurrence of a Peierls transition. In this case a site, i.e. one wavelength of the CDW, admits two electrons whose total spin is zero. In contrast to this, a $4k_F$ CDW, when it condenses, locates one electron on a site. There then remains a spin degree of freedom. The system can form an antiferromagnetic state with further decreasing temperature. When the lattice is allowed to distort, however, electron pairs made by a lattice distortion with wave number $2k_F$, may produce a state with a total spin =0. This is called a spin-Peierls transition. Crudely speaking the transition occurs when a gain in magnetic energy overcomes an energy loss due to the lattice distortion. This explanation parallels that of the normal Peierls transition. The wave number for the spin-Peierls transition is again $2k_F$.

A spin Hamiltonian for the exchange interaction is given by the Heisenberg model for a linear chain of spins [2.29]

$$H = -J \sum_{i=1}^{N/2} (s_{2i}s_{2i-1} + \alpha s_{2i}s_{2i+1}) \quad , \quad 0 \leq \alpha \leq 1 \quad . \tag{2.93}$$

The case with $\alpha = 1$ describes the state in which the $4k_F$ CDW is present and electrons are separated by an equal distance of $2\pi/4k_F$. A negative J value gives an antiferromagnetic ground state. When $\alpha \neq 1$, the separation of electrons alternates between two values (i.e. they are paired) and a magnetic phase transition called a spin-Peierls transition can take place. We consider the case of $J<0$ in the following.

Excitations in the spin system are well described by antiferromagnetic spin waves [2.30]. The dispersion relation for $\alpha = 1$ is shown in Fig. 2.25a.

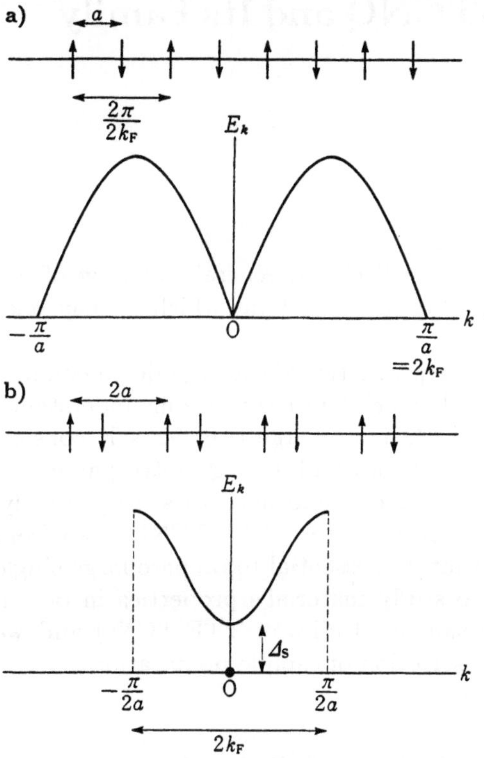

Fig. 2.25a,b. Spin-Peierls transition. **(a)** Excitation energy in a Heisenberg-spin chain with an antiferromagnetic interaction. **(b)** Excitation energy for a distorted lattice

When the lattice is distorted with the wavelength $2\pi/2k_F$, α will become less than one. Various kinds of approximate calculations and exact soutions for small N have given a dispersion relation for this case of the form shown in Fig. 2.25b [2.29]. Note the presence of an energy gap Δ_s at $k = 0$ between the ground state, $E = 0$, and the excited state, $E>0$.

The free energy of the system at $T = 0$ is lower for $\alpha<1$ than for $\alpha = 1$. The reason is that excitations can occur near $E = 0$ when $\alpha = 1$ due to the zero-point energy. When a gain in the free energy for $\alpha<1$ overcomes an energy loss due to the lattice distortion with the wave number $2k_F$, the lattice distortion takes place spontaneously. Spins form pairs and give rise to a state having zero total spin. This is the spin-Peierls transition. The excited states are triplet states with $s = 1$, separated from the ground state, $s = 0$. The magnetic susceptibility below the spin-Peierls transition temperature has a temperature dependence specified by an activation energy Δ_s.

The spin-Peierls transition gives rise to a periodic lattice distortion with the wave number $2k_F$. When we look at eperimental results, therefore, we must be careful to ascertain whether a $2k_F$ lattice distortion originates from a Peierls transition or from a spin-Peierls transition.

3. Properties of TTF-TCNQ and Its Family

In this and the following chapters we will give a general overview of experimental results and their interpretation for systems which display one-dimensional electronic conduction.

This chapter describes mainly the properties of the organic 1d conductor TTF-TCNQ, whose study has led to great progress in our understanding of 1d electron systems. TTF-TCNQ is a charge transfer salt consisting of molecules of TTF (tetrathiafulvalene) and TCNQ (tetracyanoquinodimethane), which serve as electron donors and acceptors, respectively. More experimental data have been obtained from TTF-TCNQ than from any other 1d conductor. We know that it is essential to prepare large single crystals of good quality in order to study material's properties in detail. This has again been proved important in studies of TTF-TCNQ and we shall begin our discussion with a description of sample preparation.

3.1 Molecular Synthesis and Crystal Growth

Organic materials are usually good insulators. We know, however, that there is a group of such materials in which carriers created, for example, by optical excitation have a particularly large mobility. These are the organic semiconductors; anthracene is an example. Studies of organic semiconductors have suggested that charge transfer salts have a high conductivity.

TCNE (tetracyanoethylene) was the first material to be investigated in detail as a potential acceptor of electrons [3.1]. In complexes of TCNE however, no compound was found to show a metallic conduction. In 1960 a research group in Du Pont produced TCNQ for the first time by modifying TCNE [3.2]. The maximum conductivities in the family of organic conductors in those days, of the order of $10^4\ \Omega^{-1}\ \mathrm{m}^{-1}$ were found in Q(quinolinium)-TCNQ_2, NMQ (N-methylquinolinium)-TCNQ, etc.

On the other hand little effort was made at that time to produce new electron donors. In 1970, however, TTF was synthesized by *Wudl* et al. [3.3] and the complex TTF-TNCQ was found to show a very high electrical conductivity. Studies up to this stage had usually been made on polycrystalline samples. In 1973 *Garito, Heeger* and their co-workers [3.4] investigated properties of single-crystal samples. They found a high anisotropy in the electrical conductivity and an unusually rapid increase of the conductivity with decreasing temperature. They proposed a model of fluctuating high-temperature superconductivity for TTF-TCNQ, which attracted much attention from many researchers. Several hundreds of papers have now been published about TTF-TCNQ and related compounds. The possibility of high temperature superconductivity in TTF-TCNQ has subsequently been discarded, but many other interesting phenomena, such as those described in Chap. 2, have been discovered in TTF-TCNQ.

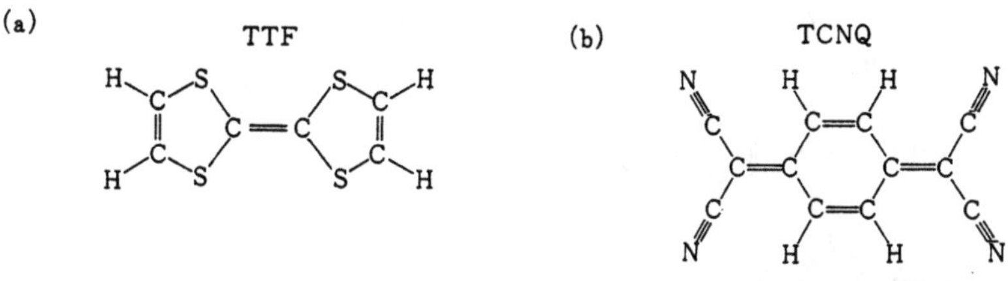

Fig. 3.1. Structure of TTF and TCNQ molecules

First, we shall describe the method of synthesizing TTF molecules. Figure 3.1a shows the molecular structure of TTF. Several methods have been used in synthesizing TTF molecules [3.5, 6]. Figure 3.2 shows one of them as an example. This process was employed by *H. Anzai* and its various steps were reported in a series of articles [3.5, 6]. The starting materials, NaCN and CS_2, are converted into the final TTF molecules after eight steps of chemical reaction. A particular advantage of this method is that the necessary reagents are readily available; it also gives a good yield. If the average yield of each process of the chemical reaction is α, the total yield after eight stages is α^8. Therefore it is crucially important to reduce the number of stages and to increase the yield of each one. As a typical example of the total yield, 13.2 g of TTF was obtained from 735 g of NaCN and 1142 g of CS_2.

TTF molecules thus synthesized[1] may be used in crystal growth after having been purified by sublimation and recrystallization. A commonly used solvent is acetonitrile.

Figure 3.1b shows the molecular structure of TCNQ. This compound may be synthesized by a process such as that shown in Fig. 3.3. In studies of TTF-TCNQ chemists have usually used commercially available TCNQ molecules. In purification the same method of recrystallization is employed, again using acetonitrile as a solvent.

Five methods are popular for the growth of single crystals: 1) a slow cooling method, 2) an evaporation method, 3) a diffusion method, 4) an

[1] Recently TTF has become commercially available.

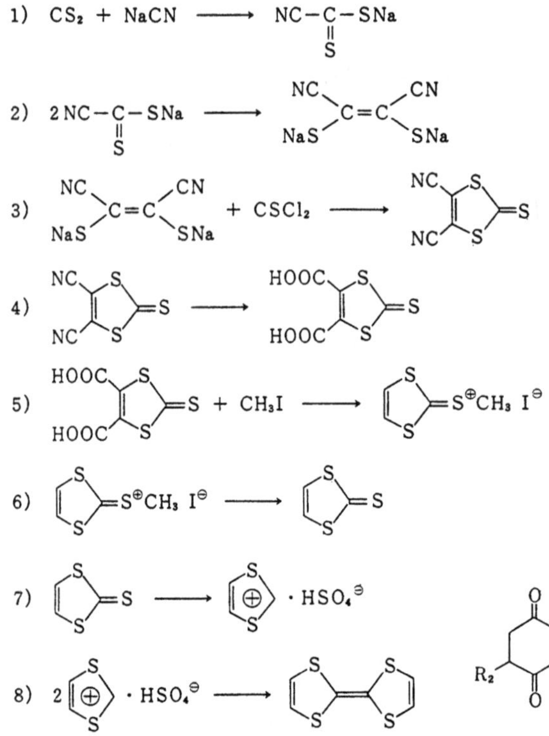

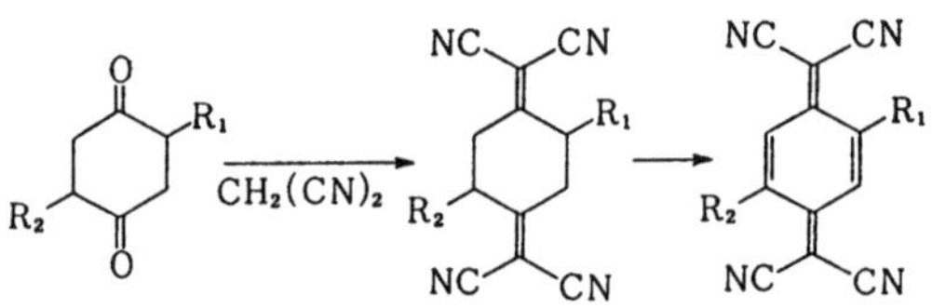

Fig. 3.2. An example of the synthesizing process of TTF molecules

Fig. 3.3. Outline of process of synthesizing TCNQ molecules

electrochemical method, and 5) a vapour-phase epitaxial method. Methods 1) and 2) are similar to each other: A mixture of TTF and TCNQ dissolved in acetonitrile reacts and gives rise to a green solution of TTF-TCNQ. Single crystals of TTF-TCNQ appear when the solution becomes saturated by slow cooling. The second method utilizes an evaporation of the solvent to saturate the solution. This is frequently used because of simplicity. Although crystals thus obtained are not big, their typical size is sufficient for resistance measurements etc. to be made. A method to obtain big crystals is 3) the diffusion method.

Figure 3.4a shows schematically the design of the apparatus used by *Anzai* [3.7] for his crystal growth. A colour photograph of the real apparatus is also shown in the frontis of this volume. Molecules of TTF and TCNQ are added to spatially separated parts of the solvent. The molecules diffuse into the midway region where they react to form TTF-TCNQ. A density gradient of TTF-TCNQ is thus present in the solution and when the region of highest density reaches saturation, crystal growth will take place selectively in this part. Therefore only a small number of crystalline nuclei are formed and

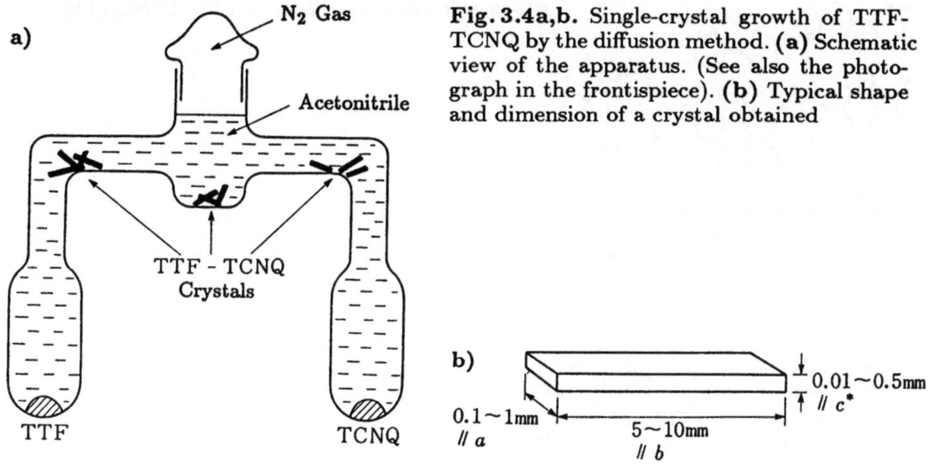

Fig. 3.4a,b. Single-crystal growth of TTF-TCNQ by the diffusion method. (**a**) Schematic view of the apparatus. (See also the photograph in the frontispiece). (**b**) Typical shape and dimension of a crystal obtained

this is expected to encourage the growth of big crystals although the total number of crystals thus obtained may be small. A typical shape and size of a crystal grown by this method are shown in Fig. 3.4b.

In the electrochemical method, 4), TTF and NMQ-TCNQ are dissolved in the acetonitrile. A small electrical current through platinum electrodes induces the reaction $\mathrm{TCNQ}^- \rightarrow \mathrm{TCNQ} + \mathrm{e}^-$ and $\mathrm{TCNQ} + \mathrm{TTF} \rightarrow$ TTF-TCNQ. This method does not on the whole produce very big crystals. The vapour-phase epitaxial growth, 5), starts from polycrystals which are sublimed at about 80° C in vacuum. A film of TTF-TCNQ then grows epitaxially on the cleaved surfaces of rock salts which are maintained at a temperature of about 70° C. The films are usually used for optical measurements although their crystallinity is not very good [3.8].

Figure 3.5 shows the crystal structure of TTF-TCNQ [3.9]. The crystal belongs to the monoclinic space group, $P2_1/c$. Its lattice parameters at room temperature are $a = 12.298$ Å, $b = 3.819$ Å, $c = 18.468$ Å, and $\beta = 104.46°$. The density is $1.62\,\mathrm{g\,cm^{-3}}$ and each unit cell contains two formula units of TTF-TCNQ. Figure 3.5 clearly shows that the two species TTF and TCNQ are segregated to form separate columns along the b axis. (This structure is sometimes called a chain structure.) The molecules stack with a definite tilt angle, the existence and value of which seem to be determined by the van der Waals force. In fact readers may make a simulation of the stacking using two molecular models. The stacking will prefer just the tilting shown in Fig. 3.5. A structure characteristic of TTF-TCNQ is also found in the alignment of the same kind of molecules along the c axis. The b axis is the one displaying 1d electronic behaviour. An electrical anisotropy is more or less expected to be observed in many kinds of organic crystals since constituent molecules usually have a geometrical anisotropy.

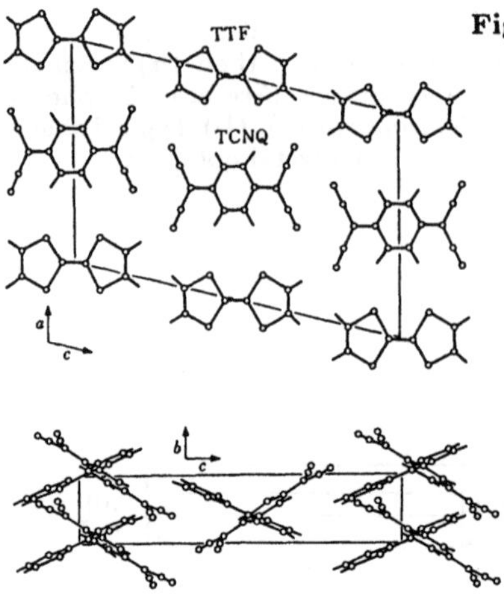

Fig. 3.5. Crystal structure of TTF-TCNQ [3.9]

3.2 Electrical Properties

3.2.1 The One-Dimensional Conduction Band and the Electrical Conductivity

The conduction band giving rise to the 1d conduction in TTF-TCNQ is formed by the wave functions of π-electrons on TTF and TCNQ molecules. Figure 3.6 shows the form of atomic bonding in neutral TCNQ and TTF. Those electrons participating in the bonding are the $2s^1 2p^3$ electrons of carbon atoms, $2s^2 2p^3$ of nitrogen, and $3s^2 3p^4$ of sulphur which now enter new orbitals made by s-p mixing. The π-orbitals of each atom are elongated perpendicular to the molecular plane and overlap with others of neighbouring molecules stacked above and below. It is unnecessary to say that these π-orbitals extend over the entire size of each molecule within the molecular plane. Figure 3.7 depicts the overlap of π-orbitals in a TCNQ stack. Electrons donated to neutral TCNQ molecules are expected to move freely via the orbital overlap along the 1d axis (the b-axis) thus giving rise to a conduction band.

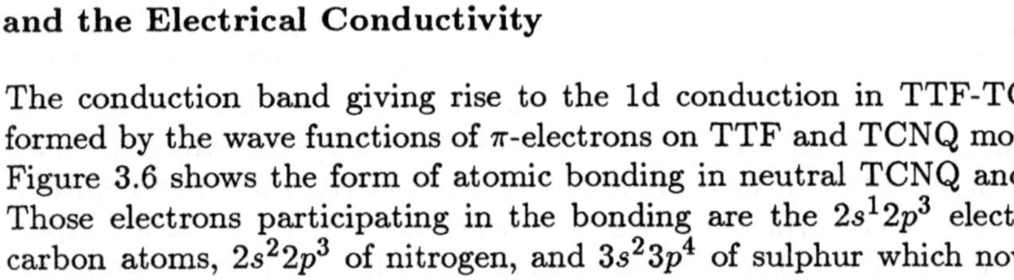

Fig. 3.6a,b. Bonding in molecules of (a) TCNQ and (b) TTF

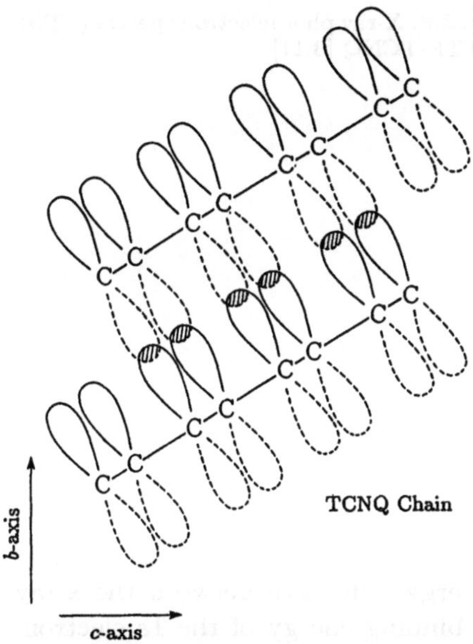

Fig. 3.7. Overlap of π-orbitals in a molecular stack of TCNQ

A similar situation is realized in the molecular stacks of TTF. A difference here however, is that vacancies in the π-orbitals of sulphur atoms due to the electron donation to TCNQ, contribute to the formation of the conduction band (hole band) in TTF stacks.

The band width is dominated by a transfer integral of electron wave functions. The band width $4t_{||}$ of TTF-TCNQ becomes large in the direction of the b-axis due not only to the short intermolecular distance in this direction but also because of the large spread of the wave functions. Although we can formally define a "band width" $4t_{\perp}$ perpendicular to the 1d direction, it is expected to be very small. In fact the conduction band model is meaningless along those directions in which conduction occurs via hopping.

The Fermi wave number in conduction bands of TCNQ and TTF stacks is determined by the number of electrons, N', transferred from TTF to TCNQ. Therefore it is important for discussions of the properties of CDWs that we can ascertain the valence of TTF and TCNQ molecules. This can be done by two methods: 1) via measurements of the density distribution of electrons in molecules by precise x-ray diffraction studies [3.10] and 2) by x-ray photoelectron spectroscopies (XPS) [3.11].

The first method determines the number of electrons on each kind of molecule by measuring the electron density distribution in a unit cell and integrating it over the TTF and TCNQ molecules. These experiments have given the number $N' = (0.48\text{–}0.60) \pm 0.15$. The second method mesures the kinetic energy of photoelectrons emitted by a crystal of TTF-TCNQ when, for example, the $1s$ electrons of nitrogen atoms are excited by monochro-

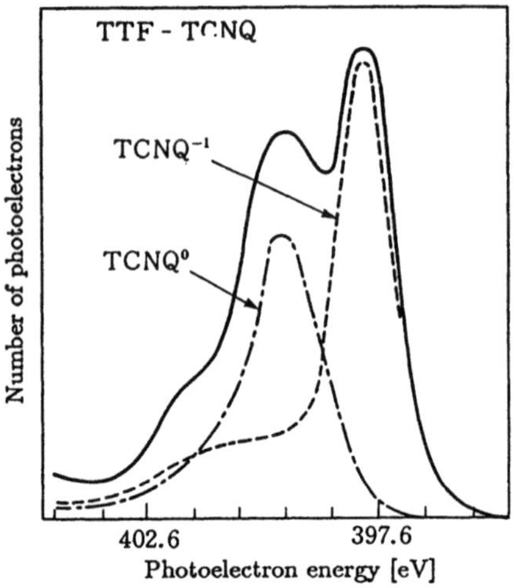

Fig. 3.8. X-ray photoelectron spectra (XPS) of TTF-TCNQ [3.11]

matic x-rays of a known energy. The energy difference between the x-ray photon and the photoelectron gives the binding energy of the $1s$ electron. (The work function must of course be taken into account.) The energy level of the $1s$ electron can then be related to the valence of the TCNQ molecules.

Figure 3.8 shows a spectrum of photoelectrons excited from the $1s$ state of nitrogen in TTF-TCNQ [3.11]. Roughly speaking it can be reproduced by a weighted sum of two spectra obtained from K-TCNQ, whose valence is known to be $TCNQ^{-1}$, and from neutral TCNQ crystals. The valence of TCNQ is obtained from the weight used for the best fit as $N' = 0.56$–0.67.[2] These results may appear to be incompatible with the metallic property of TTF-TCNQ because they mean that the apparent valence N' is essentially a mixture of those of $TCNQ^{-1}$ and $TCNQ^{0}$. However, when we consider the time scales relevant to the experiments, this removes the apparent contradiction: The band width $4t_{\|}$ is estimated to be about 0.5 eV as shown later. An electron will stay on a molecule for a time of the order of $\hbar/4t_{\|} \sim 10^{-15}$ s since the electron if assumed to be localized on the molecule, must have a kinetic energy of $\sim 4t_{\|}$. On the other hand, in the XPS measurements, the time necessary for the x-ray photon to discriminate two valence states, -1 and 0, is of the order of $\hbar/\Delta E$, where ΔE is the difference in the energy levels of valence electrons in the two states. $\hbar/\Delta E$ is estimated to be $\sim 6 \times 10^{-16}$ s because ΔE reads about 1 eV (1.6×10^{-19} J) in Fig. 3.8. Thus the measurement time in XPS is shorter than the residence time of conduction electrons on a molecule; XPS should therefore observe a mix-

[2] It should be pointed out that the spectra from TCNQ are not pefectly reproduced by the weighted sum of two spectra $TCNQ^{0}$ and $TCNQ^{-1}$. Therefore the evaluated value N' must not be taken too seriously.

ture of two valence states, 0 and −1, in spite of the metallic properties of TTF-TCNQ.

Similar measurements on $2p$ electrons of sulphur atoms in TTF have given the value $N' = 0.56 \pm 0.05$. As we will see in a later section, measurements of the values of $2k_F$ and $4k_F$ of CDWs provide $N' = 0.55$ (room temperature) and $N' = 0.59 \pm 0.01$ ($T \leq 150$ K). These values are consistent with each other.

The conduction band accepts two electrons per molecule when the Coulomb interaction among electrons is negligible compared to the band width $4t_{\|}$. Therefore, electrons with the density $N' = 0.59$ fill the band up to 0.59/2. Since $N' < 1$, an electron conduction and a hole conduction are expected to occur in the TCNQ and in the TTF bands, respectively. Figure 3.9 shows the temperature dependence of the dc electrical conductivity measured along each axis of single crystals of TTF-TCNQ [3.12]. The conductivity along the c-axis, σ_c, is not shown because of ambiguities due to a small sample size in this direction. However, σ_c seems to be a little larger than σ_a. Anisotropy of the conductivity, σ_b/σ_a and σ_b/σ_c, are of the order of 100. The b-axis is the 1d axis as expected from the crystal structure. The absolute

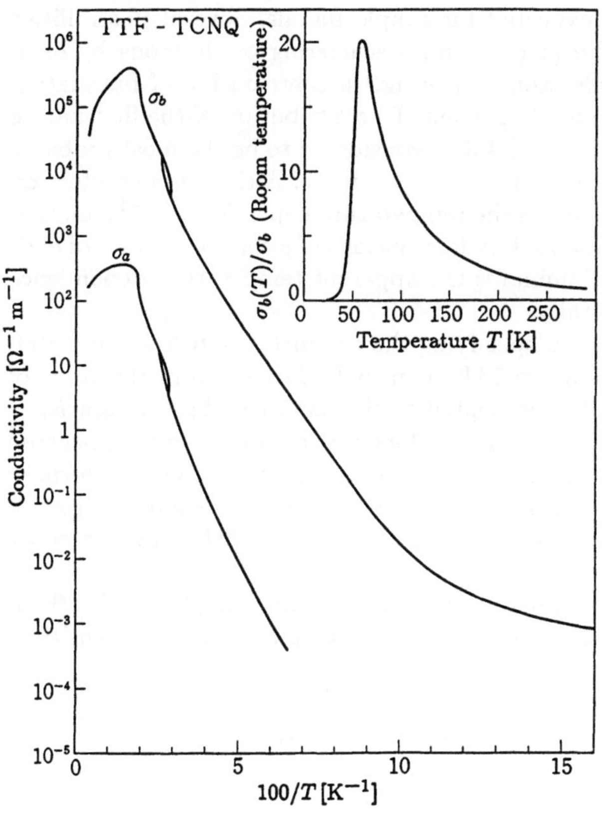

Fig. 3.9. Temperature dependence of the *dc* electrical conductivity of TTF-TCNQ [3.12]

value of σ_b at room temperature is (5–19) $\times 10^4\,\Omega^{-1}\,\mathrm{m}^{-1}$. With decreasing temperature σ_b increases, reaching a maximum value of $\sim 10^6\,\Omega^{-1}\,\mathrm{m}^{-1}$ at about 58 K, below which it decreases rapidly. We may say that a metal-insulator transition occurs at the temperature where the conductivity curve of Fig. 3.9 goes down most rapidly, namely where $-d(\log\sigma)/d(1/T)$ has the maximum value. In fact $-d(\log\sigma)/d(1/T)$ of TTF-TCNQ reaches the maximum at 53 K, which agrees with the Peierls transition temperature estimated from the behaviour of the CDW. The conductivity below 53 K is described in terms of an activation energy. A hysteresis found at 38 K corresponds to a structural change in the CDW to be discussed in Sect. 3.4.

The temperature dependence of σ_b above 58 K cannot be explained in terms of the simplest model for metallic conduction, namely a scattering mechanism by phonons. In TTF-TCNQ σ_b shows a stronger temperature dependence than that expected from the scattering by phonons, $\sigma \propto T^{-1}$. It is found that the temperature dependence of σ_b above 80 K is described by an empirical formula $\sigma_b \propto T^{-x}$ where $x = 2.3$.

We may say that TTF-TCNQ has a metallic character because of the large absolute value of the conductivity σ_b and of its increase with decreasing temperature. However there remains a problem about the scattering mechanism which brings about the strong temperature dependence of σ_b – stronger than the T^{-1} expected for simple metals. Several possibilities have been proposed for the mechanism: a scattering of electrons by optical phonons, an electron-electron scattering,[3] a contribution of fluctuating CDWs to the conduction, etc. At present the contribution of the fluctuating CDW (discussed further in Sect. 3.4.3) is considered to be the most probable mechanism for the strong temperature dependence. Although a conduction due to single particle process has the temperature dependence T^{-1}, another type of conduction brought about by fluctuations of a collective state of electrons, namely the CDW, dominates the apparent temperature dependence $T^{-2.3}$ observed experimentally.

The activation energy, Δ, specifying the conductivity below the metal-insulator transition temperature 53 K, is usually derived from the slope of the curve shown in Fig. 3.9. It is related to the band gap E_g through $E_g = 2\Delta$. In the present case, however, Δ must be temperature dependent because the gap should begin opening at 53 K. With an assumed functional form of $\sigma_b = \sigma_b^0 \exp(-\Delta(T)/T)$, where $\Delta(53\,\mathrm{K}) = 0$, the activation energy, Δ, derived from Fig. 3.9 is extrapolated to 0 K as $\Delta(0)$=200 K and, therefore, $E_g = 400$ K [3.13].

The band gap $E_g(T)$ should satisfy a condition $dE_g(T)/dT \to 0$ for $T \to 0$ since E_g is considered to be an order parameter of the transition.

[3] It is known, however, that the electron-electron scattering in an exactly one-dimensional band also gives a temperature dependence, T^{-1} for the conductivity.

However, $\Delta(T)$ derived above does not satisfy this condition. One reason for this seems to be found in the anomalous temperature dependence of σ_b for $T \to 0$, which is not represented by the above exponential function. This problem will be discussed again in Sect. 3.2.4.

The transition at 53 K is considered to be the Peierls transition characteristic of electron systems in 1d. The strongest evidence is found in the observation of the $2k_F$ CDW as described in detail in Sect. 3.4. Here one should note the following: When the $2k_F$ CDW appears, a bandgap called a Peierls gap opens in the conduction band. Since fluctuations in quasi 1d systems are expected to persist up to temperatures significantly above the transition temperature, TTF-TCNQ will have a fluctuating CDW and a pseudogap even in the phase above 53 K. Nevertheless it can show metallic properties due either to a finite density of states for single particles in the pseudogap, or due to the fluctuating CDW, or both.

3.2.2 One-Dimensionality of the Metallic Property

The high anisotropy in the electrical conductivity does not necessarily lead to the conclusion that TTF-TCNQ is a 1d metal. It must be proved that the conduction by the electrons is metallic along the *b*-axis, but is non-metallic along perpendicular directions. This has been done by measurements of the optical reflectance and thermopower.

Figure 3.10 shows reflection spectra of single crystals of TTF-TCNQ for linearly polarized infrared light [3.14]. Maxwell's equation for electromagnetic waves propagating into a material having a dielectric constant ε and an electrical conductivity σ is given by

$$\nabla^2 E = \frac{\varepsilon}{c^2}\frac{\partial^2 E}{\partial t^2} + \frac{\sigma}{c^2 \varepsilon_0}\frac{\partial E}{\partial t} \quad . \tag{3.1}$$

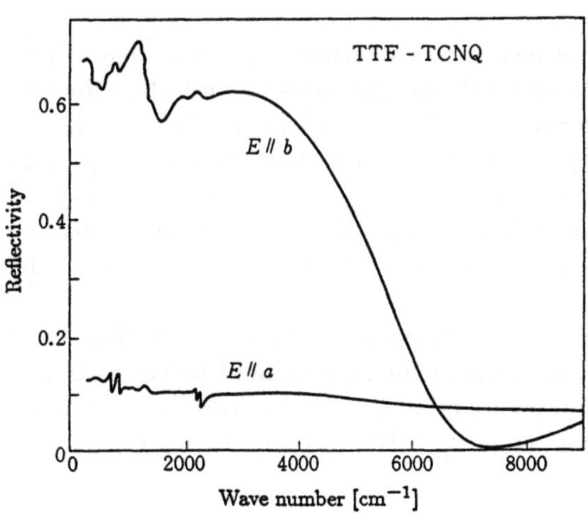

Fig. 3.10. Optical reflection spectra of TTF-TCNQ [3.14]

Substituting a solution of the form $E = E_0 \exp\{i(\boldsymbol{K}\cdot\boldsymbol{r} - \omega t)\}$, we obtain the following relation,

$$K = \frac{\omega}{c}\left(\varepsilon + \frac{i\sigma}{\varepsilon_0\omega}\right)^{1/2} \equiv K_0 N \tag{3.2}$$

where $K_0 = (\omega/c)$ is the wave number in vacuum, and N is the complex refractive index. The reflectivity is given by $R = |(1-N)/(1+N)|^2$. N^2 is approximated as $N^2 = i\sigma/\varepsilon_0\omega$ because ε, in comparison with σ, will be negligibly small in metals. Thus the reflectivity becomes

$$R = 1 - 2(2\varepsilon_0\omega/\sigma)^{1/2} \quad . \tag{3.3}$$

When $\varepsilon_0\omega/\sigma \ll 1$, total reflection occurs. On the other hand, the electromagnetic wave transmits through materials when $\varepsilon_0\omega/\sigma \gg 1$. Electrical insulators have a frequency independent reflectivity R which is less than one since their σ is equal to zero and, therefore, $N^2 = \varepsilon$.

The low reflectivity for the light whose electric vector is perpendicular to the b-axis, shown in Fig. 3.10, leads to the conclusion that TTF-TCNQ is not metallic in directions perpendicular to the b-axis. In contrast to this, the light polarized parallel to the b-axis has a high reflectivity in the long-wavelength regime and a low one in the short-wavelength regime. Such a reflection spectrum is typical of metallic materials. The rapid decrease in the reflectivity around $6000\,\mathrm{cm}^{-1}$ $(1.2\times10^{-19}\,\mathrm{J})$ corresponds to the situation where $\varepsilon_0\omega/\sigma$ reaches 1 in the above equation.

In order to handle the problem more accurately in this regime, we must take account of the frequency dependence of σ. Drude's theory gives [3.15]

$$\mathrm{Re}\{N^2\} = 1 - \frac{\omega_p^2\tau^2}{1+\omega^2\tau^2} \quad , \qquad \mathrm{Im}\{N^2\} = \frac{\omega_p^2\tau^2}{\omega\tau(1+\omega^2\tau^2)} \tag{3.4}$$

where τ is the relaxation time of electron scattering, $\omega_p = (ne^2/\varepsilon_0 m^*)^{1/2}$ is the plasma frequency, and n and m^* are the density and the effective mass of the electrons. These equations lead to the following: 1) For $\omega \ll 1/\tau$, we retrieve (3.3) since $\mathrm{Im}\{N^2\} \gg \mathrm{Re}\{N^2\}$. (Note that $\sigma(0) = \varepsilon_0\omega_p^2\tau$.) 2) When $1/\tau \ll \omega \ll \omega_p$, the reflectivity, $R = 1 - 2/\omega_p\tau$, is not small in spite of a decrease in the absorption. 3) When $\omega_p \ll \omega$, $\mathrm{Re}\{N^2\}$ becomes positive and the electromagnetic wave is transmitted through the materials; the reflectivity is small.

It is to be noted that $\omega = \omega_p$ is a characteristic frequency. The total reflection for $\omega \ll \omega_p$ is due to a screening of the electric field by the polarization of the electronic charge whose collective motion follows the oscillating field of the electromagnetic wave. Since ω_p is the eigenfrequency of the elec-

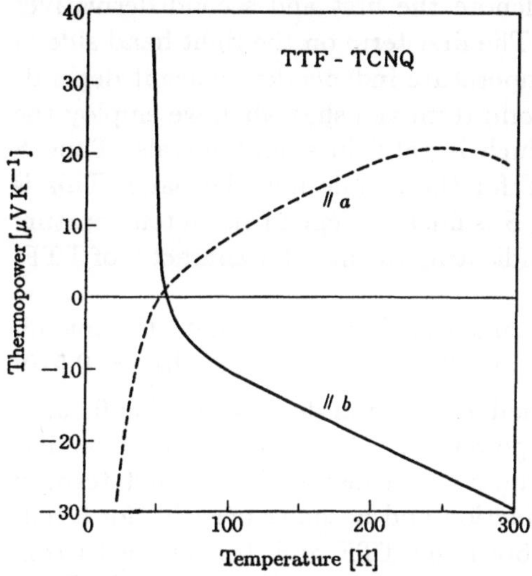

Fig. 3.11. Thermoelectric power of TTF-TCNQ [3.16]

trons' collective motion, electrons cannot be polarized by electromagnetic waves with $\omega \gg \omega_p$. Thus such waves are transmitted.

An analysis using the Drude theory of the data shown in Fig. 3.10 gives the following: $\omega_p = 9500\,\text{cm}^{-1}$ $(1.9 \times 10^{-19}\,\text{J})$, $\tau = 3 \times 10^{-15}\,\text{s}$, and $m^* = 3m_e$. One should note that the frequency dependence of the electrical conduction can be derived from the reflection spectra[4] because the real and the imaginary part of the (squared) complex refractive index are related to each other by the Kramers-Kronig relation.

Further evidence for the metallic character of TTF-TCNQ is found in the data for thermoelectric power. Figure 3.11 shows the temperature dependence of the thermoelectric power measured along the 1d axis and along the a-axis perpendicular to it [3.16]. The thermoelectric power S is defined as $S = E/\nabla T$, where E is the electric field produced by a temperature gradient ∇T. It can be interpreted physically as the ratio of the quantity of heat to that of electric charge carried by electrons.

For a simple 1d band model, S is given by [3.16]

$$S = -\frac{\pi^2 k_B^2 T}{3|e|}\left[\frac{\varepsilon_k''}{(\varepsilon_k)^2} + \frac{\tau'(\varepsilon)}{\tau(\varepsilon)}\right]_{\varepsilon = E_F} \tag{3.5}$$

where E_F is assumed to be much larger than $k_B T$, and ε_k and τ are the band energy and the relaxation time for electron scattering, respectively.

[4] Although a direct derivation from absorption spectra is possible on the basis of the Kramers-Kronig relation, we need, for reflection spectra, information about the phase shift between the incident and the reflected waves. A particular model must be chosen for this.

The prime and the double prime denote the first and second derivatives with respect to the wave number k. The first term on the right hand side of (3.5) may be taken to be nearly temperature independent since it depends on the band structure only. The second term vanishes when we employ the approximation $\{\tau'(\varepsilon)\}_{\varepsilon=\varepsilon_F} = 0$, which is valid in simple metals. Thus S is expected to be proportional to T for the simple metallic band. This is found to be the case in TTF-TCNQ; S is linearly dependent on temperature along the b-axis above about 150 K indicating the metallic character of TTF-TCNQ.

Comparing (3.5) with the experimental data we estimate the magnitude of E_F and $4t_{\|}$ as $E_F = 0.14\,\mathrm{eV}$ $(2.2 \times 10^{-20}\,\mathrm{J})$ and $4t_{\|} = 0.5\,\mathrm{eV}$ $(8 \times 10^{-20}\,\mathrm{J})$, where we have assumed an electron density of $N' = 0.59$.

The sign of the thermoelectric power, S, tells us what kind of carriers are concerned in the electrical conduction because ε_k'' in the first term of (3.5) is positive for an electron conduction and becomes negative for a hole conduction. Conduction bands on both the TTF and TCNQ stacks contribute to the thermoelectric power in TTF-TCNQ. For two bands S can be expressed as [3.16]

$$S = \frac{\sigma_Q S_Q + \sigma_F S_F}{\sigma_Q + \sigma_F} \tag{3.6}$$

where the suffixes Q and F denote the contribution from TCNQ and TTF stacks, respectively. The TCNQ band is expected to show electron conduction and the TTF hole conduction due to the degree of band filling 0.59/2 in TTF-TCNQ. Thus it is expected that $S_Q<0$ and $S_F>0$, and the observation of negative thermoelectric power along the b-axis in the metallic phase implies that $\sigma_Q>\sigma_F$.

Below about 150 K, the proportionality between the thermoelectric power and the temperature is broken. Two mechanisms have been proposed to account for this [3.16]: 1) $\tau'(\varepsilon)$ departs from zero towards negative values. This arises when a scattering by optical phonons occurs or when a gap opens in the conduction band. 2) The increase of σ is more rapid in TTF stacks than in TCNQ ones. [See (3.6).]

Below the Peierls transition temperature 53 K, the thermoelectric power along the b-axis becomes positive showing a strong temperature dependence. This fact indicates that the electrical conduction is dominated by holes which are no longer in the Fermi condensed state. TTF-TCNQ is not metallic below the Peierls transition temperature.

Thus we have learned that TTF-TCNQ has a 1d metallic character above the Peierls transition temperature 53 K. A question may arise in looking again at the *dc* conductivity data of Fig. 3.9. The perpendicular conductivity $\sigma_\perp$ also increases like a metal with decreasing temperature from room temperature although its absolute value is certainly much lower. The temperature dependence of $\sigma_\perp$ is approximately represented by $\sigma_\perp \sim T^{-1}$. Theories [3.17] have shown, however, that $\sigma_\perp$ must be proportional to the conductivity σ_b along the 1d axis even when the perpendicular conduction is diffusive but not metallic. The essence of these theories is the following. Those electrons participating in the 1d conduction along the b-axis are in states near the Fermi level having wave numbers close to k_F. When they hop onto the neighbouring stacks, their wave numbers after hoppig are again close to k_F because they must enter vacant states near the Fermi level of the neighbouring stacks. That is to say, they preserve the b-component of their wave vectors in the hopping process. When this restriction is taken into account, a comparison between theoretical formulae for σ_b and $\sigma_\perp$ gives the following [3.17]

$$\frac{\sigma_\perp}{\sigma_b} = \frac{2a_\perp^2 t_\perp^2}{\hbar^2 v_F^2} \tag{3.7}$$

where $a_\perp$, $t_\perp$, and v_F are the interstack distance, the interstack overlap integral, and the Fermi velocity, respectively. Thus, when σ_b increases with decreasing temperature, $\sigma_\perp$ must have a similar temperature dependence even though it arises from a diffusive conduction.

We recall however that $\sigma_b \sim T^{-2.3}$ in the metallic phase although $\sigma_\perp \sim T^{-1}$. This discrepancy will be shown later in Sect. 3.4.4 to be due to the sliding motion of CDWs along the b-axis.

3.2.3 High Frequency Conductivity

The electrical conductivity as a function of frequency, $\sigma(\omega)$, is given by the Kramers-Kronig transformation of the reflection spectra of infrared light or of absorption spectra obtained in thin films of TTF-TCNQ. Figure 3.12 shows $\sigma(\omega)$ along the b-axis measured below 3000 cm^{-1} (6×10^{-20} J) [3.18]. The functional form of $\sigma(\omega)$ appears to be characteristic of semiconductors except for the value of $\sigma(0)$. The large peak at around 1000 cm^{-1} is ascribed to carriers excited across a bandgap.

Here the question arises as to why $\sigma(\omega)$ shows a semiconducting character at high temperatures, 100 K and 300 K, where TTF-TCNQ is thought to be metallic. This question is answered when we assume that a fluctuating CDW is already present in the metallic phase and a Peierls gap has opened as a fluctuation. As described in Sect. 2.1.2, a fluctuating CDW must be present in a wide range of temperature above the transition temperature 53 K because of the high one-dimensionality of the system. The phase of the CDW will be fluctuating, although the amplitude is expected to have grown already. Therefore, a gap having a similar magnitude to that at 0 K may

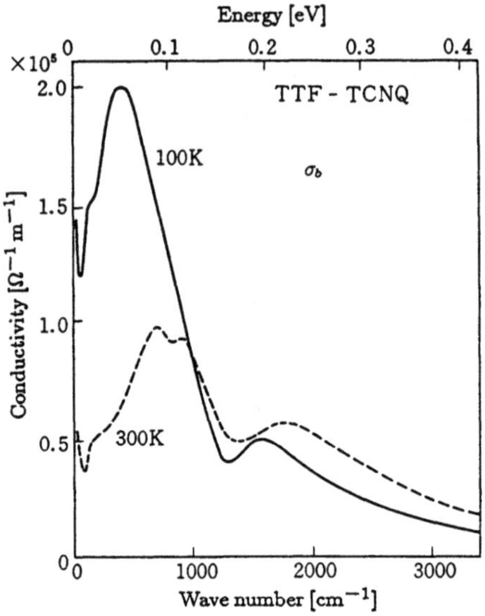

Fig. 3.12. Frequency dependence of the electrical conductivity of TTF-TCNQ [3.18]

be observed as a fluctuation. Of course the density of states in the pseudogap is not zero because of the fluctuation; $\sigma(\omega)$ below about $500\,\mathrm{cm}^{-1}$ (1×10^{-20} J) is considered to correspond to the finite density of states in the gap.

The magnitude of the fluctuating Peierls gap is estimated to be about $1000\,\mathrm{cm}^{-1}$ in Fig. 3.12. This does not agree with $E_g = 400\,\mathrm{K}$ ($=280\,\mathrm{cm}^{-1}$) obtained from dc conductivity measurements described in Sect. 3.2.1. Furthermore, the gap in Fig. 3.12 becomes 400–$500\,\mathrm{cm}^{-1}$ at about 100 K although its magnitude is predicted to increase with decreasing temperature. On the other hand a photoconductivity mesurement at low temperature by means of far infrared light, shows that the photocurrent increases rapidly when the energy of the incident light exceeds about $280\,\mathrm{cm}^{-1}$ [3.19]. The magnitude of the gap is thus estimated in this measurement to be about $280\,\mathrm{cm}^{-1} = 400\,\mathrm{K}$. It agrees with that obtained in the dc conductivity measurements. A careful inspection of $\sigma(\omega)$ shown in Fig. 3.12 may indeed suggest that $\sigma(\omega)$ increases more rapidly near $280\,\mathrm{cm}^{-1}$. Thus the data of Fig. 3.12 may also indicate that the magnitude of the gap is about $280\,\mathrm{cm}^{-1}$.

The above discussions neglect $\sigma(\omega)$ at $\omega = 0$. At room temperature $\sigma(0)$ does not appear to be anomalous. At 100 K, however, $\sigma(0)$ is estimated from dc conductivity measurements to be $(3$–$5) \times 10^5\,\Omega^{-1}\,\mathrm{m}^{-1}$, which is much larger than $\sigma(\omega)$ for $\omega \neq 0$. We may ascribe the anomalously high $\sigma(0)$ to a collective excitation, namely the sliding motion of fluctuating CDWs, rather than to single particle excitations. Later, in Sect. 3.4.4 we will see that such a conduction mechanism is really present in the metallic phase of TTF-TCNQ; it gives rise to a considerable dc conductivity.

3.2.4 Nonlinear Conduction in the Insulating Phase

We shall examine here the dc conductivity along the 1d axis at low temperatures. Figure 3.9 indicates a rapid decrease of the activation energy below about 10 K. This fact means a dominance of some conduction mechanism which has an activation energy much smaller than that of the Peierls gap. Such a conductivity is clearly observed in samples whose quality of structure and of chemical composition is expected to be high. Therefore the conduction with the low activation energy cannot be ascribed to the so-called impurity conduction due to impurity levels in the Peierls gap. Another possibility would be conduction due to the sliding motion of CDWs described in Sect. 2.2.2 or phase solitons, or both.

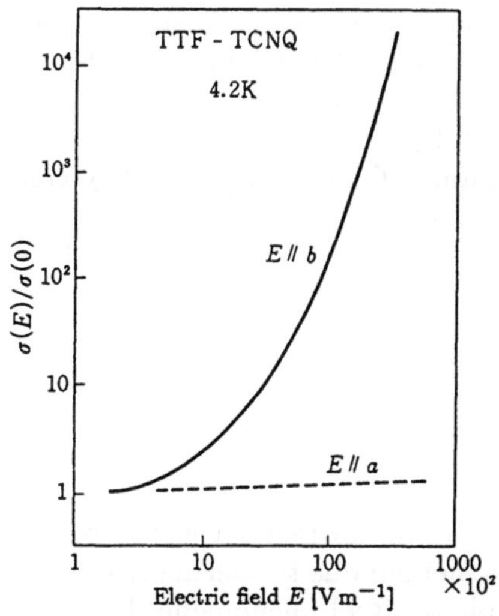

Fig. 3.13. Electric-field dependence of the dc electrical conductivity of TTF-TCNQ [3.20]

Figure 3.13 shows the dc conductivity mesured at 4.2 K as a function of the applied electric field [3.20]. The conductivity along the a-axis which lies perpendicular to the 1d axis obeys Ohm's law irrespective of the field strength. In contrast to this the conductivity along the 1d axis increases rapidly with field strength above a few V cm^{-1}; a highly non-ohmic conduction takes place. This result suggests the presence of some conduction mechanism that is activated by the electric field. It is natural to ascribe the origin of this to the CDWs since such a phenomenon is observed only along the 1d axis. We must check, however, whether the non-ohmic conduction might not arise from other causes. The first possibility, that bad contacts between measuring leads and samples cause a Joule heating due to the large current has been ruled out. Second, intrinsically nonlinear mechanisms such as field ionization and impact ionization by accelerated electrons have been examined and found not to be important. The following is some evidence for the second point above: 1) For field ionization to occur, electron wave functions of impurities must extend to a size of the order of 10^3 Å which is not realistic. 2) Intentional creation of charged impurities by deuteron bombardment does not change to nonlinear conductivity although, of course, the ohmic conductivity increases [3.20]. Thus some mechanism originating from CDWs must be the cause of the non-ohmic conduction.

The activation energy below 4.2 K decreases with increasing field strength and reaches zero at a field of $1.5 \times 10^4\,\mathrm{V\,m^{-1}}$. It is believed that the field which depins the CDWs giving rise to the sliding motion is higher than the field necessary to create phase solitons in the CDW. Accordingly, we may speculate that the nonlinear conduction observed experimentally is caused by phase solitons whose critical field for creation is $1.5 \times 10^4\,\mathrm{V\,m^{-1}}$. A determination, based on this interpretation, of several parameters which describe the dynamics of CDWs gives an estimate of the field strength necessary to depin the CDW as about $1 \times 10^5\,\mathrm{V\,m^{-1}}$ [3.20].

Thus we may say that it is experimentally verified that phase solitons are excited thermally and electrically in pinned CDWs and that they contribute to the electrical conduction.

3.3 Magnetic Properties

3.3.1 Spin Susceptibility

If TTF-TCNQ were a simple metal, a Pauli paramagnetism would be expected in the metallic phase and a paramagnetism due to thermally excited electrons would appear in the insulating phase. Real phenomena, however, are found to be more complicated as is described below.

Figure 3.14 shows the temperature dependence of the spin susceptibility of TTF-TCNQ derived by subtraction of a diamagnetic component from the total magnetic susceptibility measured by the so-called magnetic-balance method [3.21, 22]. The diamagnetic component is estimated from the sum of susceptibilities measured at room temperature in crystals of TTF and TCNQ. This estimate must give the diamagnetic component itself since these crystals have no spin. The spin susceptibility thus derived is in good agreement with that obtained from measurements of electron paramagnetic resonance.

We now examine the spin susceptibility in the metallic phase of TTF-TCNQ above 53 K. Remarkable features of this are its strong temperature dependence and very large absolute value. Is it possible to explain these in terms of the Pauli paramagnetism? First we shall consider the absolute value

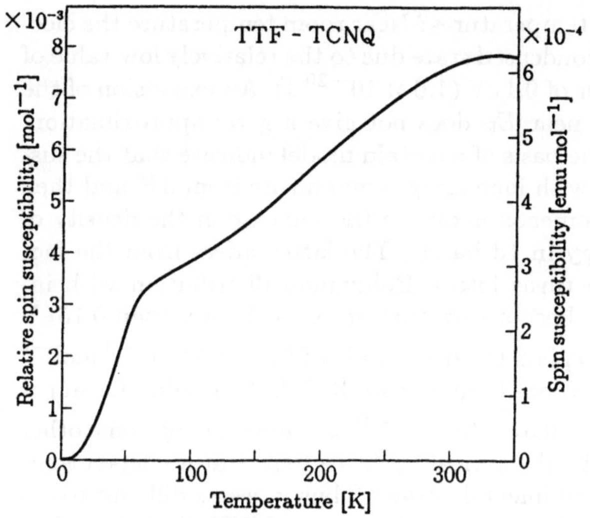

Fig. 3.14. Spin-magnetic susceptibility of TTF-TCNQ [3.21, 22]

of the susceptibility.[5] The relative susceptibility of the Pauli paramagnetism at 0 K is given by [3.23]

$$\overline{\chi}_{\mathrm{P}} = \mu_0 \mu_{\mathrm{B}}^2 D(E_{\mathrm{F}}) \quad , \tag{3.8}$$

where μ_{B} is the Bohr magneton, μ_0 the vacuum permeability, and $D(E_{\mathrm{F}})$ the density of states at the Fermi level of the conduction band. Assuming a 1d cosine band with a width $4t_{\|}$ and a degree of charge transfer N' electrons/molecule from TTF to TCNQ, we obtain the following [3.24],

$$\begin{aligned} \overline{\chi}_{\mathrm{P}} &= \overline{\chi}_{\mathrm{P}}(\mathrm{TTF}) + \overline{\chi}_{\mathrm{P}}(\mathrm{TCNQ}) \\ &= \frac{\mu_0 \mu_{\mathrm{B}} N_0^2}{\pi \sin(\pi N'/2)} \left(\frac{1}{t_{\|}(\mathrm{TTF})} + \frac{1}{t_{\|}(\mathrm{TCNQ})} \right) \quad , \end{aligned} \tag{3.9}$$

where the relation $k_{\mathrm{F}} = b^* N'/4$ is used; N_0 is Avogadro's number and $t_{\|}$ is measured in units of eV. The band width $4t_{\|}$ is estimated as $4t_{\|}(\mathrm{TCNQ}) = 0.5\,\mathrm{eV}$ (8×10^{-20} J) from measurements of the thermoelectric power and as $4t_{\|}(\mathrm{TTF}) = 0.2$–$0.4\,\mathrm{eV}$ [$(3.2$–$6.4) \times 10^{-20}$ J] from simplified band calculations. Using the value $N' = 0.59$, we obtain $\overline{\chi}_{\mathrm{P}} = 2.5 \times 10^{-3}\,\mathrm{mol}^{-1}$ ($2 \times 10^{-4}\,\mathrm{emu\,mol}^{-1}$) [3.24]. This estimate does not appear to contradict the experimental result.

[5] In this volume the magnetic susceptibility is defined as $\overline{\chi}_{\mathrm{M}} = M/H$, $B = \mu_0 H + \mu_0 M$, where $\overline{\chi}_{\mathrm{m}}$ is a relative magnetic susceptibility. A conversion from C.G.S. units to S.I. is given by $1\,\mathrm{emu\,mol}^{-1}$ (C.G.S.) $= 4\pi\,\mathrm{mol}^{-1}$ (S.I.). The Bohr magneton is defined as $\mu_{\mathrm{B}} = e\hbar/2m$. Thus $\mu_0 \mu_{\mathrm{B}} H$ has the dimension of energy.

What happens at finite temperatures? Near room temperature the electron system is not in a well-condensed state due to the relatively low value of the Fermi energy of the order of 0.1 eV (1.6×10^{-20} J). An expansion of the Fermi distribution function near E_{F} does not give a good approximation. Numerical calculations on the basis of a certain model indicate that the susceptibility at first increases with increasing temperature from 0 K and then decreases. The former phenomenon is due to the increase in the density of states with decreasing energy in 1d bands. The latter arises from the fact that the Fermi distribution turns into a Boltzmann distribution with increasing temperature. It has been shown that $4t_{\|}$ must be less than 0.15 eV (2.4×10^{-20} J) in order to explain the magnitude of $\overline{\chi}_{\mathrm{P}} \sim 7.5 \times 10^{-3}\,\mathrm{mol}^{-1}$ (6×10^{-4} emu mol^{-1}) near room temperature [3.24]. This value for $4t_{\|}$ is too small compared to that, ~ 0.5 eV (8×10^{-20} J), indicated by some other experiments. Furthermore the theoretical $\overline{\chi}_{\mathrm{P}}$ has a temperature dependence which disagrees with the experimental curve.[6] Thus it seems difficult to explain the spin susceptibility in the metallic phase of TTF-TCNQ and in particular its temperature dependence in terms of Pauli paramagnetism.

Two models have been proposed for the spin susceptibility in the metallic phase. The first model assumes that the Peierls gap is already present as a fluctuation in the metallic phase [3.22]. A decrease of the density of states at the Fermi level due to the gap fluctuation will become more pronounced with decreasing temperature giving rise to the decrease of the magnetic susceptibility. It is shown that the temperature dependence of the susceptibility can be explained by this model if the susceptibility observed experimentally is assumed to consist of a sum comprising this mechanism for the TCNQ stacks and a simple Pauli paramagnetism for the TTF stacks. The absolute value of the susceptibility, however, is not very satisfactorily explained.

The second model assumes the presence of a Coulomb repulsion U between two conduction electrons on the same molecule in at least one kind of molecular stack, TCNQ or TTF [3.24]. This is the so-called Hubbard model. Since this Coulomb interaction increases the magnetic susceptibility, the magnitude of the experimental value may be explained in this way. When U is much smaller than $4t_{\|}$ (i.e. $U/4t_{\|} \ll 1$), the susceptibility at the temperature T is given by [3.24]

$$\overline{\chi}(T) = \frac{\overline{\chi}_{\mathrm{P}}(t_{\|}, T)}{1 - U/4\pi t_{\|}} \quad , \tag{3.10}$$

where $\overline{\chi}_{\mathrm{P}}$ is the relative susceptibility of the Pauli paramagnetism. At 300 K, we can obtain a value for $\overline{\chi}$ consistent with the experimental one,

[6] However, a susceptibility which decreases with decreasing temperature may be derived theoretically when we take account of spin-spin interactions and fluctuation effects as for a spin-Peierls transition.

$7.5 \times 10^{-3}\,\mathrm{mol}^{-1}$ ($6 \times 10^{-4}\,\mathrm{emu\,mol}^{-1}$), when we assume $U = 1\,\mathrm{eV}$ and $4t_{\parallel} = 0.5\,\mathrm{eV}$ ($8 \times 10^{-20}\,\mathrm{J}$), or $U = 0.25\,\mathrm{eV}$ and $4t_{\parallel} = 0.25\,\mathrm{eV}$.

Similar results are also obtained in theories using the renormalization group [3.25]. The case $U/4t_{\parallel} \gg 1$ makes the Hubbard model equivalent to a Heisenberg model with an exchange interaction J for a 1d linear chain of spins. Its susceptibility has been examined in detail. In order to reproduce the susceptibility of TTF-TCNQ at room temperature, the values of J are set to be about 1.2 eV and 0.25 eV for $4t_{\parallel} = 0.5\,\mathrm{eV}$ and 0.25 eV, respectively [3.24]. Although the results obtained in two approximations, $U/4t_{\parallel} \gg 1$ and $U/4t_{\parallel} \ll 1$, are inconsistent with the original approximations, the conclusion derived above may be roughly correct if the results obtained under the assumptions $U/4t_{\parallel} \gg 1$ and $U/4t_{\parallel} \ll 1$ can be interpolated into a regime of intermediate U.

In reality it is believed that $U/4t_{\parallel}$ is of the order of 1 and, therefore, the Coulomb repulsion U cannot be neglected. When we employ this model, a spin density wave (SDW) is expected to appear with the wave number $2k_{\mathrm{F}}$. Neutron scattering experiments described in 3.4.1, however, failed to detect the SDW. A lack of satisfactory theories in the regime $U/4t_{\parallel} \sim 1$ makes it difficult to check whether or not J is large enough for the SDW to be observed experimentally.

The decrease of the susceptibility with decreasing temperature is qualitatively explained by taking into account the increase of the CDW's amplitude below 150 K. The reason is that two spins in a wavelength of the $2k_{\mathrm{F}}$ lattice distortion will couple into a state of total spin $= 0$ thus giving rise to the decrease of the susceptibility.

Although a complete explanation has not been given for the magnetic state of TTF-TCNQ, the importance of the Coulomb repulsion is also suggested by magnetic resonance measurements and by structural studies of CDWs described in Sect. 3.4. It is a common understanding at the present stage that on at least one type of molecular stack, the Coulomb repuslion between electrons cannot be neglected.

3.3.2 Spin Susceptibility of Each Kind of Molecular Stack

Magnetic resonance makes it possible to measure separately the susceptibility of TCNQ stacks and of TTF stacks. In this respect the method of magnetic resonance is very important for the study of TTF-TCNQ in which TTF and TCNQ stacks are expected to play different roles.

Measurements have been made of electron paramagnetic resonance (EPR), and nuclear magnetic resonance (NMR) of protons and ^{13}C. In EPR measurements one observes a process in which electrons, by absorbing microwave photons, make transitions between two levels separated in energy by the Zeeman energy $g\mu_0\mu_{\mathrm{B}}H$, where H, μ_{B} and g are the applied magnetic field, the Bohr magneton and the so-called g-factor, respectively. Usually the microwave frequency is kept constant and the applied field strength is varied. A peak in the microwave absorption is observed for an H satisfying the relation $\hbar\omega = g\mu_0\mu_B H$. The g-value usually differs slightly from 2 due to the spin-orbit interaction. Therefore each electron spin on TCNQ and TTF stacks is separately observed due to their differing g-values. The magnitude of the microwave absorption gives information about the spin susceptibility because it depends on the number of spins aligned parallel to the applied field (i.e. the magnetization) under thermal agitation at a particular temperature. Furthermore when various kinds of interactions make finite the life time of each spin state, the Zeeman level has some finite width giving rise to a resonance absorption in a certain range of the applied field H; the peak width in an absorption vs. field curve provides the spin relaxation time.

Only one resonance peak of a Lorentzian shape is observed in TTF-TCNQ [3.21]. Its width is about 6×10^{-4} T (6 G) which is rather sharp for a metal. Although the g-value is anisotropic, its average value is obtained as 2.0059 at room temperature and 2.0082 at 20 K. On the other hand g-values for TTF^{+1} and $TCNQ^{-1}$ are known to be $g_F = 2.00838$ and $g_Q = 2.0025$, respectively, from measurements in compounds having either TTF^{+1} or $TCNQ^{-1}$. The narrow resonance peak has been found to arise in a situation characteristic of 1d; the transition of electrons being inhibited for usual processes [3.26] except that from $-k_F$ to $+k_F$ or vice versa. With regard to the presence of only one resonance peak, present interpretations are the following. Spin-spin couplings are strong between TTF stacks and TCNQ ones. The coupling energy becomes larger than the difference of the Zeeman energy, $g_F\mu_0\mu_B H - g_Q\mu_0\mu_B H$, giving rise to only one resonance peak. The composite g-value is given by [3.27],

$$g = \frac{\chi_Q g_Q + \chi_F g_F}{\chi_Q + \chi_F} \tag{3.11}$$

where χ_Q and χ_F denote spin susceptibilities of TCNQ and TTF stacks, respectively. This equation serves to determine the ratio of χ_Q to χ_F at each temperature using the observed g-value. The ratio is estimated as $\chi_Q/\chi_F = 4/6$ in the metallic phase above 53 K and $\chi_Q/\chi_F \ll 1$ below 53 K [3.27].

Figure 3.15 shows the temperature dependences of χ_Q and χ_F derived from EPR measurements, and NMR described later [3.31]. The absolute value of χ_Q is not far from that of the Pauli paramagnetism discussed already, and its temperature dependence appears to be explained in terms of the fluctuating Peierls gap. In contrast to this, explanations for χ_F may necessitate the model of $U/4t_{\parallel} \sim 1$ described in the preceding section. Therefore the Coulomb repulsion among conduction electrons will be more important on TTF stacks than on TCNQ.

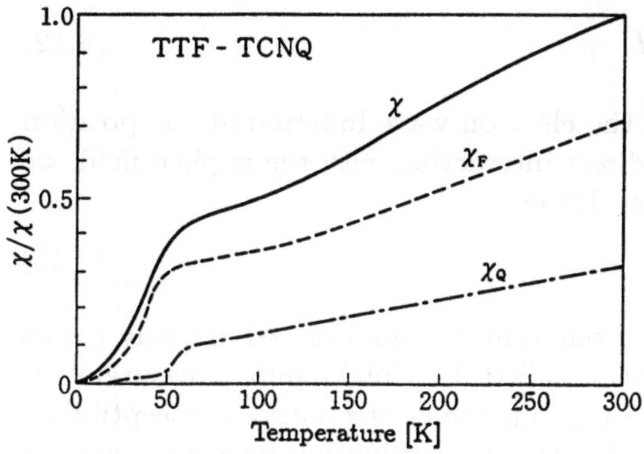

Fig. 3.15. Spin-magnetic susceptibility of TTF-TCNQ selectively measured by magnetic resonance [3.31]. The susceptibilities χ_Q and χ_F refer to the TCNQ and TTF stacks respectively

The susceptibility below the Peierls transition temperature 53 K is specified by an activation energy Δ. The activation energy is estimated as $\Delta_Q \sim 420$ K for TCNQ stacks. At 49 K χ_F begins decreasing, and it decreases more rapidly below 38 K. Δ_F below 38 K is evaluated as $\Delta_F \sim 125$ K [3.28].

These facts lead to two conclusions: 1) The Peierls transition at 53 K is predominantly due to TCNQ stacks. Temperatures at which TTF stacks undergo some change are 49 K and 38 K. (This is closely related to structural changes and the CDW as described in Sect. 3.4.) 2) Neither Δ_Q nor Δ_F is related to the Peierls gap, $E_g = 400$ K, derived from conductivity measurements discussed in Sect. 3.2 as $E_g = 2\Delta$.

Therefore the susceptibility can hardly be ascribed only to electron spins activated thermally across the Peierls gap. It may still be possible to explain that a true bandgap is $2\Delta_F = E_g - E_\mu$, where E_μ denotes the mobility gap for carriers on TTF stacks, as long as TTF stacks are responsible for both the conductivity and the susceptibiliy. However, the following interpretation is also possible. Since $U/4t_{\parallel}$ is considered to be of the order of 1, at least on TTF stacks, the decrease of χ_F may be brought about by a pair formation of two spins (i.e., the spin-Peierls transition) induced by the $2k_F$ lattice distortion. Thermal activation of singlet spins to a triplet state may dominate the temperature dependence of χ_F. If this is the case, Δ_F obtained in magnetic measurements need not to have anything to do with the Peierls gap in the conduction band.

NMR measurements reveal magnetic properties near resonanting nuclei. In a magnetic field H, electron spins are polarized giving rise to an average magnetic moment per electron $\langle\mu_e\rangle = \mu_0\overline{\chi}H/n$, where $\overline{\chi}$, H, and n are the relative magnetic susceptibility, the applied field and the number of electrons, respectively. This magnetic moment interacts with a nuclear magnetic moment μ_I via a contact interaction. The interaction energy is given by [3.29]

$$\frac{8}{3}\pi\overline{\chi}|\psi(0)|^2\frac{\mu_{\mathrm{I}}\cdot H}{n} \equiv \Delta\mu_{\mathrm{I}}\cdot H \quad , \tag{3.12}$$

where $\psi(0)$ denotes the value of the electron wave function at the position of the nucleus. Adding to this a direct interaction with the applied field, we obtain the following resonance condition

$$\hbar\omega = (\mu_{\mathrm{I}} + \Delta\mu_{\mathrm{I}})H \quad . \tag{3.13}$$

The resonance frequency shifts from that for an isolated nucleus by an amount of $K = \Delta\mu_{\mathrm{I}}H/\hbar \propto \overline{\chi}$. This is called the Knight shift. Measurements of the Knight shift give information about the electronic spin susceptibility. [it should be noted that only a part of the susceptibility is measured; namely that part which is brought about by electrons having an s-character for which $\psi(0)$ is finite.] Also the relaxation time of the nuclear spin state is measured as in the case of EPR.

Generally, different isotopes of an atomic element have a different value of nuclear spin. Therefore we can measure the local susceptibility at a certain position in TTF-TCNQ by substituting specific isotopes whose natural abundance is small for particular constituent atoms. In TTF-TCNQ, proton-NMR measurements were made in TTF(D)-TCNQ and in TTF-TCNQ(D) in which the hydrogen atoms of the TTF and TCNQ molecules respectively, were replaced with deuterons [3.30]. Also NMR of ^{13}C was measured in TTF-TCNQ(^{13}C) in which carbon atoms of TCNQ were enriched with ^{13}C, whose natural abundance is 1 %, to a level of 50 % [3.31]. Values of χ_{Q} for TCNQ stacks and of χ_{F} for TTF stacks measured by these methods almost agree with those measured by EPR. NMR measurements give the ratio $\chi_{\mathrm{Q}}/\chi_{\mathrm{F}} = 3/7$ at room temperature [3.31]. It is to be remarked that some of the changes in the properties of the TTF stacks at 38 K and 49 K already described were discovered for the first time by proton NMR measurements.

Measurements of the relaxation time in NMR have given informatin about the metallic character of conduction electrons and the electron-electron Coulomb repulsion U. It is well known that in ideal metals, the quantity K^2T_1T is a constant independent of temperature and frequency (and of magnetic field), where K, T_1 and T are the Knight shift, the spin-lattice relaxation time and the temperature, respectively. This is called the Korringa relation [3.29].

The Korringa relation can be intuitively derived as follows. The nuclear spin relaxation is caused by interactions between nuclear spins and electron spins. Electrons change their states from an initial state, i, to a final state, f with a transition probability given by

$$w = (2\pi/\hbar)|H_{\mathrm{if}}|^2\delta(E_{\mathrm{i}} - E_{\mathrm{f}}) \quad . \tag{3.14}$$

The total transition probability for all possible processes is given by integration with respect to E_i and E_f after multiplying this by the density of states for i and f and by a factor of the Fermi distribution $f(E_i)\{1-f(E_f)\}$. Using the approximation $f(E_i)\{1-f(E_f)\} = k_B T\delta(E - E_F)$ for $E_F \gg k_B T$, we obtain the following,

$$w = \text{const.}\, D(E_F)^2 T \quad . \tag{3.15}$$

As described already, the susceptibility for Pauli paramagnetism is given by $\overline{\chi}_P = \mu_0 \mu_B^2 D(E_F)$, and $\overline{\chi}_P$ is proportional to the Knight shift K. Thus the nuclear relaxation time T_1 can be expressed as

$$T_1^{-1} = w = \text{const.}\, K^2 T \quad . \tag{3.16}$$

The Korringa relation appeared to be satisfied in proton NMR measurements in the metallic phase of TTF(D)-TCNQ and TTF-TCNQ(D) [3.30]. Hence TTF-TCNQ was once considered to be a rather ideal metal. Detailed measurements revealed, however, that the Korringa relation no longer holds over a wide range of magnetic field frequency [3.32]. At present the electrical conduction in TTF-TCNQ is thought to be somewhat diffusive although it may be essentially metallic.

A detailed analysis of experimental data obtained in measurements of the relaxation time has given an estimate for the electron-electron Coulomb repulsion as $U/4t_{\parallel} = 0.8$–1 (ambient pressure, 300 K), which is consistent with that described in 3.3.1 [3.32].

3.4 Behaviour of the CDW

3.4.1 The $2k_F$ CDW and the Kohn Anomaly

The CDW is a mixed wave of charge density modulation together with a periodic lattice distortion called a Peierls distortion. Therefore a CDW can be directly observed by looking at the associated structural features.

When x-rays enter a crystal lattice, they are diffracted for a specific angle θ satisfying the well-known Bragg condition, $2d \sin\theta = n\lambda$, where d, θ, n, and λ are the distance between lattice planes, the incidence angle of the x-rays to the plane, a certain integer, and the wave length of the x-rays, respectively. The diffraction is caused by the periodic structure of the lattice specified by the repeat distance d. When a periodic lattice distortion due to the CDW is introduced, the lattice has an additional periodicity different from the original one. It results in the appearance of additional diffraction features known as satellite reflections. For example, if the original diffraction is specified by a diffraction index $(h\ k\ l)$, the satellite reflections due to the $2k_\mathrm{F}$ CDW formed along the b-axis appears at positions specified by $(h\ k{\pm}\eta\ l)$, where $\eta = 2k_\mathrm{F}/b^*$. The intensity of the satellite reflection is proportional to the squared amplitude of the atomic displacements in the periodic lattice distortion.

We now derive some formulae desribing the satellite reflections. The method of diffraction provides information about Fourier components of the periodic location of atoms in crystals. A spatial correlation function $G(r)$ describing the periodic structure is written as

$$G(r) = \langle f^*(r) f(0) \rangle \quad , \tag{3.17}$$

where $f(r)$ is the scattering factor of a scatterer at the position r, and $\langle\,\rangle$ denotes the spatial average. The scatterer is an atom for x-ray and electron beams, and a nucleus for neutrons. When a scatterer is present at $x = 0$ in a simple lattice whose lattice constant along the x-axis is a, others are located at $x = \pm a, \pm 2a, \ldots, \pm(N-1)a/2$. The correlation function $G(r)$, along the x-direction, has finite values for $x = 0, \pm a, \pm 2a, \ldots, \pm(N-1)a/2$ but it vanishes for all other x-values.

Next, the diffraction intensity I is proportional to the structure factor $S(\boldsymbol{K})$ which is a Fourier transform of $G(\boldsymbol{r})$:[7]

$$I(\boldsymbol{K}) \propto S(\boldsymbol{K}) = \int G(\boldsymbol{r}) \exp(\mathrm{i}\boldsymbol{K} \cdot \boldsymbol{r}) d\boldsymbol{r} \quad , \tag{3.18}$$

where K is called a diffraction vector (or, more generally, a scattering vector) and denotes the difference in wave vectors between the incident and the diffracted wave:

$$\boldsymbol{K} = \boldsymbol{k}_\mathrm{diff} - \boldsymbol{k}_\mathrm{inc} \quad . \tag{3.19}$$

The above example gives, for the x-component,

$$\begin{aligned} I(K_x) \propto S(K_x) &= N f^2 \sum_{n=-(N-1)/2}^{(n-1)/2} \exp(\mathrm{i}K_x \cdot na) \\ &= N^2 f^2 \Delta(K_x - G_x) \equiv I_\mathrm{B} \quad , \end{aligned} \tag{3.20}$$

[7] Usually the intensity formula is derived without using the correlation function [3.33]. However, the correlation function serves to make the intuitive meaning clear and to provide a unified understanding of diffraction and scattering phenomena of waves including light scattering. The structure factor $S(\boldsymbol{K})$ contains both the Laue function coming from the periodic structure of the lattice and the ordinary structure factor defined in a unit cell.

where N is assumed to be large, and $\boldsymbol{G}$ denotes a reciprocal lattice vector $\boldsymbol{a}^*(=2\pi/\boldsymbol{a})$, $2\boldsymbol{a}^*, \dots$. The function $\Delta(X)$ is equal to one for $X=0$ and zero otherwise. The quantity I_{B} denotes the Bragg reflection intensity. As is well known, diffraction beams with $\boldsymbol{K}=\boldsymbol{a}^*$, $2\boldsymbol{a}^*, \dots$ are called the (1 0 0), (2 0 0), ... diffraction orders.

When the lattice has a periodic lattice distortion due to a CDW, the correlation function $G(\boldsymbol{r})$ will have finite values also at $x=2\pi/Q$, $\pm 4\pi/Q$, ... in addition to the usual positions $x=0, \pm a, \dots$. Its magnitude is proportional to u_0^2, where u_0 denotes the amplitude of the distortion $\boldsymbol{u}=\boldsymbol{u}_0 \cos Qx$. $S(\boldsymbol{K})$ must then be written as

$$\begin{aligned} I(\boldsymbol{K}) \propto S(\boldsymbol{K}) &= N^2 f^2 \{\Delta(\boldsymbol{K}-\boldsymbol{G}) + (\boldsymbol{K}\cdot\boldsymbol{u}_0)^2 \Delta(\boldsymbol{K}-\boldsymbol{G}\pm\boldsymbol{Q})\} \\ &\equiv I_{\mathrm{B}} + I_1 \quad . \end{aligned} \tag{3.21}$$

This formula is an approximate one which is valid for $u_0 \ll a$. Furthermor, exact calculations tell us that $S(K)$ must also be multiplied by the Debye-Waller factor which takes account of thermal vibrations of the atoms.

Equation (3.21) demonstrates the following: 1) A diffraction intensity I_1 measured at $K \neq G$ may be brought about by the periodic lattice distortion specified by the wave vector $\boldsymbol{Q}=\pm(\boldsymbol{K}-\boldsymbol{G})$. 2) The intensity is proportional to u_0^2. 3) The factor $(\boldsymbol{K}\cdot\boldsymbol{u}_0)^2$ makes it possible to determine the direction of polarization of the lattice distortion by measuring I_1 for various directions of $\boldsymbol{K}$.

When the periodic lattice distortion is brought about by phonons, it is expressed as $\boldsymbol{u}=\boldsymbol{u}_0 \cos(\omega_Q t - Qx)$. A quantum mechanical calculation which uses statistical mechanics for the phonons gives the following,

$$\begin{aligned} I_1 \propto k_{\mathrm{B}}T/\omega_Q^2 \quad &\text{for} \quad k_{\mathrm{B}}T \gg \hbar\omega_Q \quad ; \\ \hbar/\omega_Q \quad &\text{for} \quad k_{\mathrm{B}}T \ll \hbar\omega_Q \quad . \end{aligned} \tag{3.22a}$$

When Q is parallel to the b^*-axis, the intensity I_1 is observed at points with the index $(h\ k\pm\eta\ l)$, where $\eta = Q/b^*$. As the crystal has phonons with all wave vectors less than half of the reciprocal lattice vector, the diffraction by phonons occurs for continuous indices $(h\pm\xi\ k\pm\eta\ l\pm\zeta)$, where ξ, η, and ζ take a continuous range of values less than 0.5. Taking account of this fact and noting that the "diffraction" by phonons may be considered as "scattering" by quasi-particles, we call the x-ray diffraction by lattice vibrations "x-ray scattering". Hereafter this terminology is used.

A more exact representation of (3.22a) is given by [3.33]

$$\begin{aligned} I_1 \propto (\boldsymbol{K}\cdot\boldsymbol{e}_Q)^2 \Delta(\boldsymbol{K}-\boldsymbol{G}\pm\boldsymbol{Q}) k_{\mathrm{B}}T/\omega_Q^2 \quad &\text{for} \quad k_{\mathrm{B}}T \gg \hbar\omega_Q \\ \hbar/\omega_Q \quad &\text{for} \quad k_{\mathrm{B}}T \ll \hbar\omega_Q \quad , \end{aligned} \tag{3.22b}$$

where e_Q is the polarization vector of phonons; i.e. $e_Q \| u_0$ with $|e_Q| = 1$. When phonons with $Q = 2k_F$ have an anomalously low frequency ω_{2k_F} due to the Kohn anomaly, the x-ray scattering intensity I_1 has a maximum at $Q = 2k_F$. Thus measurements of the scattering intensity provide information about the phonon frequency.

It is also well known that inelastic neutron scattering can directly measure the phonon frequency. A formula describing inelastic scattering is given by the dynamical structure factor $S(K, \Omega)$ which is the Fourier transform in space and time of the time-dependent correlation function

$$S(\boldsymbol{K}, \Omega) = \iint e^{i\boldsymbol{K} \cdot \boldsymbol{r}} e^{-i\Omega t} G(\boldsymbol{r}, t) dr\, dt \quad . \tag{3.23}$$

This result is obtained as

$$I_1 \propto S(\boldsymbol{K}, \Omega) = N^2 f^2 (\boldsymbol{K} \cdot \boldsymbol{u}_0)^2 \Delta(\boldsymbol{K} - \boldsymbol{G} \pm \boldsymbol{Q}) \Delta(\Omega \pm \omega_Q) \quad , \tag{3.24}$$

where Ω corresponds to the difference in frequency between the incident and the scattered neutron waves; namely the Brillouin shift. Neutron scattering can directly measure $S(K, \Omega)$. X-ray and electron scattering, however, provide only the integrated intensity I_1 with respect to Ω because of their poor resolution in energy. That is to say, they cannot discriminate between static lattice distortions and dynamic distortions due to phonons.

Figure 3.16 shows the wave-number dependence of the x-ray scattering intensity measured in TTF-TCNQ [3.34]. It is measured along the b^*-axis

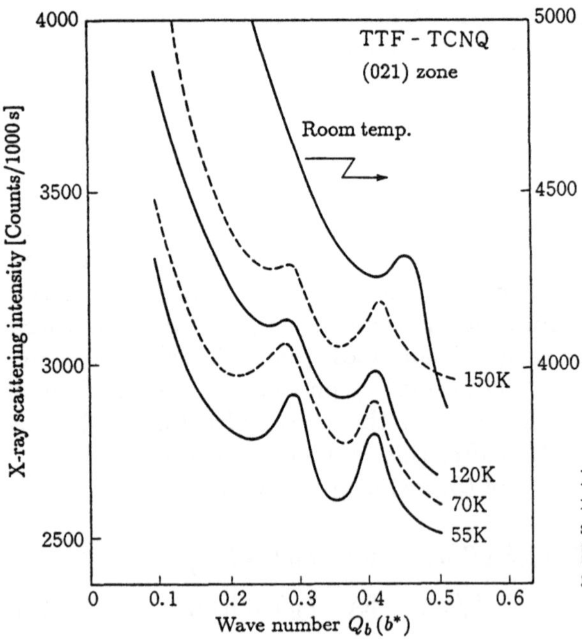

Fig. 3.16. Intensities of diffuse x-ray scattering measured in a single crystal of TTF-TCNQ [3.34]. See also the x-ray photograph in the frontispiece

from the point of the (0 2 0) Bragg reflection. Two anomalies are found in the intensity at $Q_b = 0.295\, b^*$ and $0.41\, b^*$. This result leads to the conclusion that either static lattice distortions are present having these wave numbers or that the phonons are affected by Kohn anomalies at these two points.

The value $Q_b = 0.41\, b^*$ in the reduced zone scheme is equivalent to $0.59\, b^* = b^* - 0.41\, b^*$ in the extended zone scheme. If the anomaly at $0.295\, b^*$ is due to the $2k_F$ CDW, that of $0.59\, b^*$ can clearly be ascribed to the $4k_F$ CDW. These anomalies are usually called the $2k_F$ and the $4k_F$ anomalies.

The $2k_F$ anomaly is observed below about 150 K and grows with decreasing temperature. The $4k_F$ anomaly on the other ahnd, is already present at room temperature. Its wave number changes above about 200 K reaching $0.55\, b^*$ (equivalent to $0.45\, b^*$) at room temperature. The $4k_F$ anomaly is thus not a simple second harmonic of the $2k_F$ lattice distortion. It is considered to originate in a mechanism different from that for the $2k_F$ CDW such as those described in Sect. 2.3. This conclusion is supported by the observation of the $4k_F$ CDW at higher temperatures where the $2k_F$ anomaly is not observed. The origin of the change in the $4k_F$ CDW wave number has been ascribed to a change in the charge-transfer rate from TTF to TCNQ.

The $2k_F$ and $4k_F$ anomalies are observed above 53 K and 49 K, respectively, regardless of the values of the a- and the c-components of the wave vector Q. Therefore those points at which they are observed form flat planes called diffuse sheets in reciprocal space. The data shown in Fig. 3.16 were obtained by a so-called counter method. When we employ a photographic method, a set of diffuse streaks is found on a film as shown in the frontispiece of this volume [3.35]. This fact suggests that the CDW, in the temperature range mentioned above, has a certain correlation in space along the 1d axis but has no correlation along perpendicular directions. The presence of the diffuse sheets of scattering intensity indicates the absence of any factors of the form $\Delta(K_a - G_a \pm Q_a)$ or $\Delta(K_c - G_c \pm Q_c)$ in (3.21), and accordingly, the correlation functions $G(r_a)$ and $G(r_c)$ are zero except at $r_a = r_c = 0$. Thus we may say that a CDW grows on each molecular stack and has no phase correlation with those on neighbouring stacks. This is a natural result, which reflects the one-dimensionality of the conduction electrons.

With inelastic neutron scattering it is possible to check if the $2k_F$ and the $4k_F$ anomalies are brought about by the Kohn anomaly. Figures 3.17 and 3.18 show dispersion relations of phonons measured by this method [3.36]. At $2k_F$, a clear Kohn anomaly is observed in transverse phonons with wave vector along the b-axis and polarization parallel to the c-axis. Although the frequency of soft phonons is expected to become zero at the Peierls transition temperature 53 K, direct measurements have not been made below 84 K because of the low scattering intensity here due to the thermal population factor.

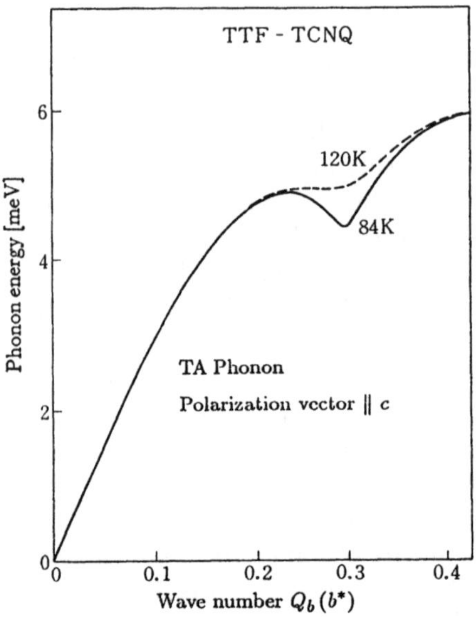

Fig. 3.17. The Kohn anomaly observed by inelastic neutron scattering in TTF-TCNQ [3.36]

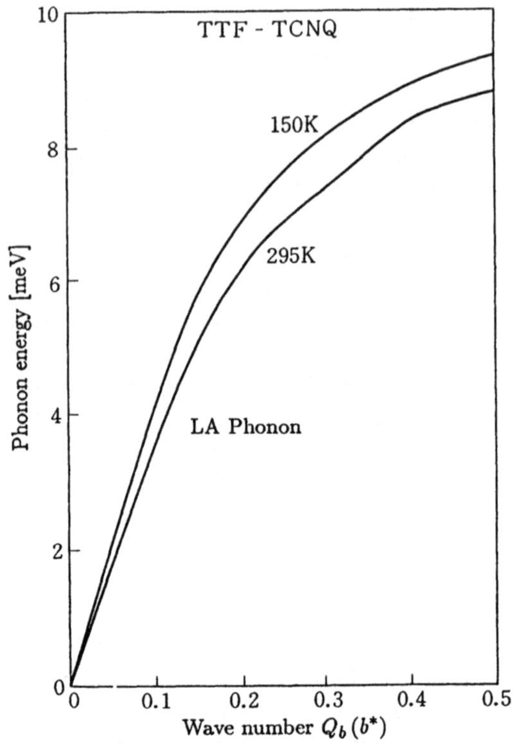

Fig. 3.18. Dispersion relations of longitudinal acoustic phonons measured by inelastic neutron scattering along the one-dimensional b-axis of TTF-TCNQ [3.36]

It is not surprising that the Kohn anomaly occurs in the transverse phonon branch. The reason is the following: The interplanar distance between neighbouring molecules in a stack is changed by a molecular displacement along the c-axis because of a tilting of the molecular plane as shown in Fig. 3.5. Thus the molecular displacement results in a change of $t_{\|}$, which is a function of the interplanar distance, giving rise to strong electron-phonon interactions. No phonon softening is observed, however, at $4k_F$, at least, not in the acoustic and the low-lying optical branches. Therefore the $4k_F$ anomaly observed by x-ray scattering must be ascribed either to a softening in other optical branches or to a static lattice distortion. At present the latter possibility is considered to be more plausible. The failure to observe the expected anomaly in the elastic intensity at $4k_F$ may be attributable to a high background intensity.

The reason why neutron scattering has not detected the anomly obtained by x-ray scattering may be the difference in the cross-section of the scatterers in the two cases. The scattering cross-section for x-rays is larger in TTF than in TCNQ, but TCNQ contributes more than TTF does in neutron scattering. If the $4k_F$ CDW is present predominantly on TTF stacks, the experimental results may be explained. This point will be examined in detail in Sect. 3.4.4.

It is to be remarked that the $4k_F$ anomaly observed only by x-ray scattering does not come from any surface mode since the x-ray beam is transmitted by the sample crystals in experiments.

3.4.2 Interchain Interactions

The intensities of the $2k_F$ and $4k_F$ anomalies increase with decreasing temperature, and their peak widths become narrow along the 1d axis. The peak width is inversely related to the spatial correlation length of the CDW lattice distortion. (When the correlation length becomes finite in (3.17), the function $\Delta(\boldsymbol{K} - \boldsymbol{G} \pm \boldsymbol{Q})$ in (3.21) has a finite width.) It is a natural result that the correlation length of the fluctuation of the order parameter, which is either the CDW itself, $\varrho = \varrho_0 \cos Q_b y$, or its structural component $u = u_0 \cos Q_b y$, increases with decreasing temperature toward the transition temperature because the Peierls transition is of the second order. The relaxation time must also increase. The correlation length and the relaxation time cannot become infinity at finite temperatures as was explained in Sect. 2.3.1, if the system is ideally one-dimensional. Real systems, however, necessarily have a three-dimensionality. When CDWs grow on each molecular stack, interchain interactions among them will make the correlation length and the relaxation time infinite at a finite temperature. This is considered to be the real nature of the Peierls transition at 53 K.

Coupled CDWs give rise to a 3d superstructure in the lattice below the transition temperature. Figure 3.19 shows the temperature dependence of the intensity of the satellite reflections due to the superstructure and also that of the wave number Q_a describing its periodicity along the a-axis [3.37]. The dash and dotted curve shows a theoretical result for the change of Q_a of the $2k_F$ superstructure between 49 K and 38 K. The theory will be discussed later. Values of the wave numbers are the following: $Q_a = x(T)a^*$, $Q_b = 0.295\, b^* = 2k_F$, and $Q_c = 0$ (or c^*) for the $2k_F$ CDW, and $Q_a = 2x(T)a^*$, $Q_b = 0.59\, b^* = 4k_F$, and $Q_c = 0$ (or c^*) for the $4k_F$ CDW, where $x(T)$ denotes a function of temperature. The temperature dependence of the intensity from the $4k_F$ CDW suggests an onset of the superstructure at 49 K. It is to be noted that another $4k_F$ CDW whose Q_a is equal to zero and temperature independent is observed below 38 K although its satellite intensity is weak compared to those shown in Fig. 3.19.

The a^*-component Q_a of the wave vector changes its value between 49 K and 38 K. That is to say, the ordered structure of CDW undergoes successive phase transitions. Although Q_b is directly related to the Fermi wave number, Q_a and Q_c must be dominated by interchain interactions.

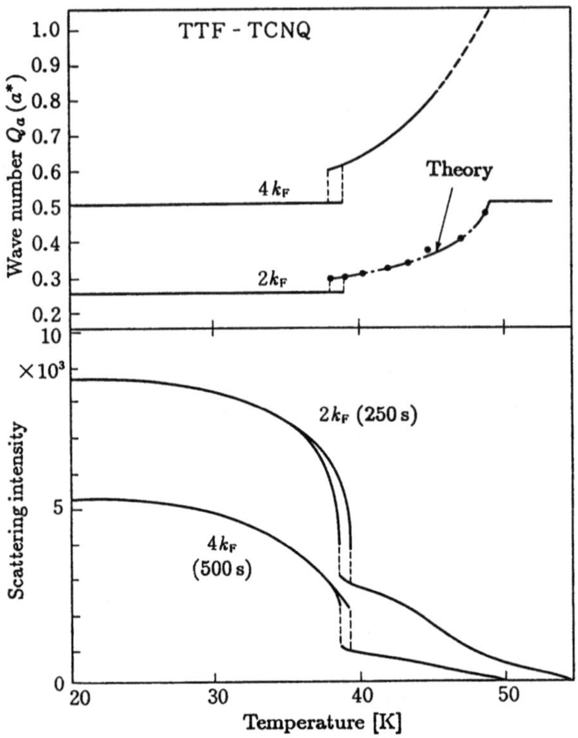

Fig. 3.19. Temperature dependence of the wave number and the intensity of the $2k_F$ and the $4k_F$ satellite reflections of TTF-TCNQ [3.37]

We shall consider now the mechanism of interactions. For the time being, discussions will focus on the $2k_F$ CDW. The important interactions are a Coulomb interaction among CDWs and the electron-phonon interaction which is related to a strain energy due to lattice distortions. Along the c-axis the Coulomb interaction energy is minimum when the phase ϕ differs by π between neighbouring CDWs; $\varrho = \varrho_0 \cos(Q_b y + \phi_i)$ on the ith stack and $\varrho = \varrho_0 \cos(Q_b y + \phi_{i+1})$ with $\phi_{i+1} - \phi_i = \pm\pi$ on the $(i+1)$th stack. The crystal structure of Fig. 3.5 shows that the period of this structure agrees with the original one, c, and, therefore, the wave number becomes 0 or c^*. In addition, from a view point of the strain energy, such a structure has here too the lowest energy. The reason is that the lattice distortion $u = u_0 \cos(Q_b y)$ has the same phase between neighbouring stacks when the CDWs on them have a phase difference of π. This arises from the tilting of planar molecules whose tilting direction is opposite on neighbouring stacks.

What happens along the a-axis? The presence of two kinds of stack, TCNQ and TTF, must be taken into account along this direction. Magnetic measurements suggest, as described in Sect. 3.3, that TCNQ stacks undergo a transition at 53 K and TTF at 49 K. Therefore we assume that CDWs on TCNQ stacks alone form the 3d structure between 53 K and 49 K. We can discuss the a-axis in a similar way to the c-axis. The phase of the CDWs will differ by π between neighbouring TCNQ stacks along the a-axis. The resultant structure has a period twice that of the original lattice thus giving rise to the wave number $0.5\,a^*$ in accordance with the experimental result. Below 49 K the $2k_F$ CDWs on TTF stacks are assumed to take part in the three-dimensional structure of the CDWs. Participation of these will modify the overall structure. This problem was phenomenologically studied by *Bak* and *Emery* [3.38] in terms of a free energy for the system. They obtained a result which agress well with the experimental result shown in Fig. 3.19.

Here we shall consider the problem using a slightly different method [3.39]. The amplitude of the $2k_F$ CDW on a TCNQ stack is set as ϱ_Q, and its phase difference from that on the neighbouring TCNQ stack along the a-axis is assumed to be ϕ. The interaction energy between the two CDWs will be given by

$$V_{QQ}\varrho_Q^2 \cos\phi \quad (V_{QQ}>0) \quad .$$

The solution for the minimum in energy, $\phi=\pi$, has already been obtained. Next, we introduce the amplitude ϱ_F and the phase α for the $2k_F$ CDW on the TTF stack which is present in between two TCNQ stacks considered above. The phase α is measured from the average of the phases of the two neighbouring CDWs on the TCNQ stacks.

The interaction energy between the CDW on the TTF stack and those on TCNQ stacks is given by

$$\begin{aligned} &V_{QF}\varrho_Q\varrho_F\{\cos[\alpha-(\phi/2)]+\cos[\alpha+(\phi/2)]\} \\ &\quad = 2V_{QF}\varrho_Q\varrho_F \cos(\alpha)\cos(\phi/2) \quad . \end{aligned}$$

Thus the total interaction energy U is

$$U = V_{QQ}\varrho_Q^2\cos\phi + 2V_{QF}\varrho_Q\varrho_F\cos\alpha\cos(\phi/2) \quad , \tag{3.25}$$

where $V_{QF}>0$. Minimizing U with respect to ϕ and α, $\partial U/\partial\phi=0$ and $\partial U/\partial\alpha=0$, we obtain solutions, $\alpha*=0$ or π. In order to explain the experimental results we have to choose the solution $\alpha=\pi$, and hence ϕ, which is equal to π for $\varrho_F=0$, decreases from π for $\varrho_F\neq 0$. In using this we obtain a solution for ϕ as $\phi=\pi-\delta$, $(\delta\ll 1)$, where $\delta=V_{QF}\varrho_F/V_{QQ}\varrho_Q$. The periodic structure having these phase differences is specified by the wave number $Q_a=0.5\,a^*-(\delta/2\pi)a^*$. If the temperature dependence of ϱ_F is described in terms of $(T_C-T)^{1/2}$ as is characteristic of the order parameter for a second-order phase transition, then Q_a can be written as $Q_a=0.5\,a^*-\text{const.}(49-T)^{1/2}a^*$.

Although the above discussion is of a largely intuitive nature, the more precise theory by Bak and Emery gives the same temperature dependence for Q_a. The first-order transition at 38 K and the value of $Q_a = 0.25\,a^*$ below it can be explained by the introduction of a commensurability energy along the a-axis.

The phases of the CDWs on both the TCNQ and the TTF stacks are obtained exactly in the above discussion; the ordered structure being definitely determined in this theoretical model. Several different models, however, have been proposed for the ordered structure [3.40]. This is because experiments have not yet revealed what type of molecular displacements occur – only the wave number and the amplitude of the sinusoidal lattice distortion have so far been measured.

It is to be noted that the 38 K transition can also be explained by a model of *Iwabuchi* et al. [3.41] which is quite different from the Bak and Emery. This model deals with both the $4k_F$ and the $2k_F$ CDW and neglects the presence of two kinds of molecular stacks. Interactions between the $2k_F$ and the $4k_F$ CDWs on one kind of stack only, can stabilize the wave number $Q_a = 0.25\,a^*$.

The ordered structure of the $4k_F$ CDW is expressed as follows. Components of the wave vector for the $4k_F$ structure are twice as large as those for the $2k_F$ one; $Q(4k_F) = 2Q(2k_F)$. The $4k_F$ CDW is not a simple second harmonic of the $2k_F$ but must have its own origin. The ordered structure will be determined by interchain interactions between $4k_F$ CDWs only. The interchain interaction may be considered to be weak because of interchain distances, 6 Å along the a-axis and 9 Å along the c, which are longer than the $4k_F$ wavelength, $\lambda = 2\pi/4k_F = 1.7\,b = 6.5$ Å. The theory by *Lee, Rice,* and *Klemm* [3.25] shows, however, that the interchain interactions among $4k_F$ CDWs are strengthened by those among $2k_F$ CDWs if the $4k_F$ CDW is formed by Coulomb repulsions among conduction electrons. Furthermore these authors point out that the $4k_F$ CDW behaves like a second harmonic of the $2k_F$ one in this case.

The other $4k_F$ satellite having a weak intensity and $Q_a = 0$ may be explained in terms of a combination of two $2k_F$ distortions; $Q(Q_a, 2k_F, 0) + Q(-Q_a, 2k_F, 0) \rightarrow Q(0, 4k_F, 0)$.

Here the model of Iwabuchi et al. must again be mentioned. This model also predicts the ordered structure of the $4k_F$ CDW [3.41]. The $4k_F$ structure with $Q_a \neq 0$, however, cannot be explained although the model can explain the presence of the weak $4k_F$ satellite with $Q_a = 0$.

3.4.3 Origins of the $4k_F$ CDW

It is believed that the $4k_F$ CDW forms a 3d structure below 49 K, i.e. at a lower temperature than the Peierls transition at 53 K, as shown in

Fig. 3.19. The magnetic measurements described in Sect. 3.3 also suggest that the TTF stacks undergo a certain transition at 49 K. A possible origin of the $4k_F$ CDW is the electron-electron Coulomb interaction, which is not negligible compared to the band energy, as described in Sect. 2.3. Data for the thermoelectric power indicate that the band width of TTF stacks is smaller than that of TCNQ stacks. Therefore the $4k_F$ CDW will appear more readily on TTF stacks, assuming that the magnitude of the Coulomb interaction is similar on each kind of stack. Furthermore neutron scattering experiments, in which the cross-section of TTF molecules is smaller than for TCNQ, cannot detect the $4k_F$ CDW, at least in the higher temperature range. These experimental and theoretical results suggest that the $4k_F$ CDW is present predominantly on TTF stacks. This idea is further supported by other experimental results. Only the $2k_F$ CDW is observed in TSeF-TCNQ [3.42], which is a selenium analogue of TTF-TCNQ, while both the $2k_F$ and the $4k_F$ CDW are observed in TTF-TCNQ. It is expected that systematic investigations of the $2k_F$ and the $4k_F$ CDWs in the alloy series $TTF_{1-x}TSeF_x$–TCNQ will clarify the kind of molecular stacks on which they are predominantly present.

Figure 3.20 shows the phase diagram obtained for the above system using x-ray scattering [3.43]. The $4k_F$ CDW loses its 3d ordering even in

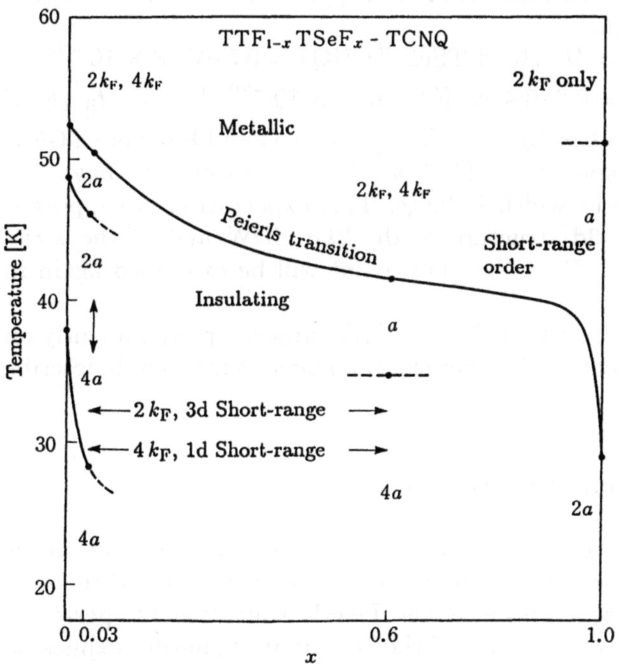

Fig. 3.20. Phase diagram of charge density waves in $TTF_{1-x}TSeF_x$-TCNQ [3.43]

the alloy with $x = 0.03$ although the $2k_F$ CDW keeps it in the highly disordered compound with $x = 0.6$.

In alloys with $x = 0.03$ and 0.6, the correlation length along the 1d axis is shorter for the $4k_F$ CDW than for the $2k_F$ CDW. For the $4k_F$ CDW the correlation length decreases with increasing x while for the $2k_F$ CDW it remains nearly constant between $x = 0.03$ and 0.6. Here it is to be noted that the disorder caused by alloying is selectively introduced into fulvalene stacks. Hence the increased sensitivity of the $4k_F$ CDW to disorder as compared to the $2k_F$ CDW, again suggests that the $4k_F$ CDW is present predominantly on fulvalene stacks. Furthermore the relative intensity of satellites is found to decrease for the $2k_F$ CDW with increasing x while it increases in the $4k_F$ case. Since the x-ray scattering factor of selenium atoms is much bigger than that of the other constituent atoms, this result also indicates that the $4k_F$ and $2k_F$ CDWs are predominantly located on TTF and TCNQ stacks, respectively.

The absence of the $4k_F$ CDW in TSeF-TCNQ would suggest an increase of the band width (or a decrease of Coulomb repulsion) in the fulvalene stacks. Intermolecular distances along the stack axis are nearly the same between TTF-TCNQ and TSeF-TCNQ, and the van der Waals radius of selenium atoms is larger than that of sulphur. Therefore the transfer integral of electron wave functions is expected to be larger in TSeF than in TTF stacks. Actually band calculations with rough approximations give following values for the band width:
$4t_{\|}$ (Q of TTF-TCNQ) = $4t_{\|}$ (Q of TSeF-TCNQ) = 0.5 eV (8×10^{-20} J), $4t_{\|}$ (F of TTF-TCNQ) = 0.2–0.4 eV [(3.2–6.4) $\times 10^{-20}$ J], and $4t_{\|}$ (F of TSeF-TCNQ) = 0.6 eV (9.6×10^{-20} J) [3.44], where Q and F denote TCNQ and fulvalene stacks, respectively. TSeF stacks are expected to have $2k_F$ CDWs because their band width is large. This expectation is supported by investigations of the 3d structure of the $2k_F$ CDW and of the x-ray scattering intensity in TSeF-TCNQ. This point will be examined again in Sect. 3.5.

Thus it is almost certain that the $4k_F$ CDW appears predominantly on fulvalene stacks. At present the Wigner crystal model seems to well describe real phenomena.

3.4.4 Dynamics of Charge Density Waves

Studies of TTF-TCNQ were much stimulated by the finding of an unexpected behaviour in the electrical conductivity; namely its rapid increase with decreasing temperature towards the Peierls transition temperature. The strong temperature dependence of the conductivity, hardly explicable by a model for simple metals, is likely to be brought about by the sliding

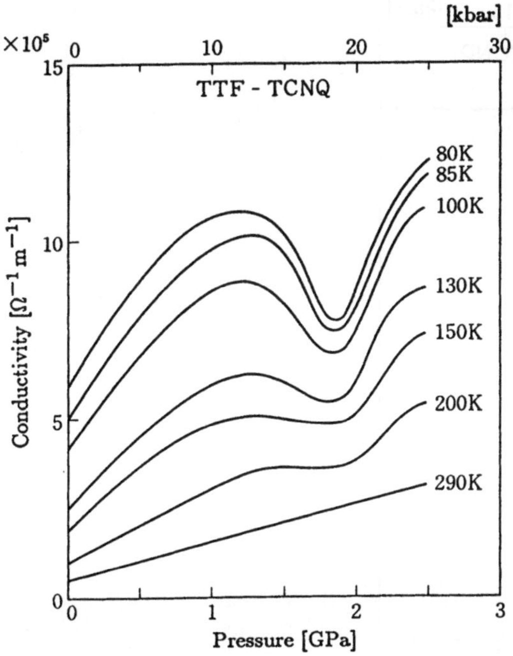

Fig. 3.21. Pressure dependence of the dc electrical conductivity measured along the *b*-axis of TTF-TCNQ [3.45]

motion of fluctuating CDWs. Also the frequency dependence of the conductivity described in Sect. 3.2.3 suggests that CDWs participate in the conduction near $\omega = 0$ since the *dc* conductivity is anomalously large compared to the high-frequency conductivity. An answer is given below to the elementary but essential problem of what dominates the conductivity in the metallic phase.

Figure 3.21 shows the conductivity along the 1d axis of TTF-TCNQ under hydrostatic pressure [3.45]. A decrease in the conductivity is observed at around 19 kbar (1.9×10^9 Pa) which becomes more pronounced with decreasing temperature below about 200 K. Such anomalies are never found along directions perpendicular to the 1d axis. Naively speaking, an application of hydrostatic pressure will reduce intermolecular distances, but is this the real mechanism which brings about the decrease of the conductivity at about 19 kbar? Some important information on this comes from neutron scattering measurements at low temperatures under pressure. Figure 3.22 shows the pressure dependence of the measured $2k_F$ value [3.46]. The $2k_F$ value increases with increasing pressure reaching a commensurable value, $1/3\, b^*$, at 14–15 kbar.

The increase of the $2k_F$ value with pressure can be understood in a simple model: We shall assume the band structure shown in Fig. 3.23 for TCNQ and TTF bands. A reduction of the intermolecular distances along the 1d axis will cause a broadening of the band. If we assume that the energy

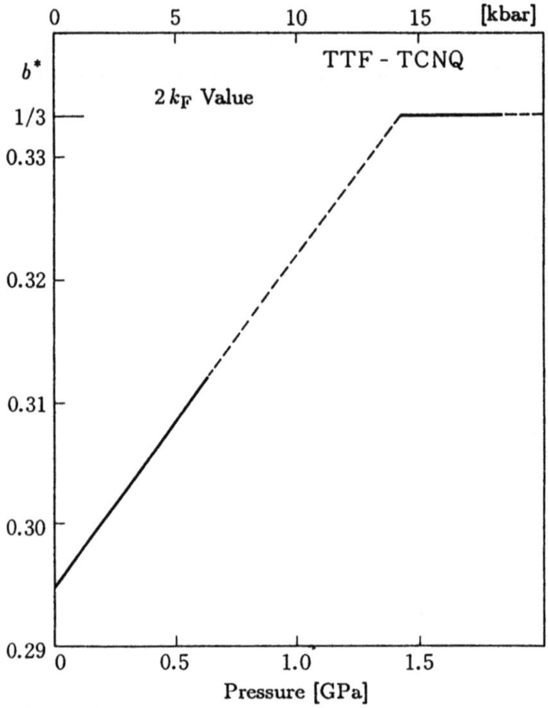

Fig. 3.22. Pressure dependence of the $2k_F$ value of TTF-TCNQ [3.46]

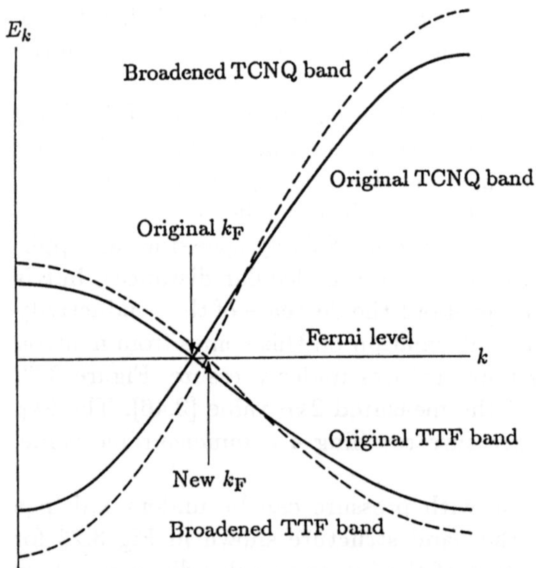

Fig. 3.23. Change of the $2k_F$ value with increasing pressure

at the band centre is independent of pressure, i.e. that the molecular energy level for electrons participating in the band formation is unchanged, band widths of both TTF and TCNQ stacks will be broadened by pressure as shown by dashed curves in Fig. 3.23. This gives rise to the increase of the $2k_F$ value.

When the value of $2k_F$ becomes commensurable, there is hardly any sliding motion of the CDWs because of the commensurability locking described in Sect. 2.2.2. Thus the decrease of the conductivity shown in Fig. 2.31 can be ascribed to a suppression of the sliding motion of fluctuating CDWs. If it is assumed that the minimum value of the conductivity, at around 19 kbar, represents the conductivity due to single particle processes, we can say that at 80 K, about half of the conductivity is due to CDWs. It is interesting to note that the temperature, 200 K, below which this phenomenon becomes more pronounced, agrees well with the onset temperature, 150 K, of the $2k_F$ CDW. The Peierls transition at 53 K under ambient pressure has been considered to be brought about by the pinning of CDWs on each molecular stack and 3d interchain coupling among them. When the commensurability locking is also present at about 19 kbar, the transition temperature must go up. In fact, conductivity measurements under pressure verify this expectation as shown in Fig. 3.24 [3.47]. The pressure dependence of the transition temperature is explained fairly well by the theory of [3.48]. The theory also predicts that the character of the transition changes into that

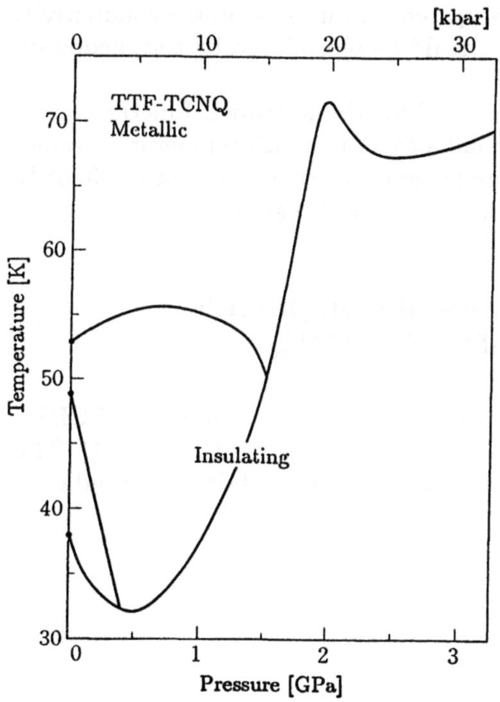

Fig. 3.24. Phase diagram of TTF-TCNQ [3.47]

of a first-order transition when the commensurability locking is present. A hysteresis in the transition temperature is really found around 19 kbar in conductivity measurements.

Similar phenomena are also expected to occur in TSeF-TCNQ. The $2k_F$ value of TSeF-TCNQ is $0.315\,b^*$ as will be described in Sect. 3.5 while that of TTF-TCNQ is $0.295\,b^*$. Therefore the $2k_F$ value of TSeF-TCNQ is expected to reach the commensurable value, $1/3\,b^*$, at a pressure lower than 19 kbar. Although its $2k_F$ value has not been measured at high pressures, data obtained for the conductivity are similar to those of Fig. 3.21 [3.49]. The conductivity minimum is observed at about 6 kbar (6×10^8 Pa), which is indeed lower than the 19 kbar for TTF-TCNQ.

The temperature dependence of the conductivity of TTF-TCNQ near the transition temperature is well explained by a pinning mechanism for CDWs. An impurity potential is given to the CDW which could otherwise slide without resistance at 0 K as discussed in Sect. 2.2.2. A solution of the Hamiltonian which is a combination of (2.63) and (2.67) gives the frequency dependence of the conductivity as shown in Fig. 2.16 [3.59]. Although the conductivity is zero at $\omega = 0$ because of the impurity pinning, for $\omega \neq 0$ it becomes finite. The energy supplied by the alternating fields, $\omega \neq 0$, excites the pinned CDWs giving rise to the finite conductivity. Here, we shall assume that the ω-axis is equivalent to a temperature axis and that the point $\omega = 0$ corresponds to the Peierls transition temperature 53 K. This assumption means that we let the phase-transition temperature, 0 K for purely 1d systems, correspond to that of a real system. If this is valid, the conductivity of TTF-TCNQ near 53 K, Fig. 3.9, should be reproduced by the theoretical curve of Fig. 2.16.

From the above discussions we conclude the following: Fluctuations of CDWs contribute to the electrical conductivity in a high temperature range, 58–150 K, and this contribution diminishes at lower temperatures, 53–58 K. The latter temperature range is a kind of critical regime.

3.5 Characteristic Properties of Compounds Belonging to the Family of TTF-TCNQ

We shall make here a general survey of the Peierls transition and of CDWs in various organic compounds belonging to the family of TTF-TCNQ. This will serve to enrich and extend our understanding obtained in studies of TTF-TCNQ.

3.5.1 TSeF-TCNQ

TSeF-TCNQ is a selenium analogue of TTF-TCNQ. Its crystal symmetry and structure are the same as those of TTF-TCNQ. The lattice parameters are slightly larger than those of TTF-TCNQ due to the larger van der Waals radius of selenium atoms compared to sulphur; $\Delta a = 0.22$ Å, $\Delta b = 0.05$ Å, $\Delta c = 0.09$ Å, and $\Delta\beta = -0.28$ Å [3.51]. Nevertheless it is considered that the transfer integral of electron wave functions is larger in TSeF stacks than in TTF ones because the difference in the van der Waals radius, 0.3 Å, is larger than the difference of the intermolecular distance between the different fulvalene stacks. The band calculation mentioned in Sect. 3.4.3 suggests that the band width of TSeF stacks is about 0.6 eV (9.6×10^{-20} J) while that of TTF ones is 0.2–0.4 eV [$(3.2\text{–}6.4) \times 10^{-20}$ J]. The band width of TCNQ stacks is estimated to be about 0.5 eV (8×10^{-20} J) in both compounds. Accordingly one can expect that the electrical conductivity of TSeF stacks slightly exceeds that of TCNQ stacks in TSeF-TCNQ. Actually the thermoelectric power of TSeF-TCNQ is positive whereas it is negative in TTF-TCNQ [3.52].

Figure 3.25 shows the temperature dependence of the *dc* conductivity measured along the *b*-axis in TSeF-TCNQ [3.13]. Data for TTF-TCNQ are also shown for comparison. TSeF-TCNQ is believed to undergo the Peierls transition at 29 K. The conductivity decreases monotonically with decreasing temperature below 29 K showing no anomalies such as the hysteresis observed at 38 K in TTF-TCNQ. TTF-TCNQ shows a wide variety of behaviour in its structural, electrical and magnetic properties since the transition temperature is different for the CDW on TCNQ stacks and that on TTF stacks. What happens in TSeF-TCNQ? If there is only one transi-

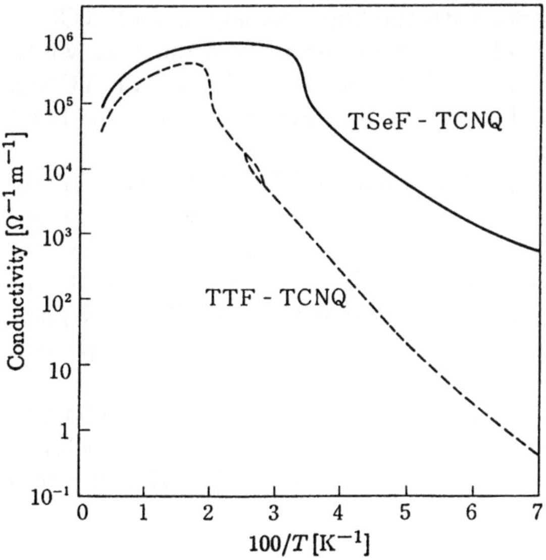

Fig. 3.25. Temperature dependence of the dc electrical conductivity measured along the one-dimensional axis of TSeF-TCNQ [3.13]

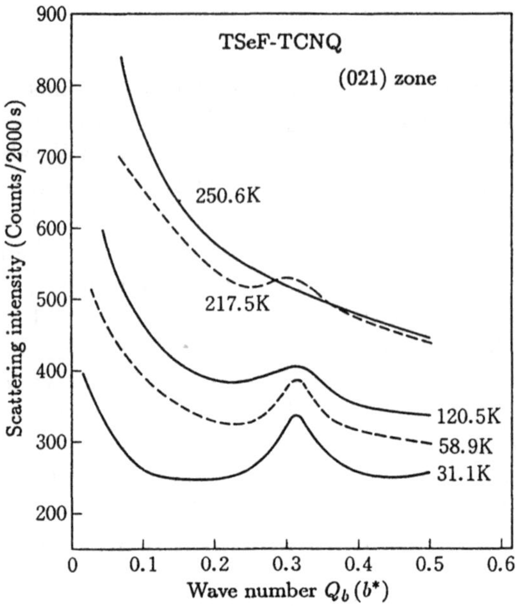

Fig. 3.26. Intensities of diffuse x-ray scattering in TSeF-TCNQ [3.53]

tion, both TCNQ and TSeF stacks must undergo the Peierls transition at the same temperature, 29 K.

From x-ray scattering data only one kind of CDW has been identified [3.42] with the wave number $Q_b = 0.315\, b^*$ below about 240 K, as shown in Fig. 3.26. It is thought that this is the $2k_\mathrm{F}$ CDW rather than the $4k_\mathrm{F}$. The reasons are the following [3.53]: 1) The factor $(\boldsymbol{K} \cdot \boldsymbol{e}_Q)^2$ in (2.21) makes it possible to determine the polarization direction of the lattice distortion. Experiments show that the lattice distortion has a transverse polarization parallel to the c-axis in accordance with the $2k_\mathrm{F}$ anomaly in TTF-TCNQ. 2) The temperature dependence of the peak intensity and the peak width at $Q_b = 0.315\, b^*$ are similar to those of the $2k_\mathrm{F}$ anomaly in TTF-TCNQ. 3) The $4k_\mathrm{F}$ CDW is not expected to appear in TSeF stacks since the band width here is larger than in TTF stacks.

It can be claimed that the $2k_\mathrm{F}$ CDW found in TSeF-TCNQ is definitely present on the TSeF stacks, because it is the selenium atoms which dominate the x-ray scattering. This claim is naturally accepted since the band width of TSeF stacks is also larger than that of TCNQ. However, it is thought that the electronic situation in TCNQ stacks is common to both TTF-TCNQ and TSeF-TCNQ. This would suggest that the $2k_\mathrm{F}$ CDW will also be present on the TCNQ stacks of TSeF-TCNQ.

This interpretation is further supported by the following fact [3.53, 54]: The $2k_\mathrm{F}$ anomaly shown in Fig. 3.26 loses its one-dimensionality below a certain temperature. It shows a 2d character below 100 K giving rise to a short-range order of CDWs with $Q_c = 0$ along the c-axis. Furthermore a 3d character appears below 55 K with $Q_a = 0$ and $Q_c = 0$ although the ordering is again only short-range. The wave number $Q_c = 0$ is explained by the same arguments as for TTF-TCNQ. Also $Q_a = 0$ can be similarly understood if we assume that the amplitude of the $2k_F$ CDW on TSeF stacks is not significantly different from that on TCNQ. The short-range order in 3d is observed in a wide range of temperature in TSeF-TCNQ although it was only found in the vicinity of the Peierls transition temperature in TTF-TCNQ. This fact suggests the presence of a strong interaction among CDWs. The $2k_F$ CDW will grow on both TSeF and TCNQ stacks.

The conductivity of TSeF-TCNQ, unlike TTF-TCNQ, appears to stop increasing below 50 K. This fact can be attributed to the onset of 3d ordering. The CDW has long-range 3d order below the Peierls transition temperature 29 K. The ordering is specified by wave numbers $Q_a = 0.5\,a^*$, $Q_b = 0.315\,b^* = 2k_F$, and $Q_c = 0$. The change of Q_a from 0 to $0.5\,a^*$ is explained by assuming that the CDW is present on both TCNQ and TSeF stacks with similar amplitudes [3.53].

3.5.2 HMTTF-TCNQ and TMTSF-TCNQ

Figure 3.27 shows the crystal structure of HMTTF (hexamethylenetetrathiafulvalene)-TCNQ. HMTSF-TCNQ is its selenium analogue [3.55, 56].

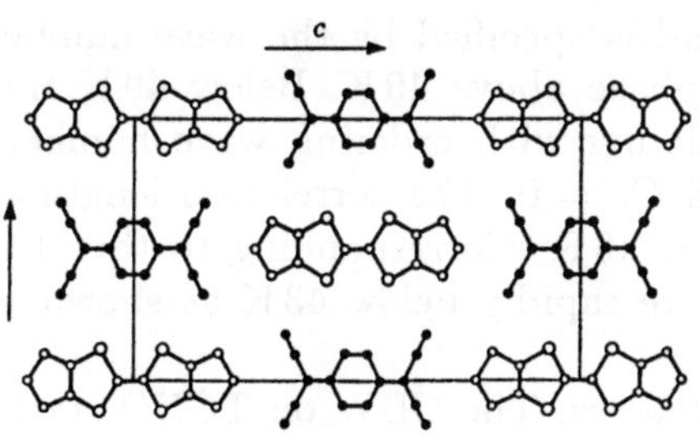

Fig. 3.27. Crystal structure of HMTTF-TCNQ. HMTSF-TCNQ is a selenium analogue of HMTTF-TCNQ [3.55, 56]

Figure 3.28 shows the temperature dependence of the *dc* electrical conductivities measured along the 1d axis in the compounds [3.55, 57]. A logarithmic derivative of the conductivity with respect to the inverse temperature indicates the presence of a metal-insulator transition at 49 K in HMTTF-TCNQ and at 24 K in HMTSF-TCNQ. The presence of another transition in HMTTF-TCNQ at 43 K is also suggested. This difference between HMTTF-TCNQ and HMTSF-TCNQ is reminiscent of that between TTF-TCNQ and TSeF-TCNQ.

Both the $2k_F$ and the $4k_F$ CDW are observed by x-ray scattering in HMTTF-TCNQ, but only the $2k_F$ CDW in HMTSF-TCNQ [3.57]. The

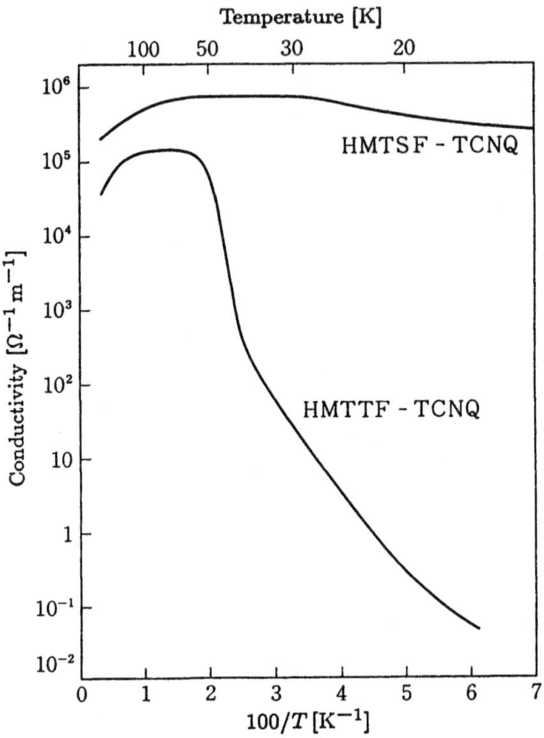

Fig. 3.28. Temperature dependence of dc electrical conductivities of HMTTF-TCNQ and HMTSF-TCNQ [3.55, 57]

$2k_F$ CDW of HMTTF-TCNQ is 1d and is specified by the wave number $Q_b = 0.36\,b^* = 2k_F$ in the metallic phase above 49 K. Below 49 K this CDW undergoes a short-range 3d ordering with ordering wave numbers $Q_a = 0.42\,a^*$, $Q_b = 0.36\,b^* = 2k_F$ and $Q_c = 0$. The correlation length of the 3d structure is infinitely long below 43 K. Corresponding to this, the satellite intensity starts to increase more rapidly below 43 K as shown in Fig. 3.29.

This result can be interpreted as follows. The CDW on TCNQ stacks undergoes a 3d ordering at 49 K whereas that on HMTTF only begins to order at 43 K. The rapid increase of the satellite intensity below 43 K is thus ascribed to the large x-ray scattering factor of the sulphur atoms in HMTTF molecules, which give rise to the 3d structure below this temperature. A similar interpretation was given for the corresponding phase transition in TTF-TCNQ. In HMTTF-TCNQ, however, the transverse wave number Q_a is commensurate along the a-axis. These phenomena are explained by taking account of the specific alignment of molecular stacks in HMTTF-TCNQ and HMTSF-TCNQ. Unlike TTF-TCNQ, here the fulvalene stacks and TCNQ stacks are alternately aligned along the a- and the c-axes giving rise to a type of interchain interaction different from that in TTF-TCNQ. The above experimental results are explained fairly well by taking account of this difference [3.58].

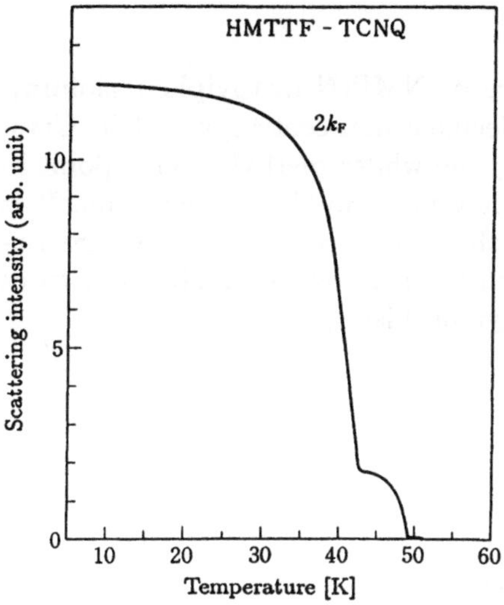

Fig. 3.29. Temperature dependence of the $2k_F$ satellite intensity of HMTTF-TCNQ [3.57]

The $4k_F$ CDW of HMTTF-TCNQ appears below 100 K. It has the wave number $Q_b = 0.28\,b^*$ along the 1d axis. The value $0.28\,b^*$ is equivalent to $0.72\,b^*$, which corresponds to $4k_F$. At sufficiently low temperatures, the $4k_F$ CDW forms a 3d structure whose wave numbers along the a- and the c-axes are just twice as large as those of the $2k_F$ CDW. However, the onset temperature of the 3d ordering has not yet been determined because of the weak scattering intensity of x-rays. We cannot discard the possibility that this $4k_F$ CDW is a second harmonic of the $2k_F$ wave since the onset temperature of the $4k_F$ CDW (100 K) is lower than that of the $2k_F$ (room temperature). For the time being, however, we shall assume that the $4k_F$ CDW has its own distinct origin as in TTF-TCNQ.

HMTSF-TCNQ, like TSeF-TCNQ, has only the $2k_F$ CDW. It is observed even at room temperature with the wave number $Q_b = 0.37\,b^* = 2k_F$ along the 1d axis. A long-range 3d ordering occurs below the transition temperature of 24 K. It is specified by wave numbers $Q_a = 0$, $Q_b = 0.37\,b^* = 2k_F$, and $Q_c = 0$. HMTSF-TCNQ undergoes only one transition unlike HMTTF-TCNQ which has two successive transitions at 49 K and 43 K. This is explained, as for TSeF-TCNQ, by assuming that the $2k_F$ CDW grows on both HMTSF and TCNQ stacks.

Thus the relation between HMTTF-TCNQ and HMTSF-TCNQ is similar to that between TTF-TCNQ and TSeF-TCNQ. Therefore we can conclude that fulvalene stacks containing sulphur atoms provide circumstances favourable for the growth of the $4k_F$ CDW, and those containing selenium atoms for that of the $2k_F$ CDW. The physical origin of this is the difference of the band width due to the difference of the van der Waals radius between sulphur and selenium atoms as discussed in detail in Sect. 3.4.3.

3.5.3 NMP-TCNQ

Figure 3.30 shows the crystal structure of NMP(N-methylphenazinium)-TCNQ [3.59]. The shape of an NMP molecule is non-centrosymmetric. Crystals can contain two types of molecules; one whose methyl groups point in the $+b$-direction and the other where they point in the $-b$-direction. It is an open question as to whether or not there is an ordering in the orientation. Furthermore NMP-TCNQ is known to have several crystal structures different from this. (This is called a polymorphism.)

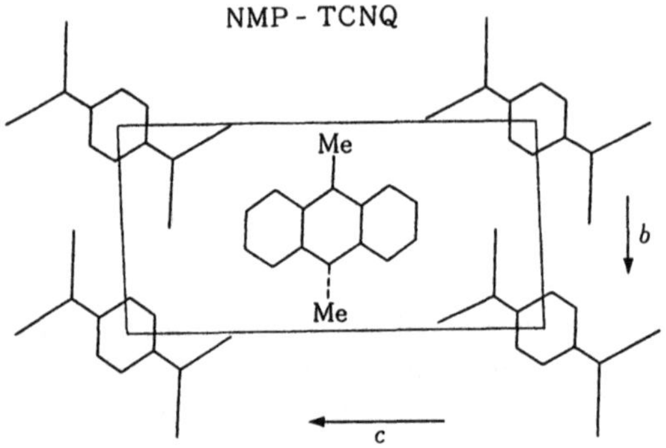

Fig. 3.30. Crystal structure of NMP-TCNQ. The solid and the broken lines for the Me-group represent two possible structures [3.59]

The properties of NMP-TCNQ were studied long before the study of TTF-TCNQ. However, some of the experimental results contradict one another. A possible reason for this lies in the different crystal structures which arise due to the polymorphism. In addition, NMP-TCNQ may contain more impurities than TTF-TCNQ.

The valence state of NMP-TCNQ had been believed to be NMP^{+1}-$TCNQ^{-1}$. NMP stacks, unlike TTF, have a filled valence band and an empty conduction band in the state of NMP^{+1}. Accordingly it was considered that conduction electrons in NMP-TCNQ were present on TCNQ stacks only. A band picture using the Hubbard model was employed because of the strong Coulomb repulsion between conduction electrons. The lower band is just filled by electrons whose density is 1 electron/molecule. NMP-TCNQ, in this case, should show insulating properties. A 1d antiferromagnetic state was expected to appear at sufficiently low temperatures [3.60].

Conductivity measurements, however, of samples of triclinic NMP-TCNQ (Fig. 3.30), whose purity and structure were carefully checked, showed a metallic character near room temperature as shown in Fig. 3.31 [3.57, 61]. Also CDWs have been discovered by x-ray measurements in these samples [3.57, 62]. Two results are of importance here. Firstly, the host lattice is found to have short-range order along the b-axis in the orientation of methyl groups although there is no ordering along the c-axis. Secondly, one

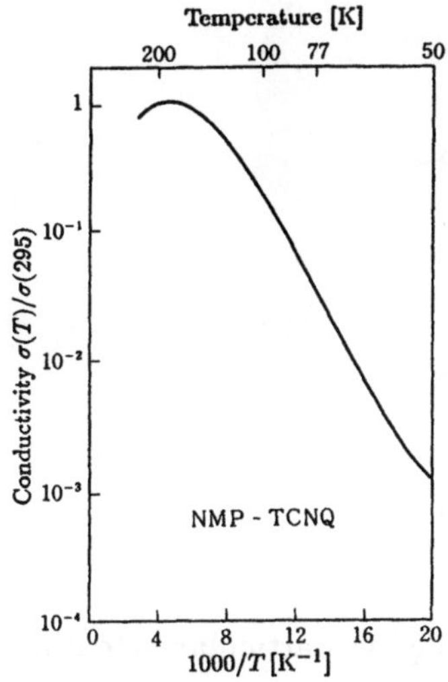

Fig. 3.31. Temperature dependence of the dc electrical conductivity measured along the one-dimensional axis of NMP-TCNQ having the triclinic symmetry [3.57, 61]

type of CDW is found below 70 K with the wave number $Q_{a1} = (1/6)\,a^*$, but another one is present even at room temperature with $Q_{a2} = (1/3)\,a^*$.

The two kinds of CDW thus found may be the $2k_F$ and the $4k_F$ CDWs since they satisfy $Q_{a2} = 2Q_{a1}$. However, a unique character of the conduction band makes this assignment uncertain. The reason is the following. A valence state of NMP$^{+\varrho}$-TCNQ$^{-\varrho}$ is formally rewritten as NMP$^{+1-(1-\varrho)}$-TCNQ$^{-\varrho}$. As the conduction band of NMP stacks is empty in a state of NMP^{+1}, the density of conduction electrons must be $(1-\varrho)$/molecule in the NMP band and ϱ/molecule in the TCNQ band. Therefore (2.8) tells us that $2k_F(\mathrm{NMP}) = (1-\varrho)/2\,a^*$ and $2k_F(\mathrm{TCNQ}) = (\varrho/2)\,a^*$. For $\varrho = 1/3$, $2k_F(\mathrm{NMP}) = Q_{a2} = 4k_F$ (NMP) in the reduced-zone scheme, $2k_F(\mathrm{TCNQ}) = Q_{a1}$ and $4k_F(\mathrm{TCNQ}) = Q_{a2}$. For $\varrho = 2/3$, on the other hand, $2k_F(\mathrm{NMP}) = Q_{a1}$, $4k_F(\mathrm{NMP}) = Q_{a2}$ and $2k_F(\mathrm{TCNQ}) = Q_{a2} = 4k_F(\mathrm{TCNQ})$. Later, CDWs were investigated in the alloy series NMP$_{1-x}$-Phen(Phenazinium)$_x$-TCNQ in order to pick the correct answer from the above two possibilities. Results obtained in alloys with $x = 0.14$ and 0.214 show that the CDW with Q_{a2} remains while the other with Q_{a1} disappears [3.63]. Therefore it is believed that the CDWs with Q_{a1} and Q_{a2} are associated with the NMP and TCNQ stacks, respectively; $Q_{a1} = 2k_F$ (NMP) and $Q_{a2} = 2k_F(\mathrm{TCNQ}) = 4k_F$ (TCNQ).[8] From the above discussion the charge transfer ϱ is determined to be 2/3, in accordance with estimates obtained from measurements of NMR and thermoelectric power [3.57].

Thus the metallic NMP-TCNQ is found to have CDWs on both TCNQ and NMP stacks. It may be necessary to reinvestigate the experimental

[8] A possibility remains that the $4k_F$ CDW may also be present on the NMP stacks because of the relation $4k_F(\mathrm{NMP}) = Q_{a1} = Q_{a2}$.

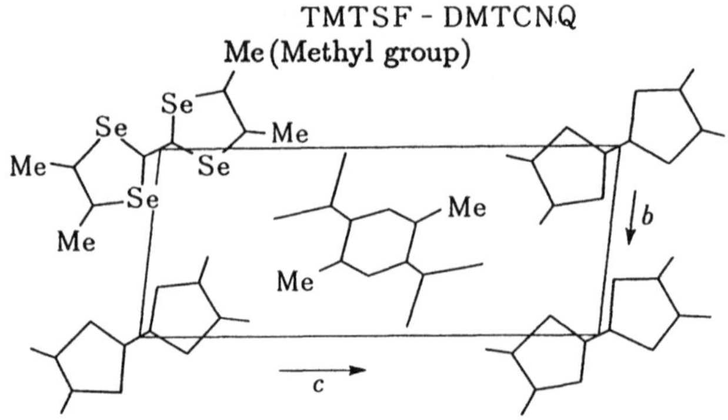

Fig. 3.32. Crystal structure of TMTSF-DMTCNQ [3.64]

results obtained in previous studies, this time paying particular attention to possible impurities and to the polymorphism.

3.5.4 TMTSF-DMTCNQ

Figure 3.32 shows the crystal structure of TMTSF (tetramethyltetraselenafulvalene)-DMTCNQ (dimethyl TCNQ) [3.64]. It has a triclinic symmetry and the a-axis, which is the stacking axis of molecules, is the electrically 1d axis. Figure 3.33 shows the temperature dependence of the conductivity [3.65]. The conductivity along the a-axis shows a metallic character. A metal-insulator transition occurs at 42 K. However, under hydrostatic pressure above about 10 kbar (1×10^9 Pa), TMTSF-DMTCNQ keeps its metallic character right down to the lowest temperature, 1.1 K, reached in experiments. A pressure-temperature phase diagram of TMTSF-DMTCNQ is shown in Fig. 3.34. It is unclear at the present stage whether the ground state continues to be metallic above about 10 kbar. Above 13 kbar (1.3×10^9 Pa) the metal-insulator transition is restored under a magnetic field applied perpendicular to the 1d axis. Furthermor a large magnetoresistance is observed in the metallic phase. Therefore it has been pointed out that a superconductivity may appear in a very low temperature range which has not yet been reached. Another possibility is that TMTSF-DMTCNQ turns into an anisotropic two-dimensional semimetal under pressure.

X-ray scattering measurements have verified that the metal-insulator transition at ambient pressure is the Peierls transition accompanied by the onset of CDWs [3.63]. A CDW with the wave number $0.5\,a^*$ is observed at room temperature, and another one with $0.25\,a^*$ is found below 225 K. These are thought to be the $4k_F$ and the $2k_F$ CDWs. The $4k_F$ CDW is believed not to be a second harmonic of the $2k_F$ wave because it is already present at room temperature where the $2k_F$ CDW is absent. The thermoelectric

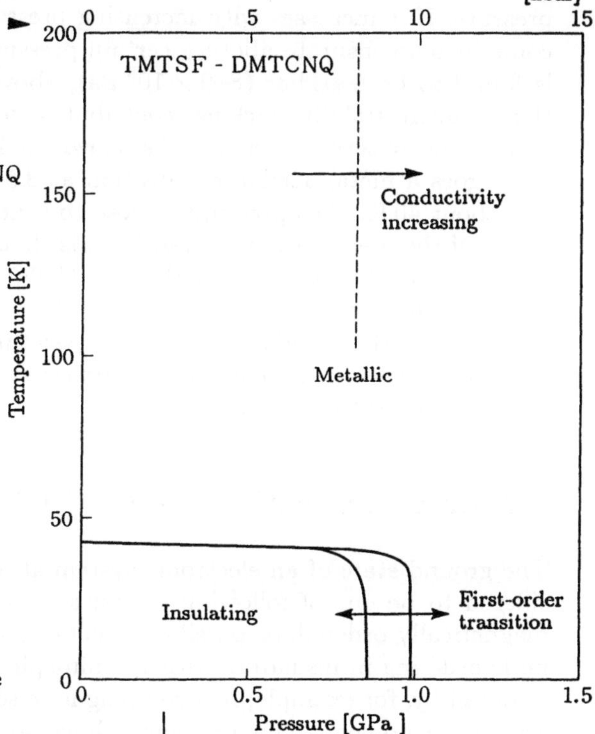

Fig. 3.34. Phase diagram of TMTSF-DMTCNQ [3.65]

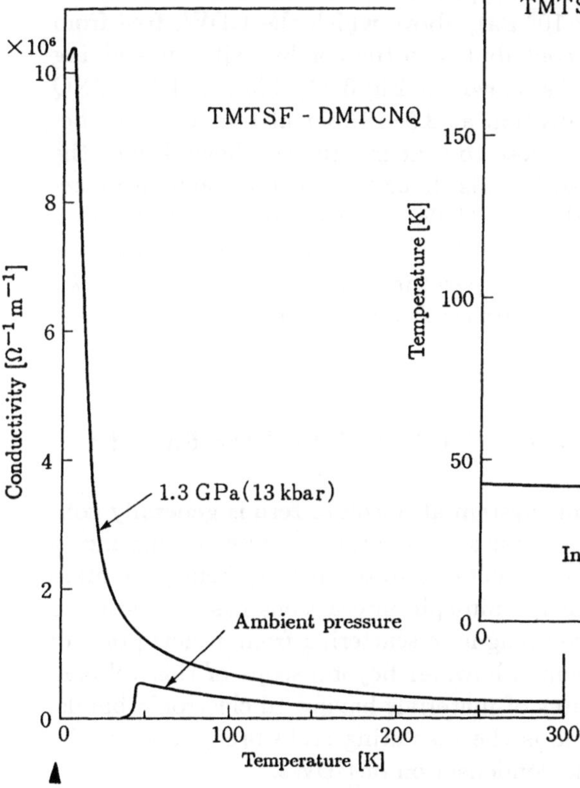

Fig. 3.33. Temperature dependence of the dc electrical conductivity measured along the one-dimensional axis of TMTSF-DMTCNQ [3.65]

power is positive in the metallic phase at ambient pressure suggesting a dominance of TMTSF stacks in the electrical conduction [3.65]. Thus if we apply the same reasoning as for TTF-TCNQ, the $2k_F$ and $4k_F$ CDWs can be ascribed to TMTSF and DMTCNQ stacks, respectively. The $2k_F$ value, $0.25\,a^*$, indicates a charge transfer of 0.5 electrons/molecule form TMTSF to DMTCNQ.

A remarkable increase is found at 8–9 kbar in measurements of the pressure dependence of the conductivity at a constant temperature in the metallic phase [3.65]. This increase, which is hardly detectable at room temperature, becomes more pronounced with decreasing temperature reaching 60 % of the original conductivity at 50 K. Such a behaviour reminds us of the commensurability locking of the $2k_F$ CDW in TTF-TCNQ.

The value of $2k_F$ in TMTSF-DMTCNQ is commensurate at ambient pressure. If it increases with increasing pressure as in Fig. 3.22, it will become incommensurate above a certain pressure. The threshold in pressure is found to be 8–9 kbar (8–9 $\times 10^8$ Pa), above which the CDW, free from the commensurability locking, contributes to the conductivity thus giving rise to the observed increase. As shown in Fig. 3.34, TMTSF-DMTCNQ undergoes a metal-insulator transition at 8.5–10 kbar at temperatures below about 40 K. This pressure is close to that mentioned above. Hence the origin of the metal-insulator transition is thought to be the commensurate-incommensurate transition of the $2k_F$ CDW. It might be expected that a Peierls transition also occurs above 10 kbar with a transition temperature below 1.1 K (the Peierls transition temperature will decrease when the $2k_F$ value becomes incommensurate). Further studies are necessary to give a definite answer to this problem.

3.6 Superconductivity in $(TMTSF)_2X$ and Its Family

The ground state of an electronic system at absolute zero is generally considered to be one of following states: an insulating, a superconducting, a magnetically ordered, or possibly, a metallic state. In 1d systems, the magnetic ordering in insulators with 1d spin-spin interactions has been studied in detail by, for example, neutron magnetic scattering from a view point of critical phenomena. This problem is however beyond scope of this volume. Among the possible ground states of materials having 1d electronic bands, our main subject in this volume is the insulating state brought about by the Peierls transition due to the condensation of CDWs.

We turn now to studies of superconductivity in 1d conductors. The superconductivity in organic compounds was really only discovered in 1980 in $(TMTSF)_2PF_6$ under high pressure [3.66, 67]. In subsequent experiments superconductivity was also found in $(TMTSF)_2$-X, where X = AsF_6, SbF_6, TaF_6, and ReO_4. Furthermore $(TMTSF)_2ClO_4$ was found to undergo a superconducting transition at ambient pressure [3.68, 69]. In 1982 superconductivity was also discovered in another kind of compound, $(BEDT\text{-}TTF)_2X$. Properties of these compounds will be briefly discussed at the end of this section.

3.6.1 Crystal Structure of $(TMTSF)_2X$

Figure 3.35 shows the crystal structure of $(TMTSF)_2PF_6$ [3.67, 70]. Its symmetry is triclinic. The *a*-axis is the stacking axis of TMTSF molecules and, at the same time, is the electrically 1d axis. Molecules of TMTSF form dimerized pairs in each stack; each unit cell contains one pair.

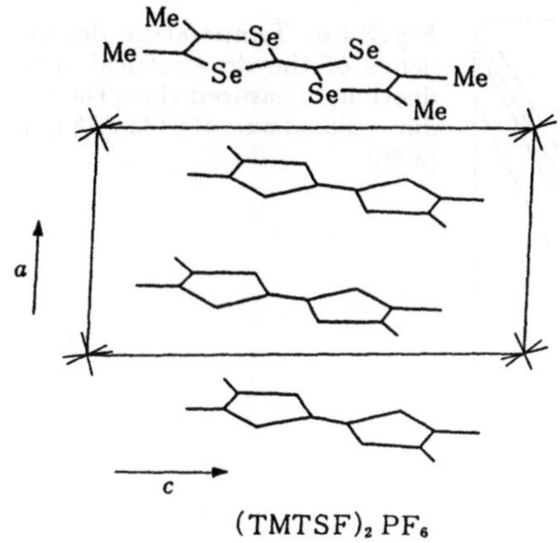

Fig. 3.35. Crystal structure of $(TMTSF)_2PF_6$ [3.67, 70]

Negatively charged groups, X, sit on sites having an inversion symmetry. They are believed to rotate at higher temperatures. When the rotational motion begins to decrease with decreasing temperature, negative groups like ClO_4 and ReO_4 which have no inversion symmetry can have two possible orientations. Therefore they may undergo an orientational ordering at a certain temperature. This problem will be discussed in Sect. 3.6.5 in relation to the interplay between superconductivity and spin density waves.

3.6.2 Electrical Properties

Figure 3.36 shows the temperature dependence of the electrical resistivity of several compounds belonging to this family [3.66]. Every compound except $(TMTSF)_2ClO_4$ undergoes a metal-insulator transition at ambient pressure. Holes are thought to be responsible for the metallic conduction for two reasons: (1) It is well known in chemistry that the anions X attain a well-defined negative charge and become X^{-1}, due to their large electron affinities. Electrons on these anions are expected to be localized. (2) The thermoelectric power is positive and nearly proportional to temperature in the metallic phase.

The carrier density is expected to be one hole per two TMTSF molecules giving a value of $0.5\,a^*$ for $2k_F$. Other compounds belonging to the family of $(TMTSF)_2PF_6$ have the same crystal structure; their electronic states are also expected to be the same.

The anisotropy in the resistivity of $(TMTSF)_2PF_6$ is estimated to be 300–1000 at room temperature. This is one of the largest anisotropies among the many kinds of conductors so far studied. $(TMTSF)_2ClO_4$, however, shows an anisotropy of only about 20. This big difference is thought to be due to the difference in the interstack distances of these compounds. Far-infrared reflectivity measurements in $(TMTSF)_2PF_6$ reveal [3.71] that a

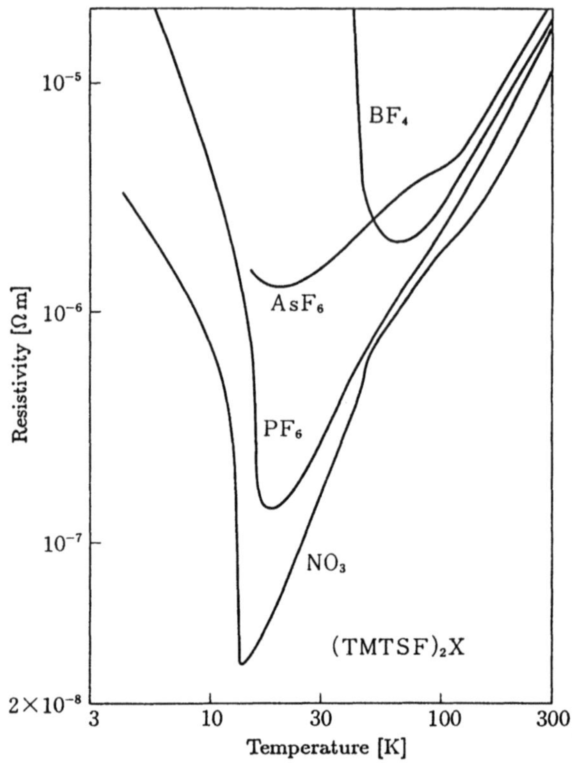

Fig. 3.36. Temperature dependence of the dc electrical conductivity measured along the one-dimensional axis of $(TMTSF)_2X$ [3.66]

plasma edge is clearly observed even for light polarized nearly parallel to the b-axis. These results suggest the presence of an at least two-dimensional character in these compounds. Band widths are estimated to be $4t_{\|} \simeq 1$ eV, $4t_{\perp}(\| b) \simeq 0.1$ eV, and $4t_{\perp}(\| c) < 0.01$ eV from band calculations and analyses of reflectivity data. The problem of dimensionality will be discussed below in relation to the nature of the superconductivity.

3.6.3 Superconductivity

The metal-insulator transition temperature of $(TMTSF)_2PF_6$ decreases with increasing pressure. Superconductivity is found to occur above about 6.5 kbar (6.5×10^8 Pa). Figure 3.37 shows the pressure dependence of the transition temperature [3.72]. Figure 3.38 shows the temperature dependence of the resistivity measured in $(TMTSF)_2PF_6$ and $(TMSTF)_2AsF_6$ under applied magnetic fields [3.67, 68]. Such curves are typical of type II superconductors. They have two kinds of critical field: H_{c1} at which the perfect diamagnetism is broken accompanied by an intrusion of magnetic flux, and H_{c2} above which the superconductivity disappears giving rise to a finite

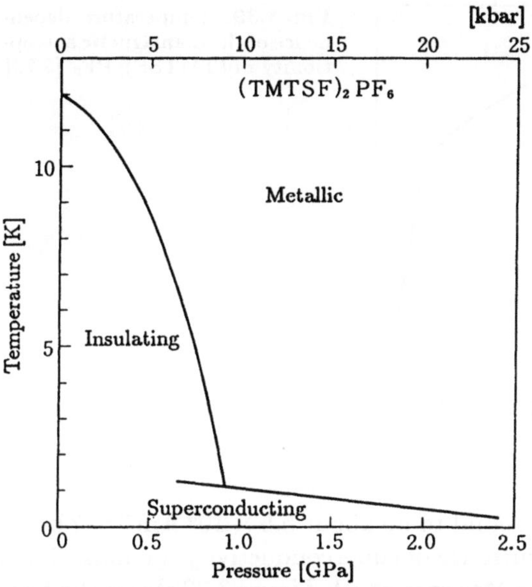

Fig. 3.37. Phase diagram of $(TMTSF)_2PF_6$ [3.71]

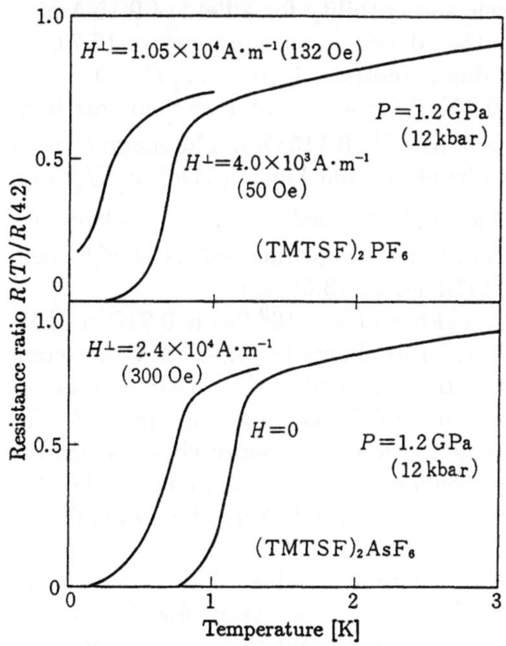

Fig. 3.38. Superconductivity transition and the magnetic field effect in $(TMTSF)_2PF_6$ and $(TMTSF)_2AsF_6$ [3.67, 68]

resistivity. Both critical field strenghts are temperature dependent. If it is assumed that the inflection point in the curve indicates a critical temperature at which $H_{c2}^{\perp}$ becomes equal to the applied field then the critical field $H_{c2}^{\perp}$ at absolute zero is estimated to be of the order of 10^4 Am^{-1} (127 Oe).

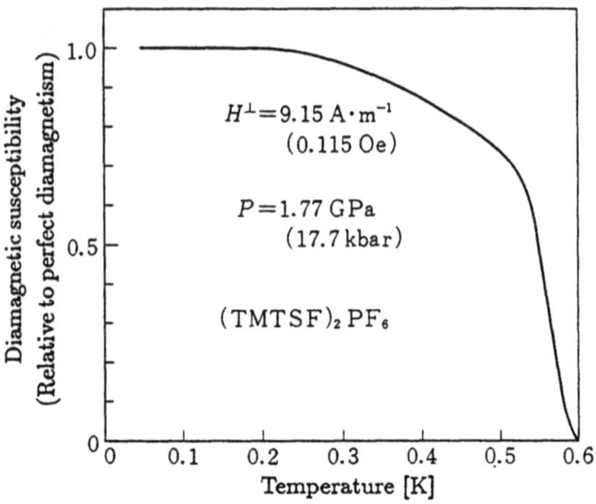

Fig. 3.39. Temperature dependence of the diamagnetic susceptibility of $(TMTSF)_2PF_6$ [3.72]

The parallel critical field $H_{c2}^{\parallel}$ is found to be about $5.5 \times 10^5\,Am^{-1}$ (7 kOe). In order to demonstrate the occurrence of superconductivity, we must find a perfect diamagnetism; this is the Meissner effect. Figure 3.39 shows the temperature dependence of the magnetic susceptbility for a field of $9.15\,Am^{-1}$ (0.115 Oe) applied perpendicular to the 1d axis under a pressure of 17.7 kbar (1.77×10^9 Pa) [3.73]. The perfect diamagnetism, $\overline{\chi}_d \equiv -M/H = 1$, is observed below about 0.2 K. Above 0.3 K $\overline{\chi}_d$ is less than 1; $H_{c1}^{\perp}$ in this temperature range must be less than $9.15\,Am^{-1}$ (0.115 Oe). The magnitude of H_{c1} at absolute zero is estimated to be about $400\,Am^{-1}$ (5 Oe) for $H_{c1}^{\perp}$ and about $40\,Am^{-1}$ (0.5 Oe) for $H_{c1}^{\parallel}$. Although detailed data have not been obtained in compounds other than $(TMTSF)_2PF_6$, other values of $H_{c2}^{\perp}$ seem to be comparable with that of $(TMTSF)_2PF_6$ [3.68, 49].

The critical temperature T_c at 12 kbar (1.2×10^9 Pa) is 0.9 K in $(TMTSF)_2PF_6$ and 1.1 K in $(TMTSF)_2AsF_6$ as shown in Fig. 3.38. Other compounds also have a critical temperature of the order of 1 K. All compounds have a very similar pressure dependence of T_c as shown in Fig. 3.37. The maximum value of T_c, namely that at the lowest pressure above which the superconductivity appears, is the following; 1.4 K (at 6.5 kbar) for the PF_6 compound, 1.4 K (at 10 kbar) for AsF_6, 1.35 K (at 11 kbar) for TaF_6, 0.38 K (at $<$ 11 kbar) for SbF_6, and 1.4 K (at 1 bar) for ClO_4.

Except for $(TMTSF)_2SbF_6$, the maximum value of T_c is nearly the same in these compounds although the pressure necessary for the onset of the superconductivity is different. This suggests that the superconductivity in them has a common character, and that its onset is dominated by lattice parameters. A comparison between lattice parameters and the critical pressure suggests that the lattice parameter c dominates the onset of the superconductivity [3.69]; the superconductivity appears when the interstack distance between TMTSF stacks along the *c*-axis becomes shorter than a critical length.

Is the superconductivity found in these compounds an ordinary one of the type observed in conventional 3d superconductors? First of all, the possibility of a sliding motion of the CDWs can be discarded because all the present compounds show the Meissner effect. It will thus be the normal superconductivity described by the BCS theory. The superconductivity has a 3d character; the transition occurs at finite temperatures and the Meissner effect is observed for magnetic fields applied parallel to the 1d axis. Nevertheless these compounds seem to have a quasi one-dimensionality in their electrical properties. Hence it is expected that the formation of Cooper pairs of electrons actually begins in each 1d stack at a temperature higher than the experimentally observed T_c; TMTSF stacks may have a flcutuating 1d superconductivity in a wide range of temperature above T_c. Its coherence length will increase anisotropically with decreasing temperature. When the coherence length along the 1d axis becomes sufficiently long, a Josephson coupling will occur between stacks giving rise to a real 3d superconductivity.

The above interpretation is based on the following experimental results: (1) In ordinary superconductors the temperature dependence of the resistivity is weak in a certain temperature range above the transition temperature. $(TMTSF)_2PF_6$ and $(TMTSF)_2AsF_6$, however, show a rather strong temperature dependence. This fact suggests the presence of a so-called paraconductivity above the critical temperature; superconducting fluctuations have become well developed along the 1d axis contributing to the electrical conduction. (2) the critical fields satisfy the following relation: $H_{c1}^{\parallel} < H_{c1}^{\perp}$ and $H_{c2}^{\parallel} > H_{c2}^{\perp}$. In order to perfectly screen the applied field $H^{\parallel}$, loop currents are required to flow through the weak couplings between molecular stacks. On the other hand $H^{\perp}$ is more easily screened by loop currents, part of which flow along the 1d axis; a large $H^{\perp}$ will therefore be needed for the magnetic flux to penetrate the samples. Thus the first relation, $H_{c1}^{\parallel} < H_{c1}^{\perp}$, may be explained. With respect to H_{c2}, the superconductivity must be broken within each molecular stack. In achieving this $H^{\parallel}$ is ineffective compared to $H^{\perp}$. The second relation is thus also understood.

It is to be noted that some objections might be raised to the above points: Metals naturally have a temperature dependence in the resistivity. In the present compounds the temperature range in which a residual resistance dominates the overall resistivity may lie lower than the critical temperature. The anisotropy in critical fields may be a natural result of the electronic anisotropy. Furthermore tunnelling spectroscopy [3.74] has revealed that the superconductivity gap can be interpreted in terms of an ordinary 3d BCS mechanism. This result would clearly seem to rule out the possibility of 1d superconductivity.

3.6.4 Spin Density Waves

In lower pressure ranges a metal-insulator transition occurs in place of the superconductivity as is shown in Fig. 3.37. Is this a Peierls transition? X-ray scattering measurements have found no CDW at temperatures down to 4.2 K [3.68]. This result does not, however, rule out the possible presence of a CDW with a very small amplitude which could not be observed experimentally. A CDW is, however, found in $(TMTTF)_2PF_6$ samples of a size similar to that of $(TMTSF)_2PF_6$ [3.75]. When we take account of the presence of selenium atoms which have a large x-ray scattering factor in $(TMTSF)_2PF_6$, we conclude that this compound has essentially no CDW of a character similar to those found in other 1d conductors.

$(TMTSF)_2AsF_6$ provides a set of data which suggests the presence of a weak x-ray scattering with the wave number $0.5\,a^*$ in a temperature range 100–50 K [3.68]. It is not thought that this scattering has any essential relation to the metal-insulator transition; its intensity is weak and it becomes hardly detectable with decreasing temperature.

Thus we conclude that the metal-insulator transition observed in the family of $(TMTSF)_2PF_6$ is not the Peierls transition although compounds of TMTTF may have this transition. Electron-phonon interactions should ideally be weak to cause the Peierls transition. At present the metal-insulator transition observed in these compounds is attributed to the development of spin density waves (SDW) due to the strong Coulomb interactions between electrons.

In the SDW state electrons have wave functions extending over the whole crystal. The density at each site, however, is different for electrons with up-spin and those with down-spin. This gives rise to spin magnetic moments. The SDW is formed by spatial waves of these moments. It is a kind of antiferromagnetic state.

The wave number specifying the SDW is $2k_F(= 0.5\,a^*)$. SDWs exert a potential with the wave number $2k_F$ on electrons through spin-spin interactions. This potential produces a gap at the Fermi level of electrons and the system turns into an insulator. The driving force for the onset of SDW is a gain in the Coulomb energy of the electrons.

In the following we discuss experimental evidence for the SDW transition. In EPR measurements the spin susceptibility of $(TMTSF)_2PF_6$ at ambient pressure is found to decrease at temperatures below the metal-insulator transition whereas the static susceptibility does not decrease [3.76, 77]. The static susceptibility below the transition temperature has an anisotropy characteristic of antiferromagnets. Its field dependence suggests the occurrence of a spin-flip process at a certain field strength [3.77]. Furthermore NMR indicates the presence of an antiferromagnetic state [3.78]. The magnitude of the effective magnetic moment is very small; it is of the order of $10^{-2}\,\mu_B$ when the moment is assumed to be present on the selenium atoms of TMTSF molecules. This result is to be expected for a magnetic moment arising from band electrons. It is now virtually certain that antiferromagnetic SDWs appear in $(TMTSF)_2PF_6$ and its family in the insulating phase.

In the temperature range below about 1 K a transition occurs with increasing pressure from the insulating SDW phase to the superconducting phase. The presence of SDWs indicates that the Coulomb repulsion between electrons is strong; the superconducting phase can have two states, either a singlet superconductivity state or a triplet state. The latter has not been found in any materials whereas the former is the usual state observed in many kinds of superconductor. (Note however, that the superfluid state in ^3He is triplet.) It is likely to be determined in the near future whether the superconductivity in the present compounds corresponds to a singlet or triplet state.

At the same time, another problem must be solved: The transition temperature for superconductivity decreases with increasing pressure. This fact contradicts the naive expectation that the lattice parameter, reduced by the increasing pressure, will provied a situation more favorable for the onset of superconductivity.

3.6.5 Orientational Ordering of Counter Anions X

As we have learned in Sect. 3.6.1, counter anions having no centre of symmetry can undergo an orientational ordering at a certain temperature. This has been observed by x-ray scattering at ambient pressure in compounds with $X = ReO_4$ (at 176 K) [3.79] and ClO_4 (at 24 K) [3.80]. It occurs also in several other compounds belonging to this family although they do not undergo the superconducting transition.

Ordered anions form a superstructure, whose period is written as $2a \times 2b \times 2c$ and $a \times 2b \times c$ in $(TMTSF)_2ReO_4$ and $(TMTSF)_2ClO_4$, respectively. $(TMTSF)_2ClO_4$ is the only compound which has the period a along the stacking axis. Other compounds, including those undergoing no superconducting transition, have the period $2a$ [3.81]. This difference in the period may be closely related to the unique character of $(TMTSF)_2ClO_4$, which becomes a superconductor at ambient pressure: Anions ordered with the period $2a$ will produce a potential with wave vector $2k_F$ giving rise to the opening of a band gap at the Fermi level. $(TMTSF)_2ReO_4$ really turns into an insulator when the anion ordering occurs. The superconductivity in $(TMTSF)_2ReO_4$ at high pressure may be brought about by a change of the superstructure or of the shape of the Fermi surface, or by a suppression of the ordering of anions. On the other hand the period, a, observed in $(TMTSF)_2ClO_4$ gives a potential with $4k_F$, which creates no gap at the Fermi level.

$(TMTSF)_2ClO_4$, when rapidly cooled from temperatures above the ordering temperature, 24 K, down to low enough temperatures, does not show the superconductivity but has instead insulating properties. Spin density waves are observed by NMR in the insulating phase [3.82] and x-ray scattering verifies directly that the anion ordering is suppressed by the rapid cooling [3.83]. Hence it is expected that the anion ordering is closely connected with the onset of superconductivity. There are several possibilities for this connection: (1) The potential produced by the ordered anions plays a crucial role in the onset of superconductivity [3.84]. (2) Reduction of interstack distances due to the anion ordering is essential for the superconductivity [3.85]. This problem requires further investigation in future studies.

3.6.6 Superconductivity in $(BEDT\text{-}TTF)_2X$

Superconductivity in organic compounds containing sulphur atoms, instead of selenium, was first discovered in 1982 in $(BEDT\text{-}TTF)_2ReO_4$ under pressure [3.86], where BEDT-TTF is the bisethylenedithiolotetrathiafulvalene molecule. Figure 3.40 shows the molecular structure of BEDT-TTF. Its transition temperature is about 2 K. In 1984 another compound $(BEDT\text{-}TTF)_2I_3$ was found to undergo the superconductivity transition also at about 2 K but this time at ambient pressure.

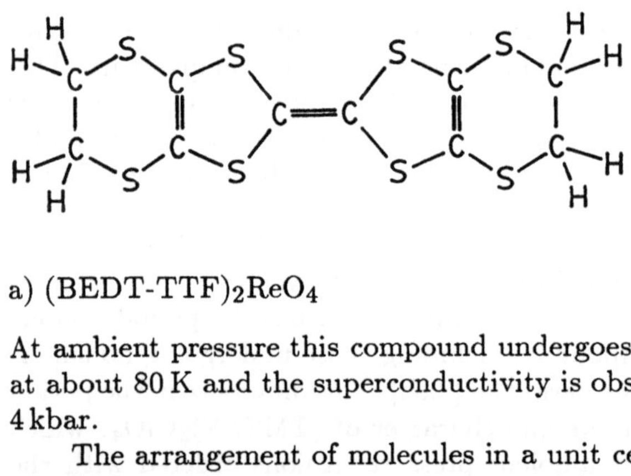

Fig. 3.40. Molecular structure of BEDT-TTF

a) $(BEDT\text{-}TTF)_2ReO_4$

At ambient pressure this compound undergoes a metal-insulator transition at about 80 K and the superconductivity is observed under pressures above 4 kbar.

The arrangement of molecules in a unit cell is similar to that of $(TMTSF)_2X$ although the present compound has two formula units in a unit cell. A difference is found in the orientation of ReO_4. The anions ReO_4 in $(BEDT\text{-}TTF)_2ReO_4$ at ambient pressure show orientational ordering even at room temperature. The metal-insulator transition observed at ambient pressure has the character of a first-order transition. The electrical resistivity increases abruptly by several orders of magnitude. Hence it is suggested that the transition is associated with the rearrangement of the ReO_4 anions.

The phase diagram of $(BEDT\text{-}TTF)_2ReO_4$ with respect to temperature and pressure is similar to those of $(TMTSF)_2X$: The metal-insulator transition temperature decreases with increasing pressure and the transition is totally suppressed above 6 kbar. Furthermore, the superconducting transition temperature at pressures above 4 kbar decreases with increasing pressure. The maximum transition temperature observed is about 2 K (at 4 kbar).

Application of a magnetic field restores the normal resistance. The critical field along the c^*-axis is about 800 G and that along the b^*-axis is twice as large. As far as these results are concerned, $(BEDT\text{-}TTF)_2ReO_4$ appears to have properties quite similar to those of $(TMTSF)_2X$.

b) $(BEDT\text{-}TTF)_2I_3$

This compound has several types of crystal structure [3.87]. One of them, called β-$(BEDT\text{-}TTF)_2I_3$, shows metallic properties down to low temperatures at ambient pressure and undergoes a superconductivity transition at about 2 K [3.88]. Interesting properties are found when hydrostatic pressure is applied. At about 1 kbar the transition temperature abruptly goes up to about 7.5 K and then decreases again with increasing pressure [3.89]. Furthermore even when the pressure is released, there remains some evidence of the superconductivity transition at about 8 K in addition to the transition at about 2 K. The transition of 8 K, although the resistivity remains finite, is considered to be of a superconducting nature due to its sensitivity to applied magnetic fields. Hence it is conjectured that a small volume of the sample becomes superconducting at about 8 K. However, it remains unclear at present what kind of change occurs under pressure.

4. Properties of MX_3

In this chapter, properties of transition metal trichalcogenides (sulfides, selenides and tellurides) are discussed. These compounds will be referred to hereafter as MX_3.[1] There are more than 10 compounds (Table 4.1) which have the triangular prismatic structure which is explained below. However, the discussion in this chapter is restricted to three compounds, which all show metallic conductivity above certain temperatures. Their crystal structures and lattice parameters are given in Table 4.2. Among these, $NbSe_3$ will be discussed in detail, partly because electrical conduction by the Peierls-Fröhlich mechanism was found, for the first time, in this material. Compared with these three compounds, other compounds have attracted far less attention, although here too the low-dimensional character of the electronic band structure and of the phonon dispersion relation is becoming clearer.

A similar family of compounds, are the transition metal chalcogenides MX_2 which have a layered structure. They are well known as typical 2d

Table 4.1. Trichalcogenides with the triangular-prismatic structure (M: metallic, I: nonmetallic)

	Ti	Zr	Hf	Nb	Ta
S	I	I	I	(I, M)	M
Se		I	I	M	M
Te		M	?		

Table 4.2. Crystal structures of MX_3

	$NbSe_3$	TaS_3		$TaSe_3$
Crystal System	monoclinic	orthorhombic	monoclinic	monoclinic
Space group	$P2_1/m$	$Pmn2_1$	$P2_1/m$	$P2_1/m$
	$a =$ 10.009	$a =$ 36.804	$a =$ 9.515	$a =$ 10.402
Parameters	$b =$ 3.480	$b =$ 15.177	$b =$ 3.3412	$b =$ 3.495
(lengths in Å)	$c =$ 15.629	$c =$ 3.340	$c =$ 14.912	$c =$ 9.829
	$\beta =$ 109.47°		$\beta =$ 109.99°	$\beta =$ 106.26°

[1] Names of MX_n are used widely. Sometimes these compounds are called as TX_3 (T: transition metals) also.

conductors or semiconductors. Remarkable progress is still being made in the study of these materials [4.1, 2].

4.1 Preparation and Crystal Structure

4.1.1 Preparation of Single Crystals

Single crystals of MX_3 compounds can be prepared by sealing the starting materials with the mole ratio of 1 : 3 in a quartz tube in vacuum and heating it at 500°–700° C. After a few weeks, many fibrous crystals can be found within the tube. Usually these crystals are very thin; typical thickness is $10 \sim 100\,\mu$m. Each crystal has the shape of rectangular parallellepiped: the longest dimension (~ 1 cm) is parallel to the 1d axis, or more exactly, to the axis along which one-dimensionality is expected. Attempts (e.g. HVT[2]) to grow larger crystals have not been successful. Sometimes, larger crystals thicker than 100 μm can be obtained accidentally. However, most of these are not single crystalline but agglomerates of twinned single crystals; several crystal orientations grow together to form an apparently single crystal, but only one or two crystal axes are parallel to each other.

4.1.2 Crystal Structure

The characteristic features of MX_3 structures will be explained with reference to the example of $ZrSe_3$, whose structure is the most simple among them. As shown in Fig. 4.1, six Se atoms form a triangular prism and Zr atom is located at the centre of the prism. The interatomic distances between Zr and Se are about 2.73 Å. $ZrSe_6$ prisms are stacked successively along the b-axis to form $ZrSe_3$ columns. There are two $ZrSe_3$ columns within the unit cell, as shown in Fig. 4.2. Points marked by crosses in the figure have inversion symmetry, so the two columns within the unit cell are crystallographically equivalent. The distance between columns is expressed by, for example, the distance betwen Zr(M) and Se(I') of 2.88 Å. (Note that these two atoms are at the same height along the b-axis.) The triangle formed

[2] HVT (halogen vapour transport method) has been used successfully for the purification of metals and the preparation of single crystals of various materials [4.3]. In the case of $NbSe_3$ for example, powders of $NbSe_3$, or mixed starting materials (niobium and selenium) are sealed in a quartz tube under vacuum. A suitable amount of iodine is added. (Other halogens, Br and Cl, can also be used.) The quartz capsule is heated in a furnace with a temperature gradient. Initially, the starting materials are placed in the hotter zone of the furnace. The chemical reaction $NbSe_3 + 2I_2 \rightleftarrows NbI_4 + 3Se$ occurs. NbI_4 and Se have higher vapour pressures, so they vapourize and move to the colder zone. There, NbI_4 and Se decompose into $NbSe_3$ and I_2. The iodine vapour moves back to the hotter zone to form NbI_4 again. If the inside wall of the capsule in the colder zone is clean and if there are a small number of nuclei for crystal growth, larger crystals can be grown there. This HVT method has been used successfully to grow large crystals of various materials. We must be careful, however, that iodine is not incorporated into the crystal as an impurity. The method explained in the text is called as DVT (Direct Vapour Transport) method, because here the material is transported by its own vapour pressure and not that of a gaseous halogen [4.3].

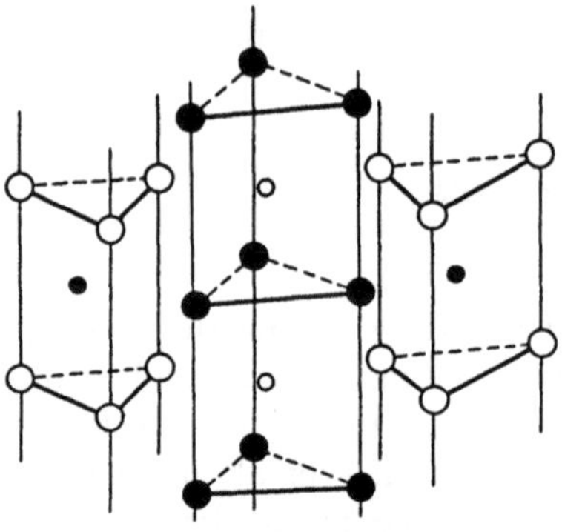

ZrSe₃

Fig. 4.1. Crystal structure of $ZrSe_3$. (Hereafter, open and filled circles denote the height along the **b**-axis of 0 and 1/2, respectively)

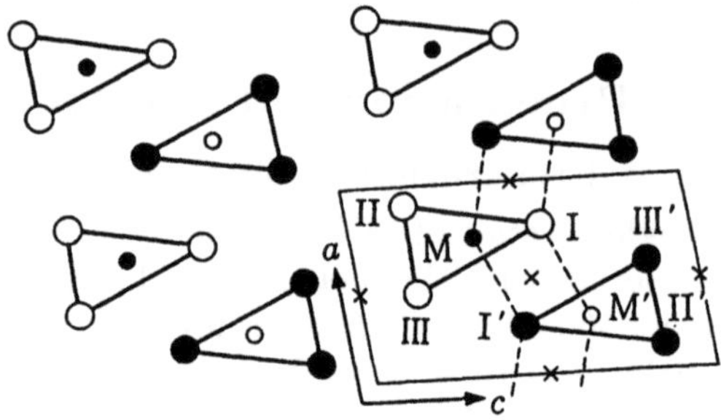

ZrSe₃

Fig. 4.2. Projection on the *a*-*c* plane of the $ZrSe_3$ structure shown in Fig. 4.1

by three Se atoms (I, II and III) is approximately isosceles; the distance Se(II)-Se(III) is the shortest.

From the shape of the triangular column, the electronic configuration of $ZrSe_3$ is estimated as follows. In general Se atoms tend to form Se^{2-} ions, but the Se(II) and Se(III) atoms should be regarded as forming a $(Se_2)^{2-}$ molecule, because the Se(II)-Se(III) separation is shorter than the van der Waals distance. In fact, the optical absorption corresponding to the vibrational mode of "Se_2" molecules has been obseved. Zirconium atoms have the outer electron configuration: $4d^2 5s^2$. Thus, $ZrSe_3$ can in fact be represented as $Zr^{4+}Se^{2-}(Se_2)^{2-}$; there is no unpaired electron and the $ZrSe_3$ crystal is expected to be insulating. In fact, it is black-purple coloured insulator with a band gap of 1.2 eV (1.9×10^{-19} J).

Next we consider the reason why $NbSe_3$ is metallic. Its crystal structure is rather complicated as shown in Fig. 4.3. Nevertheless, an obvious similarity to the structure of $ZrSe_3$ can be found. $NbSe_6$ prisms are again stacked to form $NbSe_3$ columns which lie parallel to each other and to the monoclinic *b*-axis. The height along the *b*-axis of neighbouring columns is

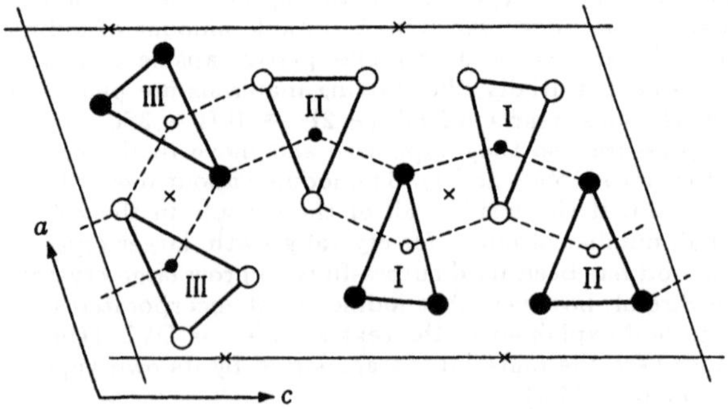

Fig. 4.3. Crystal structure of $NbSe_3$ [4.5]

shifted by half a lattice constant. Therefore, the coordination number of the Nb atoms is eight rather than six. There are six columns within the unit cell, but due to the inversion symmetry at points marked by × in the figure, there are three distinct kinds of $NbSe_3$ columns. A van der Waals gap is found in both the $ZrSe_3$ and $NbSe_3$ parallel to the *b-c* plane. No chemical bond extends across those planes. This is the reason why these crystals can be easily cleaved. The characteristic feature of the $NbSe_3$ structure is the shape of the cross sections of the triangular columns. The shortest sides of the approximately isosceles triangles are 2.49 Å and 2.37 Å in columns (I) and (III) respectively, and 2.91 Å in (II). (Compare these with the distances of other edges: 3.7 ~ 3.8 Å). Therefore it is not likely that there is a "Se_2 molecule" in column (II). Bearing in mind that a Nb atom has the outer electronic structure ($4d^4 5s^1$), the three non-equivalent columns out of six within the unit cell should be regarded as

$$3NbSe_3 \rightarrow (Nb^{5+})_3(Se^{2-})_5(Se_2)_2^{2-} + e^- \quad .$$

There is one electron which remains unpaired [4.6], and which thus partially fills the conduction band of crystalline $NbSe_3$.

For the compound TaS_3 two polytypes have been found. One has been designated as orthorhombic [4.7], although its structure has not been determined yet in spite of many trials. The other is monoclinic [4.8] with a structure quite similar to that of $NbSe_3$.

Figure 4.4 shows the crystal structure of $TaSe_3$. There are four triangular columns (of two kinds), within the unit cell, each with a cross section that is an isosceles triangle. The interatomic distance Se(5)-Se(6) is 2.5 Å, shorter than the 2.90 Å of Se(2)-Se(3). Two molecules of $TaSe_3$, from two non-equivalent columns, can be represented as $(Ta^{5+})_2(Se^{2-})_4(Se_2)^{2-}$.

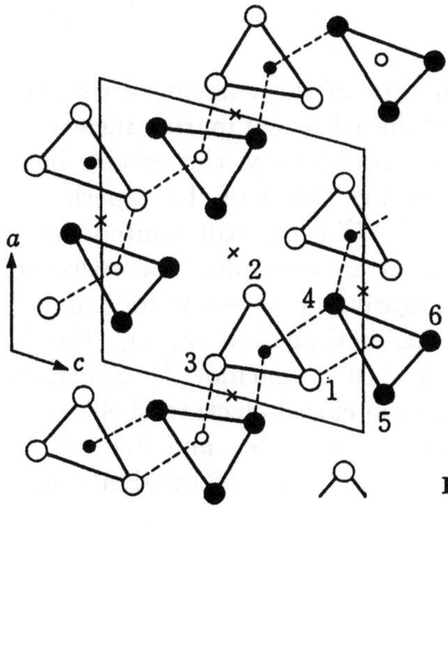

Fig. 4.4. Crystal structure of $TaSe_3$ [4.9]

Therefore, $TaSe_3$ might be expected to be insulating. However, $TaSe_3$ in fact shows metallic conductivity and the Shubnikov-de Haas effect (cf. [4.4]) has been observed at low temperature. The latter is the clearest evidence that $TaSe_3$ has conduction electrons. Thus, we must be careful when assigning an electronic configuration to the crystalline state from an argument such as that above. This is because we have no way of knowing the critical interatomic distance for the picture of "Se_2 molecules" to be valid. Another difficulty is that the ionic approach mentioned above is not always useful for predicting the band structure of crystalline state. For example, even if an argument of the above type does not predict the presence of unpaired electrons, the crystal can still be a semimetal when the wave functions of neighbouring columns overlap sufficiently with each other.

Recently these three compounds $NbSe_3$, TaS_3 and $TaSe_3$ have been synthesized under pressure [4.10]. All the products are in powder form, so the information gained on physical properties is limited. Nevertheless, it was found that both TaS_3 and $TaSe_3$ when prepared under pressure, have the $NbSe_3$-type structure: monoclinic with six chains in the unit cell. It was proposed that orthorhombic TaS_3 is S-deficient and similarly, that the usual $TaSe_3$ is in reality $TaSe_{2.8}$. As far as diffraction results and physical properties observed up to now are concerned however, neither compound seems to show significant structural disorder, e.g., excess Ta atoms in irregular sites.

Before turning to the interesting properties of individual materials, we discuss some properties common to all the metallic MX_3 compounds. The first is that the electrical conductivity is about $10^5\,(\Omega^{-1}\,\mathrm{m}^{-1})$ at room temperature. This value is about two orderes of magnitude smaller than that of typical pure metals, and nearly equal to that of a semimetal such as Bismuth (Bi) or Antimony (Sb). The poor conductivity may arise from a low carrier density, since the conductivity is given by

$$\sigma = ne^2\tau/m \tag{4.1}$$

where n and τ are density and relaxation time of the current carrier, respectively. Usually, the density n can be estimated by comparing measured values of the Hall coefficient, the magnetic susceptibility, the specific heat at low temperature and various other properties. Most of the experimental data on these fundamental properties of MX_3 are still limited, however. Consider the magnetic susceptibility as an example. The susceptibility of $TaSe_3$ at room temperature is negative; $\chi = -6 \times 10^{-4}\,\mathrm{mol}^{-1}$ $(-5 \times 10^{-5}\,\mathrm{emu\,mol}^{-1})$, which is much smaller than that of, e.g., the 2d conductor 2H-$TaSe_2$, $+2 \times 10^{-3}\,\mathrm{mol}^{-1}$ $(1.6 \times 10^{-4}\,\mathrm{emu\,mol}^{-1})$. Subtracting off the diamagnetic contribution due to the closed electron shells gives $+1.3 \times 10^{-3}\,\mathrm{mol}^{-1}$ $(+1.1 \times 10^{-4}\,\mathrm{emu\,mol}^{-1})$ for $TaSe_3$ and 3.4×10^{-3} mol^{-1} $(2.7 \times 10^{-4}\,\mathrm{emu\ mol}^{-1})$ for $TaSe_2$ as the sum of Landau-Peierls diamagnetic and Pauli paramagnetic susceptibilities of conduction electrons. (In some transition metal compounds, van Vleck paramagnetism is also important.) If the two contributions can be separated, we can estimate the density of carriers, because the Pauli susceptiblity is proportional to the density of states at the Fermi level, $N(E_F)$. In fact the separation is not always reliable, so we must be satisfied, at present, to find that the value of n of $TaSe_3$ is much smaller than that of $TaSe_2$. (Density of states at the Fermi level of 2H-$TaSe_2$ is $2 \sim 3\,eV^{-1}\,mol^{-1}$ ($(1.3–1.9) \times 10^{-19}\,J^{-1}\,mol^{-1}$). Other MX_3 also show small values of the magnetic susceptibility.

As shown in the following section, the electrical conductivity of MX_3 is quite anisotropic; it is largest along the axis of triangular column. If each column were isolated, the conduction electrons could only move along this axis, and MX_3 would be a purely 1d conductor. However, perpendicular to the column, the chemical bond -X-M-X can be found parallel to the van der Waals gap. So it is also possible to regard MX_3 compounds as 2d materials like MX_2 but with a strong anisotropy within the layer. When we consider various properties of MX_3, it must be borne in mind that the electronic band structure cannot be always approximated by a 1d model since 3d effects, such as the Landau-Peierls diamagnetism and the Hall effect, which should be absent in purely 1d conductors, have been observed.

4.2 Charge Density Waves in $NbSe_3$

4.2.1 Anomalies in the dc Conductivity and the Periodic Lattice Distortion

Historically, it was following the discovery of the anomalous conductivity of $NbSe_3$ that MX_3 compounds began to draw much attention. Hence, this section deals with the conductivity of $NbSe_3$. Figure 4.5 shows the tem-

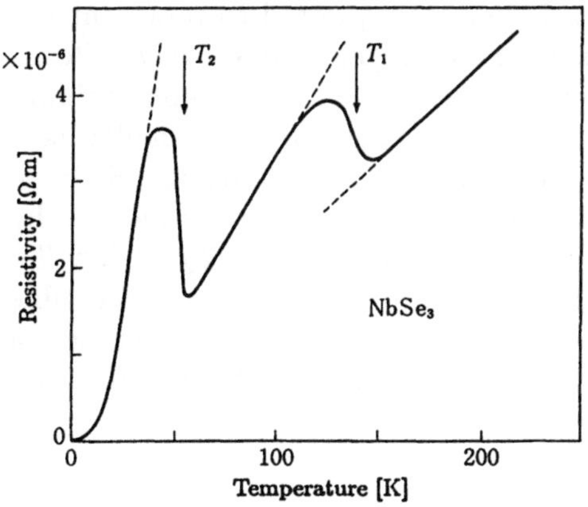

Fig. 4.5. Electrical resistivity of $NbSe_3$

perature dependence of the electrical resistivity along the b-axis. (In this chapter, both the terms conductivity, and resistivity, which is the reciprocal of conductivity, are used. This is because the term resistivity, rather than conductivity, has been traditionally used for metals.) The resistivity decreases as the temperature is lowered, but two maxima appear at about 140 K and 60 K. The value of the residual resistivity, which is obtained by extrapolating to $T = 0$, increases with increasing impurity or defect concentration in the crystal. Shubnikov-de Haas oscillations have been observed [4.11, 12] at liquid helium temperature, which demonstrates definitely the presence of degenerate electrons. Therefore we may claim that $NbSe_3$ is a metal.

Each of the two maxima in the resistivity is caused by ordering of the charge density waves (CDWs), an interpretation also supported by the observation of a periodic lattice distortion associated with the CDWs. As explained in Sect. 3.2.1, the maximum of $|d \ln \sigma(T)/dT|$ is assumed to correspond to the transition temperature. The transition temperatures are given as 142 K and 58 K. Hereafter, these temperatures are called as T_1 and T_2, respectively.

When the CDW with the wave vector $\boldsymbol{Q}$ is ordered, a new periodicity with the same wave vector $\boldsymbol{Q}$ appears in the lattice (cf. Sect. 2.1.2). At the same time the conduction electron bands show a discontinuity at $\boldsymbol{k} = \pm\boldsymbol{Q}/2$. If $\boldsymbol{Q}$ is commensurate with the lattice periodicity, then $\boldsymbol{k} = \pm\boldsymbol{Q}/2$ corresponds to the first Brillouin zone boundaries of the superlattice. In a purely 1d metal, where the Fermi surface (FS) is composed of a pair of parallel planes, a Peierls gap opens over the entire area of FS and there is no remaining FS in the CDW state at all. In $NbSe_3$, however, there must be a remaining section of FS even below T_2. It is quite probable that FS of the high temperature phase (above T_1) is partially curved or it is composed of several pairs of planar sheets. In fact, band structure calculations have revealed that the FS is composed of both planar and more curved sheets [4.13]. In either case, a certain area of FS is vanishing below T_1 and the electrical conductivity decreases. Below T_2, a part of the remaining sheets vanishes again. In Fig. 4.5 three dotted curves are plotted to represent the resistivity in the phases below T_2, between T_2 and T_1, and above T_1. Resistivity peaks appear as one moves from a curve to the next. From the figure one can understand that the lower-temperature phases have a smaller electron density n [cf. (4.1)], because of the successive vanishing of FS. If we assume that only n changes at the transitions, the area of FS is reduced by $\sim$30 % at T_1 and $\sim$70 % of the remaining area vanishes at T_2. Below T_2, only $\sim$20 % of FS of the phase at $T > T_1$ is surviving; a quite remarkable change is occurring in the electron system. Conversely, if the geometry of FS is such that a large change cannot be expected, a CDW transition is not likely to occur (cf. Sect. 2.1.4).

It should be emphasized that the above estimate is only qualitative, because any change in the relaxation time is neglected. To make the situation clearer, let us consider how (4.1) is derived. Under the electric field E_x along the x-direction, the electric current flows in the same direction (cf. [4.14]). The current density, J_x, is given by

$$J_x = \sum ev_x$$

or, by replacing the summation in the $\boldsymbol{k}$-space by integration,

$$J_x = \frac{2}{(2\pi)^3} e \int f(\boldsymbol{k}) v_x d\boldsymbol{k}$$

where v_x is the electron velocity along x. We assume that the distribution function $f(\boldsymbol{k})$ deviates from the equilibrium distribution $f_0[\varepsilon(\boldsymbol{k})]$ by $(df_0/d\varepsilon)(-eE_x)[v_x\tau(\boldsymbol{k})]$. The quantity $-(df_0/d\varepsilon)$ is approximated by the delta function $\delta(\varepsilon - \varepsilon_F)$. The volume element $d\boldsymbol{k}$ is written as $(\hbar v)^{-1} dS\, d\varepsilon$, where dS is the element of a constant energy surface and v is the group velocity at $\boldsymbol{k}$. Integration over ε gives

$$\sigma = \frac{J_x}{E_x} = \frac{e^2}{4\pi^3\hbar} \int \tau \frac{v_x^2}{v} dS \ .$$

In general τ depends on $\boldsymbol{k}$, but if we assume that it depends only on ε, (4.1) can be obtained in the case of a 3d free electron system. The same relation holds for free electrons in 1d, where the FS is a pair of planes, because $v = v_x$. Therefore, the conductivity decrease is proportional to the fractional decrease of the FS area, if τ is assumed to be constant.

The relaxation time $\tau(\boldsymbol{k})$ is the effective length of time for which an electron stays in the state $\boldsymbol{k}$ without scattering. The average of $\tau(\boldsymbol{k})$ is given by

$$\frac{1}{\tau} = \int P(\theta) dS_\varepsilon$$

where $p(\theta)$ is the probability of elastic scattering $\boldsymbol{k} \rightarrow \boldsymbol{k}'$, and θ is the angle between $\boldsymbol{k}$ and $\boldsymbol{k}'$. The integration is over the constant energy surface $\varepsilon = \varepsilon_{\boldsymbol{k}}$. Even if we do not know the functional form of $P(\theta)$, we see that τ increases when a part of the FS vanishes because the integration range decreases. The change in the conductivity at the transition thus comes both from the decrease of n and from the increase in τ. So it is not easy to estimate more accurately the relative area of the vanishing FS sections from the resistivity increase.

The result, that a large fraction of the FS vanishes at the T_1 and T_2 transitions, is not contradictory with the result of the Shubnikov-de Haas oscillation measurements. Under a strong magnetic field, the Landau levels of conduction electrons shift in proportion to the magnetic induction B and, as the result, the resistivity oscillates with $1/B$ [4.15]. From the oscillation period $\Delta(1/B)$ the extremal cross-sectional area A_0 of the FS perpendicular to the magnetic field can be obtained:

$$\Delta\left(\frac{1}{B}\right) = \frac{e}{h} A_0^{-1} \quad . \tag{4.2}$$

By measuring A_0 in various directions, the shape of the FS can be determined, at least in simple cases. Quite clear oscillations have been observed [4.11, 12] in $NbSe_3$ as shown in Fig. 4.6. The FS below T_2 is ellipsoidal and its volume is only a few percent of the Brillouin zone. The longest axis of the ellipsoid, parallel to the b^*-axis, is approximately three times longer than the other; the axial ratio is not large or, in other words, the remaining FS is not strongly anisotropic. This might seem in contradiction with the one-dimensionality of the system. In reality, however, the FS observed at low temperature is the section remaining after the two successive CDW transitions. Planar sections of the FS (of the high temperature phase above T_1) have vanished at the transitions; the remaining sections must be necessarily more or less isotropic. The more one-dimensional the high temperature phase is, the smaller the remaining sections. The curvature of the FS in the low temperature phase is determined by reconstruction of the remaining FS pieces. To understand this point more clearly, assume a FS with a small curvature and a commensurate Q-value and reconstruct the new FS, of the low temperature phase, within the new Brillouin zone.

It has been confirmed, by observing the periodic lattice distortion (PLD), that the conductivity anomalies are due to the long-range order of

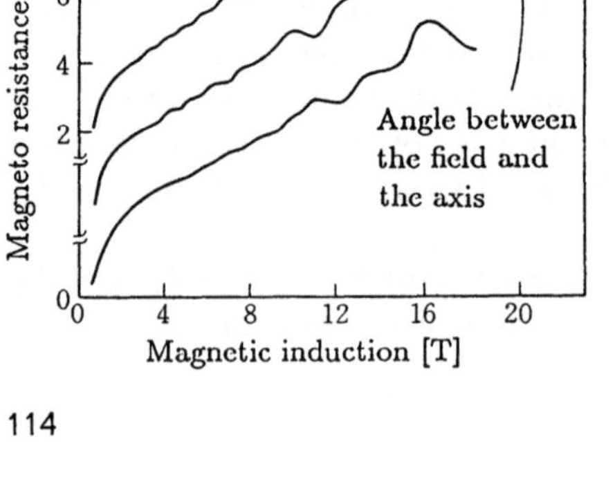

Fig. 4.6. Shubnikov-de Haas oscillation [4.12] in $NbSe_3$

the CDW. Neutron diffraction experiments are extremely difficult because of the small size of the crystals obtained up to now. By x-ray and electron diffraction [4.16, 17], a PLD with the wave vector $\boldsymbol{Q}_1 = (0, 0.24, 0)$ is observed below T_1, and another one with $\boldsymbol{Q}_2 = (0.5, 0.26, 0.5)$ occurs below T_2 which coexists with $\boldsymbol{Q}_1$. (Compared with x-rays, the electron diffraction method has the advantage of being able to detect weaker reflections due to the stronger interaction of electrons with the materal. On the other hand, it has the drawbacks of low accuracy in determining crystal orientation and the sample temperature. Also we must be careful that the sample is not damaged by electron irradiation.)

It should be noted that both $\boldsymbol{Q}_1$ and $\boldsymbol{Q}_2$ have incommensurate b^*-components, which are nearly equal to 0.25. If these values are determined by the FS nesting (cf. Sect. 2.1.4), the volume of the FS is approximately equal to a quarter of the Brillouin zone. From the local approach explained in Sect. 4.1.2, one unpaired electron is shared by the two triangular prisms which contain the "Se_2" molecules. Since each branch of electron energy bands can accommodate up to 2 electrons per unit cell, the conduction band is one quarter filled, in good agreement with the measured b^*-components of $\boldsymbol{Q}_1$ and $\boldsymbol{Q}_2$. Here, we have assumed that electrons are localized in each column, as in a hypothetical unit cell with one column within it. Of course, it is not clear to what extent this assumption of independent columns, is justified.

The b^*-components of $\boldsymbol{Q}_1$ and $\boldsymbol{Q}_2$ can be viewed from a different approach [4.6]. If an uncertainty of $\lesssim 2\,\%$ is allowed in the experimentally determined b^*-components, 0.24 can be equal to 9/37 (0.243) or to 10/41 (0.244). The former value means that the lattice has a long-range modulation with the period of $(37/9)b$. Taking into consideration the commensurability with the lattice, the basic modulation should have the period of $4b$ (wave vector of 0.25). The period of $(37/9)b$ corresponds to a structure of 9 periods of $4b$ plus one $1b$, that is, the superstructure of $4b$ but with interstitial defects periodically arranged in every 9 repeats of $4b$ [Fig. 4.7a]. In the same way, $(10/41)b^*$ corresponds to the structure with an interstitial defect in every 10 beats of $4b$ [Fig. 4.7b]. The b^*-component of Q_2, 0.26, can be viewed similarly. It is approximately equal to 5/19, which corresponds to the structure with vacant sites periodically arranged in every 5 beats of $4b$ [Fig. 3.7c][3].

[3] It is not clear whether these models can explain the results of diffraction experiments. In the case of $9/37b^*$ for example, the unit cell is 37 times larger than that of the high temperature phase. Bragg reflections should be observed at (h, k, l), $(h, k+(1/37), l)$, , $(h, k+(n/37), l)$, but only at (h, k, l) at high temperature, where h, k and l are integers. Therefore effects of the extinction rule are expected, obtainable by calculating the structure factor of the low temperature phase, in which atoms are displaced from the equilibrium (high temperature phase) positions.

Fig. 4.7. Discommensurations in $NbSe_3$. (–•) denotes the unit cell of $4b$ superlattice. (After [4.6])

Such a superstructure can be regarded as a kind of regular arrangement of discommensurations (DC). The CDW whose wave vector is incommesurate but nearly commensurate can gain commensurability energy by making locally commensurate domains [4.18, 19]. On the other hand, there is an energy loss due to the deformation of the CDW. Therefore, under certain conditions, a structure with commensurate domains and regularly arranged boundaries is possible. In such a modulated structure the averaged wave vector is incommensurate. The domain boundary is called DC. Consider the $\boldsymbol{Q}_1$ structure of $NbSe_3$ as an example. The charge density is $\varrho_0 + \Delta\varrho \cos(Q_1 x + \phi_0)$, where the x-axis is chosen to lie parallel to $\boldsymbol{Q}_1$. If the modulated structure is incommensurate and homogeneous, the phase ϕ_0 is constant. Using the commensurate wave vector $Q_1' = (1/4)b^*$, $Q_1 x + \phi_0$ is equal to $Q_1' x + \phi_0 + (Q_1 - Q_1')x$. In this expression the phase is not constant but varies in proportion to x [Fig. 4.8a]. On the other hand, if the wave vector is equal to Q_1', the phase is constant but with a multiple of $\frac{\pi}{2}$ [Fig. 4.8b]. In the discommensurate structure, DC's are arranged regularly as in Fig. 4.8c. The distance between DC's can be found from Fig. 4.8b as $(\frac{\pi}{2})|Q_1 - Q_1'|^{-1}$, since the gradient of Fig. 4.8a is equal to $|(Q_1 - Q_1')|$. [Compare with the soliton – a kind of excitation in the CDW state, explained in Sect. 2.2.2.)] As shown in (2.81), there is an electric charge piled up where the phase varies gradually in space. Therefore, DC's are charged; this picture has a similarity with the atomic picture shown in Fig. 4.7. Notice that (2.81) is valid only if the phase varies so slowly in space that the local charge density can be defined: this is different from the local picture.

The lattice distortion associated with the discommensurate CDW is periodic but not sinusoidal in space; it cannot be expressed by a single sinusoidal function but contains higher harmonics. Therefore satellite reflections corresponding to the higher harmonics can be observed in careful diffraction

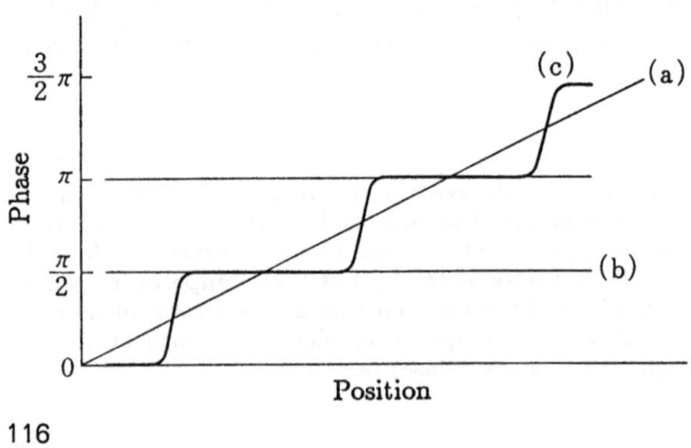

Fig. 4.8. Phase of the charge density wave

experiment. Note that higher order satellite reflections of this kind are different in origin from the $4k_F$-reflection in TTF-TCNQ and other compounds, explained in Sect. 3.4.1.

Consider now the transverse components of Q_1 and Q_2. Both transverse components of Q_1 are equal to 0; one may expect the related sheets of the FS to be planar. On the other hand, both the a^*- and c^*-components of Q_2 are equal to 0.5. There are two possibilities. The first is that the nesting vector is not parallel to the 1d axis, i.e. the related sheets of the FS have a curvature [cf. Fig. 2.8d]. The other is that, though the nesting vector has only a b^*-component, long-range order perpendicular to the b^*-axis is favoured by the electrostatic interaction between CDWs on neighbouring chains. It is reasonable to assume that the Q_1-CDW is on a certain kind of triangular chain and the Q_2-CDW on another, but nevertheless no unique explanation can be found as to why the transverse periods are twice the lattice constants (wave numbers of 0.5). As explained in Sect. 4.1.2 each column in the unit cell is connected to another by inversion symmetry, but it is not self-evident that the inversion symmetry still holds in the CDW state. Similarly, no explanation has been found to the problem of why the transverse components of Q_1 are equal to 0.

Another question concerns the possibility of a mutual relation between the Q_1 and Q_2-CDWs. Intensities of the Q_1 and Q_2 satellite reflections are shown in Fig. 4.9. The intensity of the Q_1 satellite does not show an anomaly at T_2, below which the Q_2 satellite begins to grow. The Q_2-CDW has no effect on the Q_1-CDW; the two CDWs are independent of each other below T_2, at least apparently. This insensitivity of the Q_2-CDW can be explained by a model in which the Q_1-CDW lies in a certain type of triangular column and the Q_2-CDW in another. When there is no special relation between Q_1 and Q_2, the three dimensional arrangement to the Q_1-CDW is not affected by the Q_2-CDW as far as averaged structure is concerned, even if

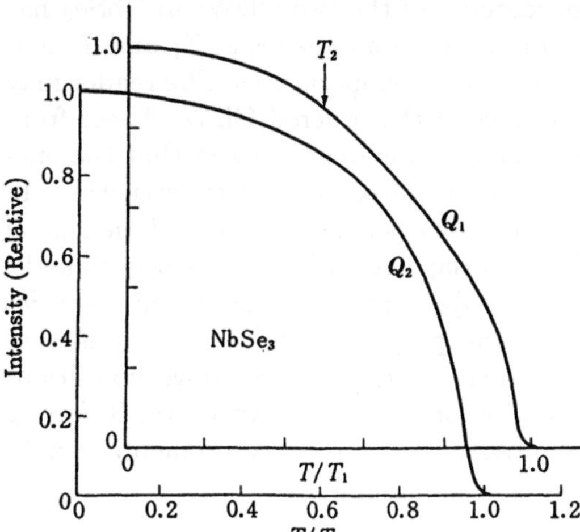

Fig. 4.9. Intensities of x-ray reflection [4.17] of Q_1 and Q_2 satellites in $NbSe_3$

the columns bearing the $\boldsymbol{Q}_1$-CDW lie between $\boldsymbol{Q}_2$ columns. Another idea is that a pair of planar sheets of FS contributes the $\boldsymbol{Q}_1$-CDW and that the reconstructed FS (below T_1) forms the $\boldsymbol{Q}_2$-CDW. These two models are essentially identical if the flat sheet of FS comes from wave functions with large amplitudes on a particular kind of triangular columns and if more curved FS sections correspond to wave functions approximately localized on the other type of column.

Several theoretical arguments have been presented on the mutual relation between $\boldsymbol{Q}_1$- and $\boldsymbol{Q}_2$-CDWs. It can be noticed that $2(\boldsymbol{Q}_1 + \boldsymbol{Q}_2)$ is approximately equal to the vector $\boldsymbol{G}(1, 1, 1)$ in the reciprocal lattice. When a PLD with the wave vector $\boldsymbol{Q}$ is not purely sinusoidal in space, the electronic system feels a potential with wave vectors $\pm 2\boldsymbol{Q}$, $\pm 3\boldsymbol{Q}$, ... , as well as the fundamental component $\pm\boldsymbol{Q}$. Energy gaps appear in the conduction band at $(\pm 2\boldsymbol{Q})/2$, $(\pm 3\boldsymbol{Q})/2$, ... , as well as at $\pm\boldsymbol{Q}/2$. Usually only the fundamental component is taken into consideration, but if $2\boldsymbol{Q}_2$ is equal to $-2\boldsymbol{Q}_1$ (plus a reciprocal lattice vector) the effect is not necessarily small, because they are mutually locked. In such a case, it may be that the wave vector $\boldsymbol{Q}_2$ is not determined solely by the nesting condition but is slightly modified by the periodic lattice modulation of $\boldsymbol{Q}_1$ below T_1. The situation is quite similar to the usual commensurability locking. Recent high resolution experiment have revealed, however, that the above relation does not hold exactly; $\boldsymbol{Q}_1 = (0, 0.24117\pm 0.0002, 0)$ and $\boldsymbol{Q}_2 = (0.5, 0.26038\pm 0.0002, 0.5)$, at low temperature [4.20]; the two wave vectors are determined independently of one another from the size of the FS.

4.2.2 Conductivity Anisotropy, Hall Effect and Thermoelectric Power

As we have already explained, the existence of the two CDWs in $NbSe_3$ has been established by observing the conductivity anomalies at T_1 and T_2 and the satellite reflections below these transition temperatures. The reader may wonder if we can predict the appearance of the ordered CDW phases from the material's high temperature properties. One might expect that the one-dimensionality of the system can be inferred from its crystal structure. In reality, however, such an attempt is not always fruitful. In one-dimensional conductors, electronic wave functions overlap each other sufficiently to make the band picture valid, but only in one direction, and the Fermi level lies within a band, while the overlap in the perpendicular directions is very small. Usually, we cannot predict a priori whether the above conditions are fulfilled, even by careful examination of the crystal structure. Referring to the formula (4.1) for the conductivity, the above requirements can be approximately expressed as

$$l_{\parallel} = v_{\parallel}\tau \gg a_{\parallel}$$
$$l_{\perp} \lesssim a_{\perp} \tag{4.4}$$

where $l_{\parallel}(l_{\perp})$ and $a_{\parallel}(a_{\perp})$ are the mean free path and the lattice constant parallel (perpendicular) to the 1d axis, respectively. The following relation can then be derived from (4.4):

$$\sigma_{\parallel}/\sigma_{\perp} \gg 1 \quad . \tag{4.5}$$

If sufficiently large crystals are available, we can measure $\sigma_{\parallel}$ and $\sigma_{\perp}$ independently, as in the case of TTF-TCNQ, by the four probe configuration. If the crystals are small in width, as in cases of MX_3, it is extremely difficult to measure the conductivity perpendicular to the needle axis independently. In such a case the Montgomery method [4.21] is used. The principle is as follows. The steady current flowing in a conductor satisfies the continuity condition

$$\operatorname{div} j = 0 \quad ,$$

where j is the current density. When the conductivity tensor σ_{ij} is diagonalized,

$$\sum \frac{\partial}{\partial x_i} j_i = \sum \frac{\partial}{dx_i}(\sigma_i E_i) = -\sum \sigma_i \frac{\partial^2 V}{\partial x_i^2} \quad .$$

By the transformation $x_i' = x_i/\sqrt{\sigma_i}$, the above equation is reduced to the Laplace's equation

$$\sum \frac{\partial^2 V}{\partial x_i'^2} = 0 \quad ,$$

from which the current distribution in an isotropic conductor can be determined. Electrical contacts are put on the four corners of a parallelepiped conductor. A current I is passed from one electrode to a neighbouring one, and the voltage V is measured between the other two contacts. By comparing the ratio V/I for various sizes of parallellepiped isotropic conductor with that of an anisotropic conductor (again parallelepiped shape with four contacts on the corners), the anisotropy of the conductivity within that plane, can be obtained.

The ratio of resistance, R_2/R_1 as a function of l_2'/l_1', for an isotropic conductor (Fig. 4.10) can be found in [4.21]. Then, from the measured value of R_2/R_1 of the anisotropic sample, its effective length ratio l_2'/l_1' is found. By measuring the actual size ratio of the sample, the ratio $\sqrt{\sigma_1/\sigma_2} = (l_2'/l_1')(l_1/l_2)$ can be obtained. When the thickness of the sample is not small, or when the resistivity in that direction is large, the current density

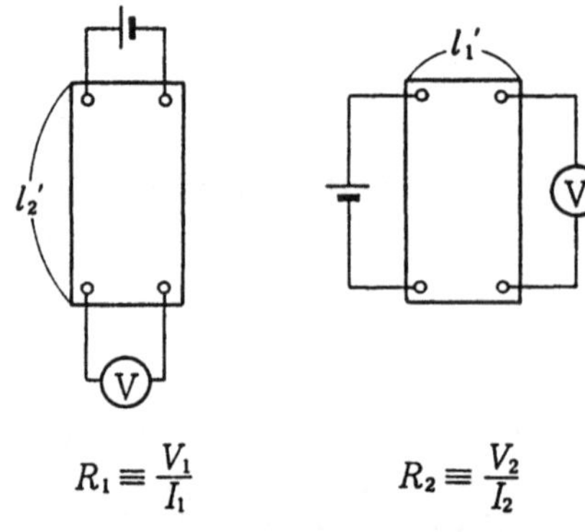

Fig. 4.10. The Montgomery method

is not uniform throughout the depth. The correction factor necessary in those cases is also given in [4.21].

Crystals of $NbSe_3$ have the shape of thin parallelepipeds, with the large face parallel to the b-c plane. The ratio σ_b/σ_c can be measured by the Montgomery method. Because the crystals are very thin, the resistivity perpendicular to the face can only be measured in a two probe configuration. In the two probe configuration, the measured resistance is the sum of those of contacts and of the crystal. The contact resistance is estimated from the measured resistance at 4 K, by assuming that the resistance of pure $NbSe_3$ at 4 K is much smaller than that at room temperature along every direction. The anisotropy at room temperature has been measured as $\sigma_b/\sigma_c = 17$ [4.22] and a value of 10 was obtained for the same ratio in another experiment [4.23]. The ratio σ_b/σ_a^* was found to be about 10^3 [4.23]. The small σ_{a^*} can be inferred from Fig. 4.3; it arises due to the presence of the van der Waals gap and the thus presumably small overlap of wave functions in that direction. On the other hand, we note that the ratio σ_b/σ_c is not very large. The relatively large σ_c should be related to the narrow temperature range of fluctuating CDWs (see Fig. 4.9), and also to the experimental finding [4.17] that the anisotropy in the width of the satellite reflections just above T_p, ~ 5, is not very large. Therefore $NbSe_3$ is sometimes called as an anisotropic two-dimensional metal.

Corresponding to the drastic change in the resistivity at the transition, anomalies have been observed in related properties. The Hall coefficient [4.24] under weak magnetic field is shown in Fig. 4.11 and the thermoelectric power [4.25] in Fig. 4.12. The former was obtained from the Hall voltage measured parallel to the c-axis with current and magnetic field parallel to the b- and a^*-axis, respectively. The conductivity tensor $\sigma_{ij}(B)$, in magnetic field B is given in footnote 4. Because $\sigma_{bc}(B) \propto B$, the Hall coefficient R_H is given by

$$R_H = \frac{\sigma_{bc}/B}{\sigma_b\sigma_c + \sigma_{bc}^2} \cong \frac{\sigma_{bc}/B}{\sigma_b^0\sigma_c^0} \tag{4.6}$$

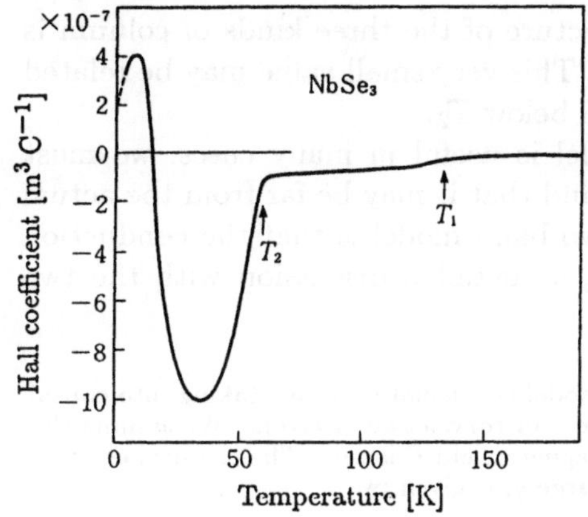

Fig. 4.11. Hall coefficient [4.24] of $NbSe_3$

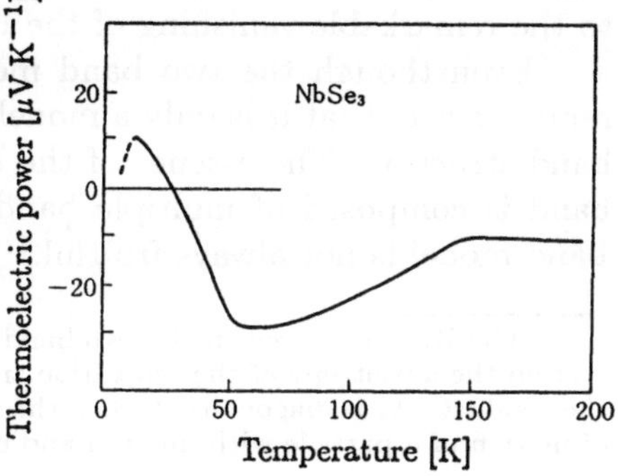

Fig. 4.12. Thermoelectric power [4.25] of $NbSe_3$

where σ_i^0 is the conductivity in zero magnetic field; $\sigma_i^0 = \sigma_i(B = 0)$. In the simplest case of a single band, the sign of charge and concentration n of current carriers can be obtained from the relation

$$R_{\mathrm{H}} = 1/ne \tag{4.7}$$

where e is negative for electrons and positive for holes. Assuming (4.7) to be valid in $NbSe_3$, n is found to be 0.05 per Nb atom at room temperature. However, it is not appropriate to use (4.7) to relate the changes in R_{H} below T_1 and T_2 to those in n, because the temperature dependence of R_{H} is too complicated especially below T_2. If we use the two band model (an electron and a hole band), the temperature dependence of R_{H} reflects the difference between the relaxation time of electrons and that of holes, therefore the densities of carriers cannot be obtained. In good crystals, carrier mobility increases considerably with decreasing temperature and the terms in R_{H} proportional to B^2 cannot be neglected even when the magnetic field is low. Under high magnetic field both σ_x and σ_y are proportional to $1/B^2$ and σ_{xy} to $1/B$. Therefore, when $n \neq p$, R_{H} becomes temperature independent: $R_{\mathrm{H}} = (\sigma_{xy}B)^{-1} = -1/e(n-p)$. From Fig. 4.13, $p - n = 2.7 \times 10^{18}\,\mathrm{cm}^{-3}$

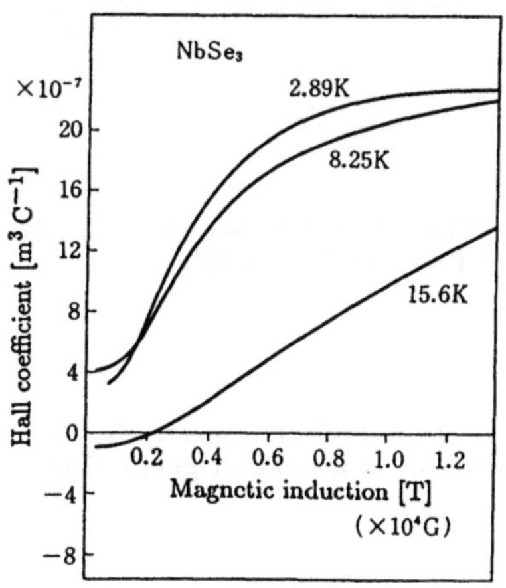

Fig. 4.13. High field Hall coefficient [4.24]

below T_2. If the difference in the structure of the three kinds of column is neglected, $p - n = 10^{-3}$ per Nb atom. This very small value may be related to the remarkable vanishing of the FS below T_2.

Even though the two band model is useful in many cases, we must remain aware that it is only a model and that it may be far from the actual band structure. The essence of the two band model is that the conduction band is composed of multiple bands, so detailed discussion with the two band model is not always fruitful.[4]

---

[4] The Hall coefficient in the two band model is calculated below, taking into consideration the anisotropy of the relaxation time τ of conduction electrons. We assume that the conductivity is diagonalized when the magnetic field B is zero. The classical equation of motion of a particle with mass m and charge q, is given by

$$m(d\boldsymbol{v}/dt) = -m\boldsymbol{v}/\tau + q(\boldsymbol{E} + \boldsymbol{v} \times \boldsymbol{B}) \quad .$$

In the steady state, $d\boldsymbol{v}/dt = 0$. Under a magnetic field parallel to the z-axis,

$$mv_x/\tau_x = q(E_x + v_y B) \quad , \quad mv_y/\tau_y = q(E_y - v_x B)$$

where τ_j is the relaxation time for the motion along the j-axis. Velocities v_x and v_y are given by

$$v_x = (\mu_x E_x + \mu_x\mu_y E_y B)/(1 + \mu_x\mu_y B^2)$$
$$v_y = (-\mu_x\mu_y E_x B + \mu_y E_y)/(1 + \mu_x\mu_y B^2)$$

where $\mu_i = \tau_i(q/m)$ is the mobility along the i-axis. Since the current density $\boldsymbol{j}$ is equal to $nq\boldsymbol{v}$, the conductivity tensor is given by

$$(\sigma_{ij}) = \frac{nq}{1 + \mu_x\mu_y B^2}\begin{pmatrix} \mu_x & \mu_x\mu_y B & * \\ -\mu_x\mu_y B & \mu_y & * \\ * & * & * \end{pmatrix} \quad .$$

To obtain the Hall coefficient, the electric field E_y is measured under the condition:

$$j_y = \sigma_{yx}E_x + \sigma_{yy}E_y = 0 \quad .$$

When the magnetic field is low,

$$E_y = (\sigma_{xy}/\sigma_{xx}\sigma_{yy} + \sigma_{xy}^2)j_x \cong (\sigma_{xy}/\sigma_{xx}^0\sigma_{yy}^0)j_x$$

so the Hall coefficient $R_{\rm H} = E_y/j_x B$ is given by (4.6). If the conduction band is composed of electron and hole bands, σ_{ij} is the sum of two contributions; for example,

$$\sigma_{xy} = \left(\frac{-n\mu_x\mu_y}{1 + \mu_x\mu_y B^2} + \frac{p\nu_x\nu_y}{1 + \nu_x\nu_y B^2}\right)|e|B \quad ,$$

where $n(p)$ and $-e(+e)$ are the density and charge of electrons (holes), respectively. Their mobilities are expressed by μ_i and ν_i. The Hall coefficient in a weak magnetic field is then given by

$$R_{\rm H} = -\frac{(n\mu_x\mu_y - p\nu_x\nu_y)|e|}{(n\mu_x + P\nu_x)(n\mu_y + P\nu_y)} \quad .$$

The thermoelectric power has been measured under a temperature gradient parallel to the b-axis. For a crystal with cubic symmetry[5] and a single band, the thermoelectric power S is given by

$$S(T) = \frac{\pi^2 k_B^2 T}{3e}\left[\frac{\partial \ln \sigma(\varepsilon)}{\partial \varepsilon}\right]_{\varepsilon_F} \quad . \tag{4.8}$$

Above T_1, $S(T)$ is proportional to T as in TTF-TCNQ. At low temperatures however, its temperature dependence is complicated. If the conduction band is composed of several bands, $S(T)$ is given by the weighted sum:

$$S = \sum \sigma_i S_i / \sum \sigma_i$$

and the actual experimental data are extremely difficult to analyze. Nevertheless, $S(T)$ is sometimes quite useful because, compared to the conductivity, it is more sensitive to small changes in the system.

Anomalies have been observed in the specific heat [4.26] and the magnetic susceptibility [4.27] of $NbSe_3$.

4.2.3 Nonlinear Electrical Conductivity: Sliding Motion of Charge Density Waves

a) The dc Conductivity

In this section the non-ohmic conductivity, the most remarkable feature of $NbSe_3$, and other related phenomena are discussed. As illustrated in Fig. 4.14, the conductivity σ_b (along the b-axis) below T_1 is constant in a weak electric field but increases sharply at a critical field E_T, and then approaches a constant in the high field limit [4.28]. If this phenomenon were caused by the temperature rise due to Joule heating, the conductivity would

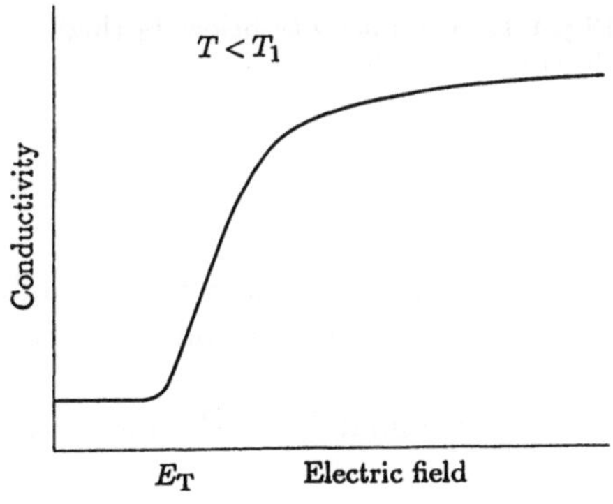

Fig. 4.14. Nonlinearity of the conductivity of $NbSe_3$ below T_1 (schematic)

[5] Equation (4.8) can be derived by assuming cubic symmetry. However, it is also often used for low symmetry crystals. This expression is equivalent to (3.5) in Chap. 3; it is reduced to (3.5) by writing explicitly the energy dependences of n and μ.

increase or decrease according to the measuring temperature (cf. Fig. 4.5); this is found not to be the case. Moreover, it has been confirmed that there is no temperature rise, by observing the voltage response for high pulse current (in a high electric field). The important features of this non-ohmic conductivity can be summarized as follows.

i) The nonlinear conductivity is observed only below T_1. Therefore, it is undoubtedly related to the long-range order of the charge density wave (CDW).
ii) It can be observed only in the b-component; the transverse conductivities remain ohmic [4.29] in electric fields at least several times higher than E_T.
iii) Another step-like increase in conductivity is observed below T_2 in higher fields [4.30] (Fig. 4.15).

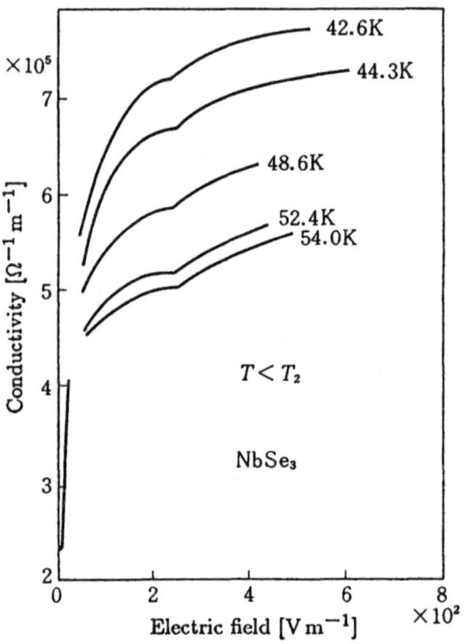

Fig. 4.15. Conductivity below T_2 (higher electric field) [4.30]

iv) In the high field limit the anomalies at T_1 and T_2 almost vanish and the conductivity varies smoothly from above the transition temperatures [4.31].
v) The conductivity σ_b can be expressed approximately by the following empirical formulae:

$$\sigma_b = \begin{cases} \sigma_1 & E<E_T \\ \sigma_1 + \sigma_2 \exp[-E_0/(E-E_T)] & E>E_T \end{cases} \qquad (4.9a)$$

or

$$\sigma_b = \sigma_1 + \sigma_2\left(\frac{E - E_T}{E}\right)\exp\left[-E_0/(E - E_T)\right] \quad E > E_T \quad . \tag{4.9b}$$

Above E_T the conductivity is the sum of the ohmic part σ_1 and the non-ohmic contribution which approaches σ_2 in the high field limit.

vi) Both E_T and E_0 are sensitive to temperature [4.32] (Fig. 4.16) and to the species and concentration of impurities or defects. The latter feature will be described in detail below.

vii) In fields higher than E_T, the electrical noise is found to be remarkably enhanced. This is explained below.

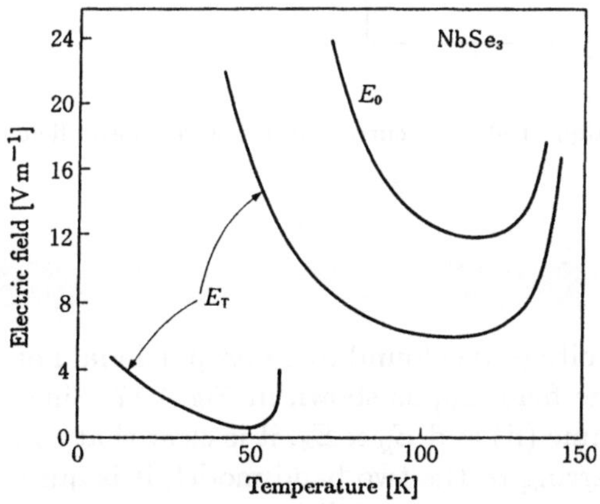

Fig. 4.16. Temperature dependences [4.32] of E_T and E_0

When nonlinear conductivity was first observed, it was suggested that the sheet of the Fermi surface, which had vanished with the onset of the CDW, reappears in high fields by some unknown mechanism and as a result the carrier density n increases. But this was proved not to be the case by x-ray diffraction from samples in high electric fields [4.17]. If the vanishing Fermi surface partially revives, the CDW amplitude would decrease and the intensity of satellite reflections would decrease. But the observed intensity shows no detectable change. It is also unlikely that the mobility (rather than the density) of carriers increases; the electric field is not high enough to modify the distribution of normal electrons significantly. Thus the increase in the conductivity cannot be attributed to a change in the motion of conduction electrons, and must therefore be due to the motion of CDWs which become depinned by the applied electric field (cf. Sect. 2.2.1).

More definite evidence for depinning of the CDWs has come from the Hall coefficient measured in low magnetic field but high electric field [4.29, 33]. As explained in Sect. 4.2.2, the Hall field (E_y) in a low magnetic field is given by

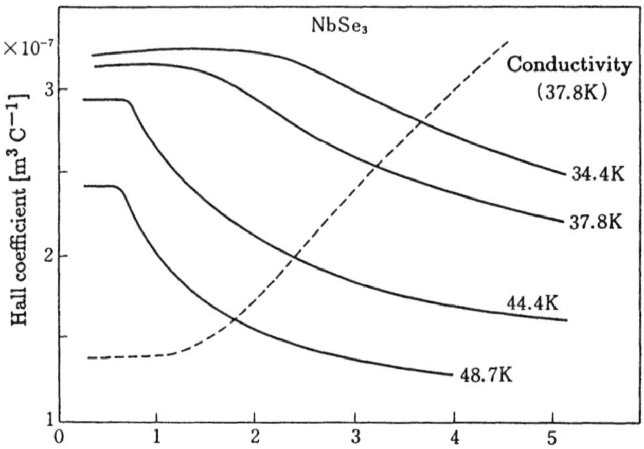

Fig. 4.17. The Hall coefficient as a function of electric current in a low magnetic field [4.29]

$$E_y = \left(\frac{\sigma_{xy}}{\sigma_x^0 \sigma_y^0}\right) j_x = \left(\frac{\sigma_{xy}}{\sigma_y^0}\right) E_x \quad .$$

In the non-ohmic regime, the Hall voltage was found to be proportional not to the current density j_x, but to the field E_x, as shown in Fig. 4.17. Since the transverse conductivity σ_y^0 is ohmic (ii) and $E_y \propto E_x$, it is clear that σ_{xy} is independent of j_x (and E_x). Referring to the two band model, it is quite probable that both carrier densities and their mobilities are independent of the applied electric field, E_x. (One may argue that the densities and/or mobilities of electrons and/or holes are field dependent but the conductivity σ_{xy} itself is field independent. But in reality it would be difficult to explain such peculiar field dependences of these quantities.)

The experimental finding that E_y is not proportional to j_x means that σ_x^0 varies with electric field. Since the single particle conductivity remains constant, it is quite natural to conclude that a new conduction mechanism arises along the chain axis. In 1d conductors such a new conduction mechanism must be the sliding of CDWs. Moreover, the nonlinear Hall voltage is observed only below T_1. Thus we can say that the existence of the sliding motion of CDWs, i.e. electrical conduction by the Peierls-Fröhlich mechanism, is definitely confirmed. The conductivity σ_1 in (4.9a, b) is the contribution of conduction electrons (single particles) while the second term is due to the sliding CDW.

The important mechanism for pinning the CDW is impurity pinning in cases where the CDW wave vector is incommensurate with the lattice[6].

[6] In the case of the discommensurate structure, described in Sect. 4.2.1, the pinning by the commensurability must be taken into consideration, as well as the pinning of discommensurations by impurities.

Therefore the threshold electric field E_T is expected to be sensitive to impurities. According to the theory of *Lee* and *Rice* [4.34], E_T is proportional to the impurity concentration c if the CDW is strongly pinned, and to c^2 if the pinning is weak (cf. Sect. 2.2.3). The electrostatic potential of an impurity is written approximately as $V(r) = V_0\delta(\boldsymbol{r} - \boldsymbol{r}_i)$, where $\boldsymbol{r}_i$ is the position of the impurity atom; the Coulomb potential is screened by conduction electrons and is of short-range. As was explained in Sect. 2.2.2, the interaction between CDWs and impurity atoms, F_{imp} is given by

$$\begin{aligned} F_{\text{imp}} &= -\int \varrho_0 \exp\{\mathrm{i}(\boldsymbol{Q}\cdot\boldsymbol{r} + \phi)\}V(\boldsymbol{r})d\boldsymbol{r} \\ &= -\varrho_0 V_0 \sum \exp\{\mathrm{i}(\boldsymbol{Q}\cdot\boldsymbol{r}_i + \phi)\} \quad . \end{aligned} \tag{4.10}$$

If there were only one impurity atom, the CDW phase ϕ would be determined uniquely so as to minimize F_{imp}. If many impurities are randomly distributed in space, the CDW must be distorted. Since r_i is random, $\sum \exp(\mathrm{i}\boldsymbol{Q}\cdot\boldsymbol{r}_i)$ can be written approximately as (number of impurities)$^{1/2}$ and F_{imp} per unit volume is equal to $-V_0\varrho_0(cL^3)^{1/2}/L^3 = -V_0\varrho_0 L^{-3/2}c^{1/2}$, where L^3 is the sample volume. Provided the interaction with each impurity atom is weak (weak pinning), the CDW phase is approximately constant over a certain volume. Neighbouring domains have different values of $\overline{\phi}$, so that near the boundary the CDW is deformed to keep its phase contiuous. The elastic energy per unit volume is proportional to the squared gradient of the order parameter, ψ (the periodic lattice distortion or the accompanying modulation of the charge density):

$$F_{\text{el}} = f_0 \int \xi^2 |\nabla\psi|^2 d\boldsymbol{r} \sim f_0\xi^2|\psi|^2/L^2 \quad , \tag{4.11}$$

where ξ is the coherence length of the CDW and the approximation $|\partial\psi/\partial x| \sim |\psi|/L$ is used. The total free energy $F_{\text{imp}} + F_{\text{el}}$ has the minimum value $f_0\xi^2|\psi|^2/L^2$, when

$$L \cong (f_0\xi^2|\psi|^2/\varrho_0 V_0)^2 c \quad . \tag{4.12}$$

The CDW can be regarded depinned in electric field if it is displaced by its period; the threshold electric field E_T is given by:

$$\varrho_0 E_T(2\pi/Q) \cong f_0\xi^2|\psi|^2/L^2 \propto c^2 \quad . \tag{4.13}$$

If the interaction between the CDW and each impurity is strong, the CDW phase tends to be fixed near an individual impurity atom even if the elastic energy cost increases. When an electric field is applied the CDW phase far from impurity atoms increases, and thus its gradient in space also increases. Since F_{imp} is given by (4.10) also in the case of strong pinning,

$F_{\mathrm{el}} \sim f_0|\psi|^2\xi^3[\phi(r\to\infty)-\phi(0)]$ per impurity atom and the relation

$$E_{\mathrm{T}} \propto c \tag{4.14}$$

is obtained. Experimentally [4.35] it was found in Ta-doped samples that E_{T} is proportional to the square of the residual resistance ratio, $\varrho(4\,\mathrm{K})/\varrho(300\,\mathrm{K})$, below T_2. Tantalum and niobium have the same number of outer electrons, so a Ta atom works as a weak impurity. The residual resistance ratio is proportional to Ta concentration (Matthiessen's rule). Thus (4.13) is verified. The linear relation (4.14) was observed in samples irradiated by high energy protons [4.36].

Feature (ii), of the non-ohmic conductivity, is consistent with the limited freedom of the CDW motion; it can slide only parallel to the wave vector. The second discontinuity in the conductivity below T_2 (iii) reflects the depinning of the Q_1 CDW. The steep increase of E_{T} near the transition temperatures has been attributed to the decreasing CDW amplitude (weakening of the effect of applied field) and to the rapidly increasing coherence length; the latter has been verified experimentally in the case of BCS superconductors. Remaining problems are: why E_{T} increases again with decreasing temperature and what is the physical picture of the electric field dependence of the nonlinear conductivity, i.e. the second term of (3.9a or b). In other words, the physical meaning of the characteristic field E_0 is not clear. It was found that E_0 increases in proportion to the square of the residual resistance ratio (impurity concentration). If the effect of dynamic friction on the CDW is entirely included in σ_2, it is possible that E_0 is a measure of some inhomogenity, for example the variation of E_{T} within the sample. In that case, the measured E_{T} is approximately equal to the minimum value of the local E_{T} and its increase with lowering temperature is the effect of thermal depinning. On the other hand, the nonlinearity of the conductivity in some samples shows a drastic change even when the temperature is changed only slightly [4.37], as illustrted in Fig. 4.18. It was proposed [4.37] that the CDW moves coherently over a certain volume, and that individual domains move independently in low fields and that the size of domains becomes larger with increasing field. Here, the physical picture of "inhomogenity" becomes important. However, the experimental results are not fully understood so the term "inhomogenity" is sometimes equivalent to "not understood yet"! Let us compare the present status of research on the sliding CDW with the past one on imperfections in crystals. Real crystals contain various defects which have an important influence on the mechanical properties of crystal. The idea of a dislocation was originally proposed to explain the mechanical weakness of certain crystals. Only later was the existence of dislocations experimentally verified. If the CDW motion is likened to the plastic deformation of the "CDW crystal", it is necessary to consider the effect of extended defects, such as dislocations, and their distribution [4.34], as well as the effects of point defects.

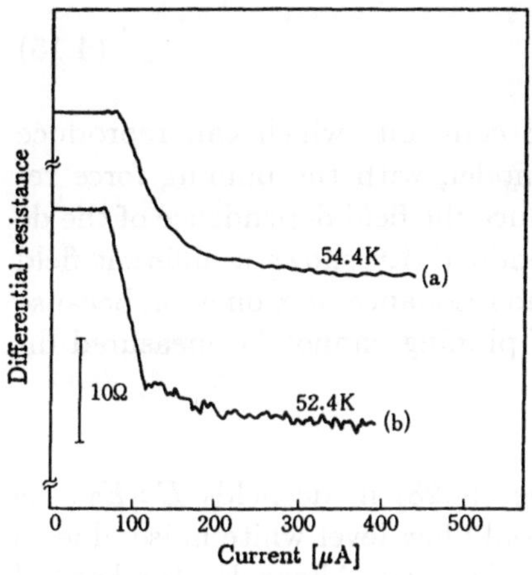

Fig. 4.18. Nonlinearity in the differential resistance [4.37]

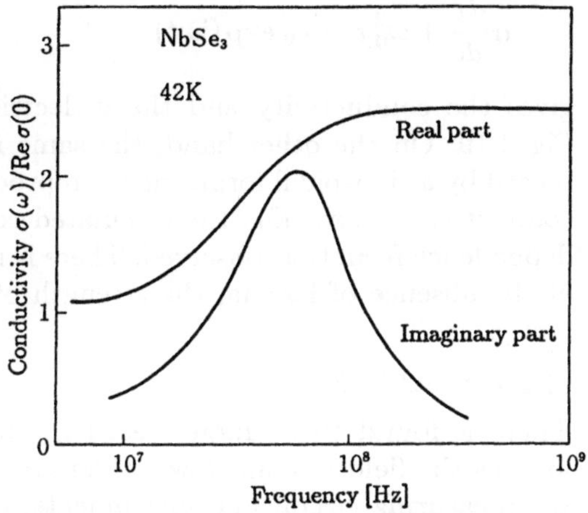

Fig. 4.19. Frequency dependence [4.38] of the conductivity of $NbSe_3$

b) High Frequency Conductivity and Dielectric Constant

Figure 4.19 shows the frequency dependence of the conductivity [4.38]. Neither real nor imaginary part shows such a dependence above T_1. This strong dependence on frequency is most clearly observed at the temperature where the dc resistivity is maximum. The dielectric costant, which is equal to the imaginary part of the conductivity divided by the angular fequency, $\mathrm{Im}\{\sigma(\omega)\}/\omega$, becomes as large as 10^8. When the limiting conductivity $\sigma(\omega \to \infty)$ is plotted against temperature, the anomalies at T_1 and T_2 are hardly observable. Qualitatively, the following relation can be found: $\sigma(\omega \to \infty) = \sigma(E_{\mathrm{dc}} \to \infty)$.

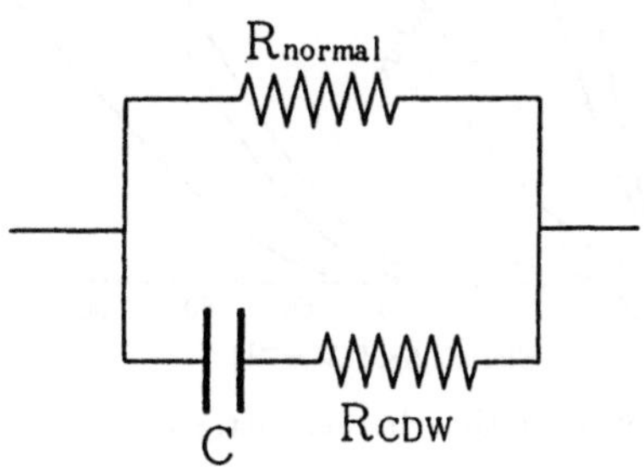

Fig. 4.20. Equivalent circuit to explain the conductivity as the function of frequency [4.39]

The equivalent circuit shown in Fig. 4.20 can reproduce the observed frequency dependence [4.39]. Here, the current path for normal electrons is represented by R_{normal} and parallel to the CDW by R_{CDW} and the capacitor C_{CDW}. In this circuit, there is no inductance coil, so there is no effect of inertia. In the classical mechanics analogue, this circuit corresponds to the overdamped oscillator, where the inertia term is neglected. If the potential is harmonic, the solution of

$$\alpha\frac{dx}{dt} + \omega_0^2 x = e_0 \exp(\mathrm{i}\omega t) \tag{4.15}$$

gives the conductivity and the dielectric constant, which can reproduce Fig. 4.19. On the other hand, the same model, with the pinning force replaced by a sinusoidal form, cannot reproduce the field dependence of the dc conductivity above E_T; the calculated conductivity shows a different field dependence from that observed. There is no resonance phenomenon because of the absence of inertia; the strength of pinning cannot be measured directly.

c) Electrical Noise

There is found to be large electrical noise [4.28] in dc fields $E > E_T$. As long as the field remains lower than E_T, only low level white noise, due to the measuring circuit and instruments, is observed. Above E_T the level of white noise increases rapidly, as can also be seen in Fig. 4.18,[7] especially in

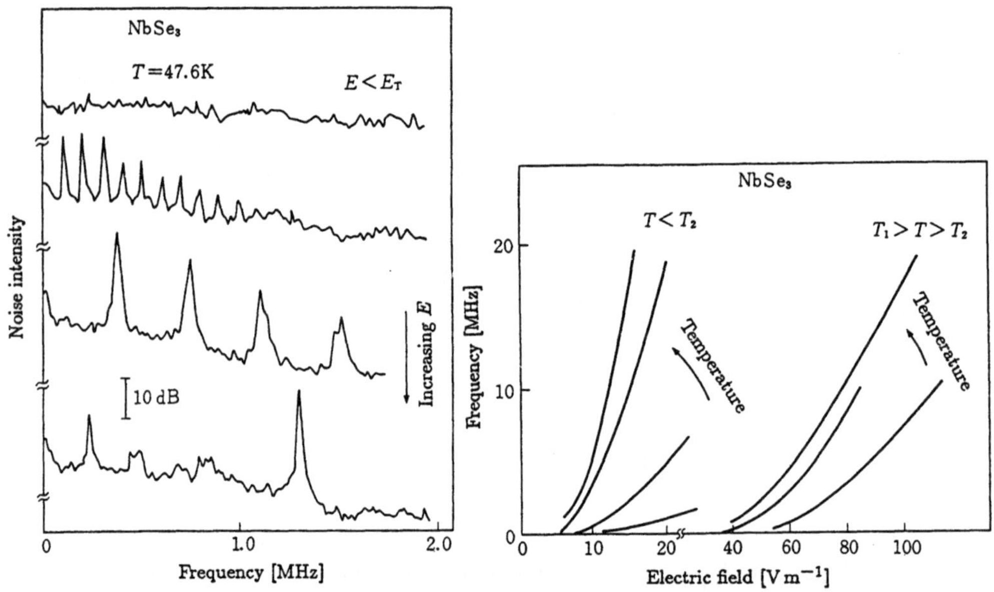

Fig. 4.21. (a) Noise spectrum. **(b)** Field dependence of the characteristic frequency [4.28]

[7] The differential resistance dV/dI is obtained by measuring the voltage under dc current I_0 superposed by a small ac current $i_0 \cos \omega t$. As far as $I_0 \ll i_0$,

$$V = V(I_0 + i_0 \cos \omega t) \cong V(I_0) + (dV/dI) i_0 \cos \omega t \quad .$$

The differntial resistance dV/dI at $I = I_0$ can be obtained by measuring the ω component of voltage selectively. From Fig. 4.18 it can be seen that the ω component of the white noise increases remarkably above E_T.

curve (b). At the same time, there appear sharp peaks in the noise spectrum as shown in Fig. 4.21a. The characteristic frequencies increase with increasing dc field (Fig. 4.21b). Some of them are higher harmonics of the others but sometimes several fundamental frequencies are observed. (Figure 4.21a shows the Fourier spectrum of the noise across the sample, when only dc current is passed. If only the fundamental component ω_0 is observed, the voltage across the sample contains the sinusoidal component of angular frequency ω_0, in addition to the dc component. The smaller width $\Delta\omega_0$ corresponds to the longer duration (slower damping) of the oscillation in real time. When a higher harmonic component, say $3\omega_0$, is also observed, the voltage oscillation in real time has the non-sinusoidal form $A_0 \sin(\omega_0 t) + A_3 \sin(3\omega_0 t + \delta)$. The relative amplitude A_3/A_0 is determined from the relative strength in the noise spectrum, but the phase difference δ cannot be determined from the frequency distribution. In short, a sharp noise spectrum with higher harmonics corresponds to a non-sinusoidal distorted voltage oscillation which decays slowly in time.) This phenomenon has been observed by many workers but the detailed results are still irreproducible. An important key to understanding this phenomenon is that the characteristic frequency is proportional to the non-ohmic current irrespective of temperature [4.40]. As shown in Fig. 4.22, a linear relation can be found between the characteristic noise frequencies and the current carried by the CDW. As the CDW amplitude is not likely to change in a strong field, the characteristic frequency ν is proportional to the drift velocity v_D of the CDW. When measured at constant voltage (constant force on the CDW), the current noise corresponds to the acceleration and slowing down of the CDW as its maxima and minima pass through the pinning sites.[8] The drift velocity is of the order of $10^{-3} \sim 10^{-2}$ m/s, since $v_D(1/\nu) = (Q/2\pi)^{-1}$.

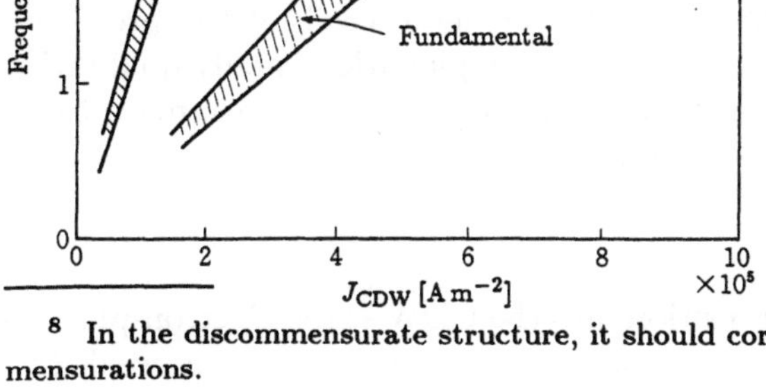

Fig. 4.22. Proportionality between noise frequencies and the nonlinear current [4.40]

[8] In the discommensurate structure, it should correspond to the passage of discommensurations.

The above interpretation is based on the assumption that the whole CDW is moving in the electric field; the field dependence of the non-ohmic conductivity comes from the field-dependent mobility, not from the variation in the volume fraction of the CDW set in motion. The solution of (4.15), with the harmonic potential replaced by a sinusoidal one and the external force by a dc field, contains harmonic components. Their dependence on electric field, however, is not in agreement with the experimental results.

Though not explained in this text, different mechanisms for the non-ohmic conductivity [4.41, 42] and for the noise [4.43] have been proposed. In particular, *Bardeen* [4.41] proposed that the process of depinning the CDW is not a classical phenomenon, as explained above, but that the CDW is depinned by the quantum-mechanical tunnelling. Extension of his theory has successfully explained a variety of experimental data. However, at present, neither classical nor quantum theory can explain all the data accumulated in this field.

d) Sliding CDW: A Rigid Body?

Some important remaining questions include the following: why are effects of the CDW mass not observed and to what extent can the CDW be regarded as the rigid body? Related to these questions, it was found that the voltage response to current pulses is strongly dependent on the electrical history of the sample [4.44]. When successive high current pulses, square in shape, are passed through the sample, the voltage response to each pulse overshoots initially and then gradually approaches the stationary value. If the direction of the current is reversed, a slow increase of the voltage is observed; the sample is in a less resistive state. At the end of each pulse, the voltage drops to zero instantaneously; the transient response to the current pulse is not attributable to a finite inertia of the CDW.

Various other similar phenomena, "memory effects", have been found, all of which are suggestive that the sample is not in configurational equilibrium. As explained above [e.g. (4.11)], the CDW, when at rest, is deformed locally to optimize the interaction with pinning sites. Because the sample dimension is larger than the coherence length, there are various metastable configurations. Once a high current pulse has bene passed through the sample, the CDW is "trained" to respond to the next pulse in a different fashion; for example, the CDW can be set in motion more easily if the preceding pulse was of the opposite direction. Such an explanation is only qualitative and oversimplified; many of the various memory phenomena remain to be fully understood in a consistent manner.

4.2.4 Superconductivity

In materials which exhibit a transition into the CDW state, electron-phonon interactions are strong while the repulsion between electrons is weak (cf.

Sect. 2.4). In general, these conditions are exactly those required for superconductivity to appear. When these two ordered phases are compared, the energy gap in the electron energy spectrum in the CDW state is in general larger than the superconducting energy gap; the CDW transition temperature is expected to be higher than that of superconductivity. Since the Fermi surface vanishes when the CDW is formed at higher temperature, superconductivity is not likely to appear at a lower temperature. In $NbSe_3$, however, the Fermi surface does not completely vanish, even below T_2, so there remains a possibility that superconductivity will appear. Under hydrostatic pressure [4.45] the CDW transition temperature T_2 decreases and approaches zero at $p \sim 6$ kbar (6×10^8 Pa). The upper CDW transition temperature T_1 also decreases under pressure. As for the superconducting transition temperature, T_c, this increases with pressure presumably because of the enhanced three-dimensionality. T_c becomes ~ 2 K at $p \sim 6$ kbar (Fig. 4.23). The difficulty is that the pressure dependence of T_c determined by resistivity measurements is different from that determined by the magnetic (Meissner effect) measurements. This could be due to a difference in the quality of the applied pressure: inhomogenity and/or deviation from hydrostatic pressure, e.g., presence of a shear component. At present, we can only conclude qualitatively that the CDW state and superconductivity are competing phenomena.

There has been some confusion as to whether $NbSe_3$ becomes superconducting at atmospheric pressure. In some samples the resistivity decreases at temperatures below 2 K, corresponding to the onset of superconductivity, but with a broad transition width. A finite resistance is often observed even at the lowest temperature, while in other samples, no superconducting

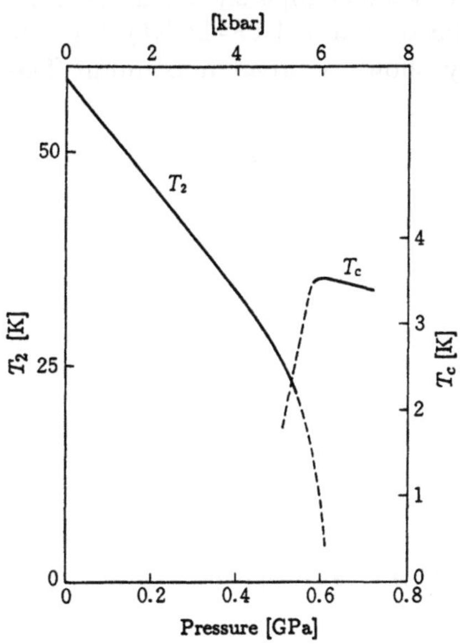

Fig. 4.23. Pressure dependences of T_2 and the superconduction transition temperature (T_c) [4.45]

transition is observed down to 0.05 K. The diamagnetic susceptibility of an agglomerate of many crystals is much smaller than that of usual superconductors and it increases gradually with decreasing temperature [4.46]. The presence or absence of the decrease of resistivity associated with superconductivity was not controlled. This is presumably one of the explanations for the poor reproducibility of T_c under pressure. It has been established that the superconducting decrease of resistivity can be observed only in twinned crystals [4.47]. In a true single crystal, in which only one set of x-ray diffraction spots appears, no superconducting transition can be observed. A small fraction of sample volume near the twin boundary is responsible to the superconductivity in "pure" $NbSe_3$; it is not a bulk property.

In Ta-doped samples, both T_1 and T_2 are lower and T_c increases but both the CDW transitions and the superconducting transition are observed in the same sample so long as the Ta concentration is low [4.48]. At a higher concentration, resistivity anomalies at T_1 and T_2 become broad and can hardly be observed while T_c becomes as high as 2 K. The superconducting upper critical field shows a large anisotropy; the electronic system is still anisotropic [4.49]. Qualitatively, superconductivity and the CDW can coexist but in a competitive way. The relation between the two types of ordering has not yet been fully understood experimentally nor from the microscopic point of view.

4.3 Charge Density Waves in TaS_3

4.3.1 "Orthorhombic" (o-)TaS_3

The resistivity of o-TaS_3 as measured in low fields [4.50] is shown in Fig. 4.24. The temperature dependence is quite similar to that of TTF-TCNQ (Fig. 3.9). Near room temperature the conductivity shows a broad maximum. The

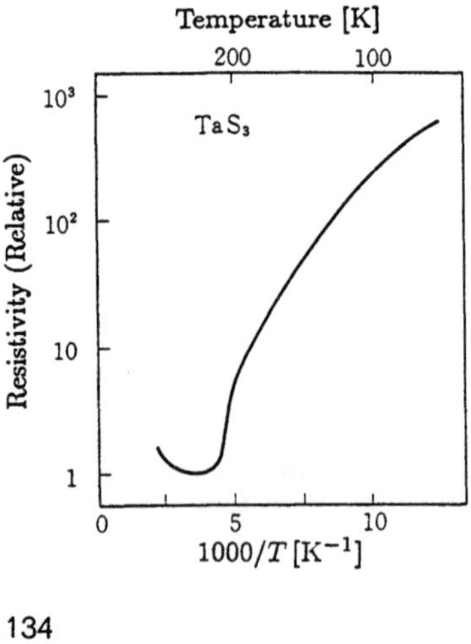

Fig. 4.24. Resistivity [4.50] of o-TaS_3

transition temperature $T_{\rm p}$ was determined from the maximum of $|d\ln\sigma_c/d\ln T|$ to be 220 K. Above room temperature the one-dimensional diffuse diffraction streaks are observed at $Q_c = 0.25c^*$; a CDW is formed on each triangular chain but their phases are not correlated. Below $T_{\rm p}$ o-TaS_3 is insulating; a nearly planar Fermi surface vanishes. At low temperature the Peierls phase shows a commensurate lattice modulation [4.51] with the wave vector $Q = (1/2)a^* + (1/8)b^* + (1/4)c^*$, where the 1d axis is parallel to the c-axis. A recent electron diffraction experiment [4.52] suggests that both the longitudinal (c^*-) and the transverse (a^*-) components of the wave vector are incommensurate below $T_{\rm p}$ and only lock into the commensurate values at $\sim$160 K. (The b^*-component was not measured, because it is parallel to the electron beam.) Both components are temperature dependent but the change at 160 K is continuous, in contrast with $NbSe_3$ and most organic conductors in which the londitudinal components are temperature independent. A re-examination by x-ray would be desirable to help to establish why the wave vector in o-TaS_3 is temperature dependent.

At room temperature the conductivity is anisotropic [4.23]; $\sigma_c/\sigma_b \sim 150\pm50$, $\sigma_c/\sigma_a \sim 5(\sigma_c/\sigma_b)$. These values are larger than those of $NbSe_3$, consistent with the non-metallic conductivity below $T_{\rm p}$ and with the weak temperature dependence of the resistivity over a wide temperature range above $T_{\rm p}$ (broader fluctuation range).

In order to estimate the Peierls gap 2Δ, the conductivity below $T_{\rm p}$ is expressed approximately by

$$\sigma_c(T) = \sigma_c^0 \exp\{-\Delta(T)/k_{\rm B}T\} \tag{4.16}$$

where the maximum of σ_c just above $T_{\rm p}$ is substituted for σ_c^0. The gap $\Delta(T)$ should be temperature dependent as expected from the mean-field theory (BCS theory of superconductivity). From Fig. 4.25, $2\Delta(0)$ is found (by ex-

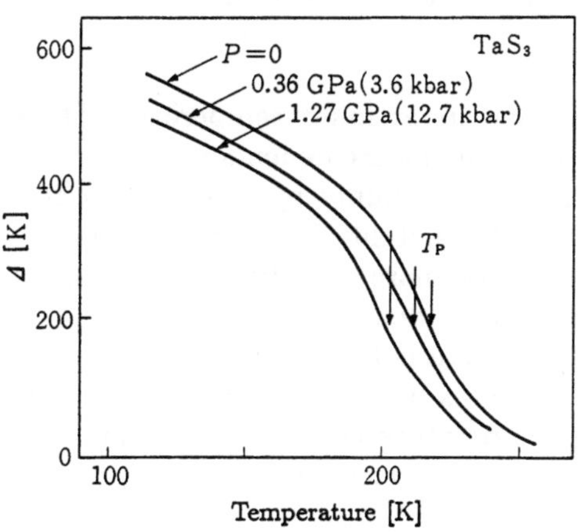

Fig. 4.25. Order parameter as functions of temperature and pressure [4.53]

trapolating to $T = 0$) to be $\sim 1200\,\mathrm{K}$ ($\sim 6k_\mathrm{B}T_\mathrm{P}$). Since the mean-field theory gives $2\Delta(0) = 3.5k_\mathrm{B}T_\mathrm{p}^\mathrm{MF}$, the low dimensionality is reflected in the low T_p of $\sim 0.6T_\mathrm{p}^\mathrm{MF}$. Under hydrostatic pressure the distance between the triangular chains would be reduced and T_p would increase due to the enhanced interchain coupling. From the measurement of conductivity under pressure [4.53], however, the following relations were obtained: $dT_\mathrm{P}/dp = -1.3\,\mathrm{K\,kbar}^{-1}$ ($-4 \times 10^8\,\mathrm{K\,Pa}^{-1}$). The ratio $T_\mathrm{P}/\Delta(0)$ or $T_\mathrm{P}/T_\mathrm{p}^\mathrm{MF}$ is pressure independent; the three dimensional ordering in not sensitive to pressure (interchain coupling), even though both T_p^MF and $2\Delta(0)$ are pressure dependent. Since the interchain Coulomb interaction is expected to be sensitive to the distance between chains, the commensurability energy is the more important mechanism of 3d ordering in o-TaS_3.

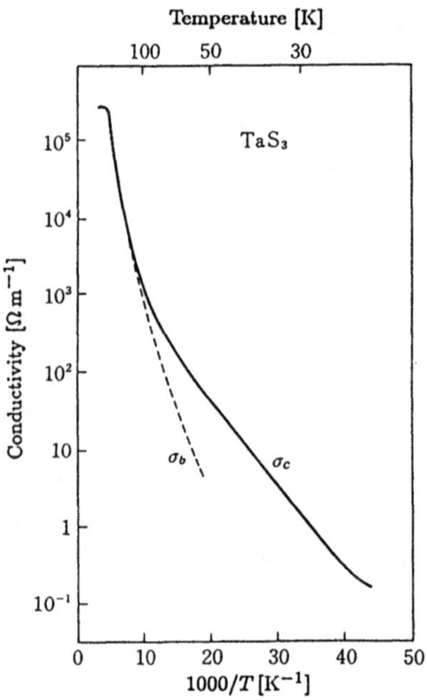

Fig. 4.26. Temperature dependence of conductivity and its anisotropy [4.54]

The conductivity components σ_c and σ_b at lower temperature [4.54] are shown in Fig. 4.26. Above 100 K the temperature dependences of σ_c and σ_b are the same; the electrical conduction is mainly by electrons excited thermally across the energy gap. Here, the mobility is assumed to be approximately independent of temperature, or more exactly, its temperature dependence is weaker than that of the carrier density.

By a careful measurement [4.55], hysteresis has been found in the resistance in low electric fields at temperatures between 205 K and 55 K; the resistivity measured with increasing temperature is larger than that measured during cooling. Since the measured wave vector is nearly commensurate [4.52], the lattice distortion is presumably discommensurate. The variation of the wave vector with temperature corresponds to a variation in the density of discommensurations (DC). But impurities in the lattice oppose the movement of DCs, and therefore the density of DCs and the local wave vector are not always equal to the thermal equilibrium values. The difficulty lies in the fact that the wave vector locks in at ~ 160 K[9], while the hysteresis of resistivity is observed down to 55 K; the exact lock-in temperature has not been established.

Below 100 K, the two curves in Fig. 4.26 deviate from each other [4.54]. The transverse component σ_b still decreases exponentially but in σ_c another activation energy of 250 K can be identified between 30 K and 70 K. It means that another thermally activated process appears, but only along the c-axis. As explained in Sect. 2.2, solitons can move only along the 1d axis and are indeed able to carry current. The commensurability energy must be important in the formation of solitons in o-TaS_3 and the potential $V(\phi)$ in (2.74) is replaced by (2.68);

$$V(\phi) \propto \cos(N\phi) \quad ,$$

where the order of commensurability N is equal to 4 since $q_c = 1/4$. The solution of the sine-Gordon equation (2.74) for $\psi = N\phi$ gives a soliton with a phase change of $2\pi/N$ and a creation energy of $(8n_s m^*/Q^2 L)/N$. Its width is equal to v/ω_F, where ω_F is the measure of strength of the commensurability pinning (cf. Sect. 2.2.2). The excitation energy was estimated as ~ 30 K in o-TaS_3; it is significantly smaller than the measured activation energy of 200 K even if one takes into account that only a soliton-antisoliton pair can be generated thermally.

It should be emphasized that the presence of solitons and their effect on various properties of 1d conductors have not been established experimentally, perhaps with the exception of $(CH)_x$. One of the best candidates for such a study is o-TaS_3.

At lower temperatures, σ_c becomes non-ohmic in high electric fields, while the transverse component remains ohmic. Because the conductivity is small and the material is semiconducting, it must be carefully examined whether the non-ohmic conduction might not be due to other mechanisms such as those explained in Sect. 3.2.4. Let us estimate the threshold field E_T of the commensurate CDW. From the theory by *Lee* et al. [4.56] the energy gained through commensurability is given by

$$v_c = (|\Delta|^2/\varepsilon_F)(e|\Delta|/w)^{N-2} \tag{4.17}$$

[9] At ~ 160 K, no discontinuous change in the wave vector is observed. It is not clear whether the observed continuous change is intrinsic or due to a high impurity level even in carefully prepared crystals.

where ε_F and w are the Fermi energy and band width, respectively. With the experimental results of $\Delta \cong 600\,K$ and $N = 4$, and by assuming one-dimensionally free electrons, v_c is of the order of 0.6 K. The threshold field E_T is given by the relation

$$eE_T(\pi/k_F N) \cong v_c \quad , \tag{4.18}$$

where $\pi/k_F N = (2\pi/2k_F)/N$ is equal to $(1/N)$ times the wave length. From the value $E_T = 6 \times 10^4\,V/m$ measured at low temperature [4.54], v_c is estimated to be 0.25 K, which is in qualitative agreement with the above estimate.

The threshold field [4.57] shows a maximum at $\sim 150\,K$ and a rapid increase near 50 K with decreasing temperature. It might be possible to attribute the maximum at $\sim 150\,K$ to the locking into the commensurate structure, but because thermal hysteresis can be observed down to 55 K the CDW must be still incommensurate, at least locally. No satisfactory explanation has been given for the maximum at $\sim 150\,K$.

4.3.2 Monoclinic (m-)TaS_3

Besides the "orthorhombic" polytype just described a monoclinic (m-) form of TaS_3 has been found recently. TaS_3 synthesized under pressure is monoclinic; it is possible that the o-TaS_3 synthesized by the usual method has a composition somewhat different from 1 : 3. Both polytypes are often found in the same stock and they can be separated only by microscopic examination. As shown in Table 4.2, the *c*-axis of o-TaS_3 has the same length as that of the *b*-axis of m-TaS_3. Therefore in some cases crystallites of both polytypes stick together and grow as an apparently single crystal. The properties of m-TaS_3 are expected to be similar to those of $NbSe_3$, because of the similar structure. As shown in Fig. 4.27, the resistivity [4.8] of m-TaS_3 displays two

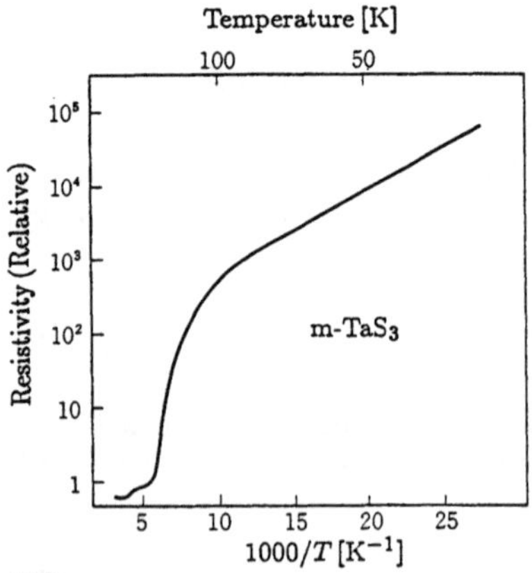

Fig. 4.27. Resistivity [4.8] of m-TaS_3

abrupt increases, at ~240 K (T_1) and ~160 K (T_2). Although not always reproducible, the resistivity decreases with decreasing temperature just below T_2; the material is metallic between T_1 and T_2. Thus, m-TaS_3 is different from $NbSe_3$ only in that its Fermi surface vanishes completely below T_2. The wave vector [4.57] of the lattice distortion below T_1 is $Q_1 = (0.253 \pm 0.003)b^*$, and another with $Q_2 = (0.5a^* + (0.245 \pm 0.003)b^* + 0.5c^*)$ is observed below T_2 in coexistence with Q_1. The features common with those of $NbSe_3$ are clear; the transverse components of Q_2 are again equal to 0.5 and the relation $2(\boldsymbol{Q}_1 + \boldsymbol{Q}_2) \simeq \boldsymbol{G}(1, 1, 1)$ also holds. Therefore, a comparison of the non-ohmic conduction properties of m-TaS_3 with those of $NbSe_3$ is quite interesting. Unfortunately, however, no significant progress has been made, because preparation of truly single crystal of m-TaS_3 is difficult. To date however, no essential difference has been found between m-TaS_3 and $NbSe_3$ in the non-ohmic conduction properties.

The Peierls-Fröhlich conduction mechanism was first observed in $NbSe_3$ and then in TaS_3. Though not covered in this text, non-ohmic conductivity, noise phenomena and other interesting properties arising from the sliding CDW have been observed in other classes of materials: blue bronze $K_{0.3}MoO_3$ and $([Nb,Ta]Se_4)_nI$. Significant progress has been made in this field through extensive research by many workers, stimulated partly by the discovery of these new materials. Nevertheless, much further work is needed to achieve a comprehensive understanding of the nature of the sliding motion of CDWs. The reader who would like further details should consult the review articles [4.58–60].

4.4 Superconductivity in $TaSe_3$

No structural phase transition has been found in $TaSe_3$; its resistivity decreases smoothly with decreasing temperature without any anomaly, and it becomes superconducting below 2 K. It must be kept in mind that the absence of any anomaly in the resistivity does not always imply the absence of e.g. a CDW transition. For example, $NbSe_2$, one of the typical MX_2 compounds, shows a CDW transition near 35 K. The Hall coefficient changes in sign at that temperature, but the small anomaly in the resistivity was not detected for long time. Because of the presence of the superconductivity and the absence of the CDW, the wave functions on the triangular prismatic chain presumably extend far enough towards the neighbouring chains such that $TaSe_3$ is a 3d metal. The superconducting transition temperature, $T_c \sim 2$ K, is rather high, probably because of a strong electron-phonon coupling. (Remember that $TaSe_3$ is a semimetal.) Another possibility is that, although $TaSe_3$ is strongly one-dimensional, its Fermi surface (FS) is closed and has a large curvature due to the small electron density and thus no suitable nesting vector can be found (cf. Fig. 2.8).

For $TaSe_3$ many different oscillation periods are found in the Shubnikov-de Haas effect [4.11]; the shape of the FS is more complicated than that of $NbSe_3$ (below T_2). If the FS consists of several closed surfaces, the volume of each sheet must be small and cannot exceed $\sim 10\,\%$ of the Brillouin zone.

Under hydrostatic pressure [4.61], T_c decreases rapidly at a rate of $-7.5 \times 10^{-5}\,\mathrm{K\,kbar^{-1}}$ ($-7.5 \times 10^{-10}\,\mathrm{K\,Pa^{-1}}$). In MX_2 compounds which show the CDW transition at higher temperature, T_c increases with pressure. Hydrostatic pressure suppresses the CDW and the increase in T_c is presumably due to the increasing three dimensionality (cf. Sect. 4.2.4). It is difficult, however, to explain quantitatively the negative pressure coefficient of T_c in $TaSe_3$; we cannot give a reliable estimate of the pressure coefficients of the quantities determining T_c.

There are many theoretical expressions for T_c. For example, *McMillan* [4.62] derived the following formula

$$T_c = (\theta_D/1.45)\exp\left(-\frac{1.04(1+\lambda)}{\lambda - \mu^*(1+0.62\lambda)}\right) \quad ,$$

where θ_D is the Debye temperature, $\lambda = N(\varepsilon_F)\langle I^2\rangle/M^2\langle\omega^2\rangle$ is the strength of electron-phonon interaction and μ^* is the effective Coulomb interaction. [$N(\varepsilon_F)$: density of states at the Fermi Level, $\langle I^2\rangle$: square of the matrix element of the electron-phonon interaction averaged over the FS, M : ionic mass, $\langle\omega^2\rangle$: mean square phonon frequency.] Usually μ^* is insensitive to pressure. Nevertheless, many parameters are left undetermined; it is difficult to estimate their dependences on pressure. For example, if it is assumed that only $\langle\omega^2\rangle$ varies under pressure, the above value of dT_c/dp gives $d\langle\omega\rangle/dp \sim 10^{-5}\,\mathrm{s^{-1}\,bar^{-1}}$ ($10^{-10}\,\mathrm{Pa^{-1}}$), which is much larger than those of most materials. Theoretical estimates of T_c are given in [4.63].

Up to now, it has been believed that the major effect of hydrostatic pressure on the properties of 1d conductors is to enhance the three-dimensionality by reducing the distances between chains, which themselves are approximately rigid. But we must be careful in making this approximation, or in other words, about the extent to which the chain can be regarded as rigid. This depends on the anisotropy of the fractional contraction of the lattice constant under pressure (compressibility), and on how the properties of interest are sensitive to the anisotropic contraction of the lattice. Data on anisotropic compressibility are essential for a discussion of physical properties under pressure. Nevertheless, such data are quite rare; the compressibility of MX_3 was measurd for the first time only recently [4.64]. Therefore, no experimental support has been accumulated to conclude that the rapid decrease of T_c is solely due to the enhanced three-dimensionality. We are frequently surprised by the lack of fundamental data. The reasons lie in the experimental difficulties in obtaining reliable data.

The magnetic field in which the bulk crystal becomes normal (upper critical field) is given by

$$H_{c2} = \phi_0/2\pi\xi^2 \tag{4.19}$$

where ξ is the coherence length and $\phi_0 = h/2e$ is the magnetic flux quantum. The physical meaning of (4.19) can be understood by rewriting it as $(\pi\xi^2)H_{c2} \sim \phi_0$. In anisotropic superconductors, the coherence length is anisotropic because it is related to the effective mass by the relation

$$\xi \propto 1/\sqrt{m} \quad . \tag{4.20}$$

Both ξ and m can be diagonalized simultaneously. The diagonal elements of ξ correspond to the axial lengths of an ellipsoid in ξ-space, therefore $\pi\xi^2$ in (4.19) corresponds to the cross sectional area of the ellipsoid viewed from the direction of magnetic field. If the effective mass components satisfy the relation $m_x \sim m_y > m_z$, then ξ can be written as

$$\xi = \begin{pmatrix} \xi & & \\ & \xi & \\ & & \varepsilon\xi \end{pmatrix} \quad , \quad \varepsilon \equiv \sqrt{m_x/m_z} \quad . \tag{4.21}$$

The cross-sectional area of the flux quantum is given by that of the ellipsoid perpendicular to the magnetic field. When the magnetic field is applied at an angle θ to the z-axis, H_{c2} is given by

$$H_{c2}(\theta) = \frac{\phi_0}{2\pi\xi^2(\cos^2\theta + \varepsilon^2 \sin^2\theta)^{1/2}} \quad . \tag{4.22}$$

This effective mass model has been successfully used to explain the experimental results of MX_2 superconductors.

The angular dependence of H_{c2} for $TaSe_3$, obtained from resistive transition curves in magnetic field [4.65], is in good agreement with the above model as shown in Fig. 4.28. It shows a maximum when the field is parallel to the b-axis; the effective masses perpendicular to the b-axis are large as expected from the triangular prismatic chain structure. The ratio of the effective masses m_x/m_z is, however, too large; it is approximately equal to 700. Such a large anisotropy is hardly expected since the van der Waals gap is not extended in the a-c plane. The critical field H_{c2}, given by the anisotropic Ginzburg-Landau theory, shows the same angular dependence as that in (4.22) but the effective mass therein is not equal to the band mass but the effect of the anisotropic superconducting energy gap is also included. Therefore it cannot be compared directly with the effective mass measured in the normal state. Figure 4.29 shows the angular dependence of

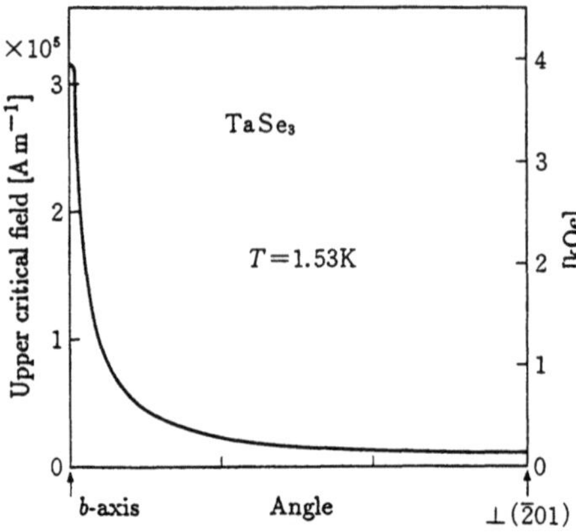

Fig. 4.28. Upper critical field (H_{c2}) for $TaSe_3$ [4.65]

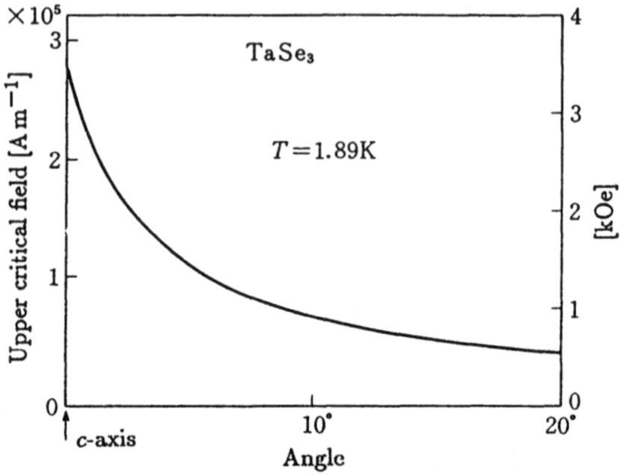

Fig. 4.29. Upper critical field (H_{c2}) perpendicular to the b-axis [4.65]

H_{c2} measured in a magnetic field perpendicular to the b-axis. In this configuration also, a large anisotropy is found, contrary to the naive expectation. In the effective mass model (4.22), H_{c2} changes smoothly with the angle θ; $dH_{c2}/d\theta = 0$ at $\theta = 0$ and $\frac{\pi}{2}$. On the other hand, in Fig. 4.29 $dH_{c2}/d\theta$ is not equal to zero when the field is parallel to the c-axis. Such a feature has been found in thin films of isotropic superconductors [4.66]. Therefore, one might suppose that each crystal of $TaSe_3$ is composed of many thin crystals stuck together parallel to the c-axis, but this idea is not compatible with the cleavage habit parallel to ($\overline{2}$, 0, 1). This difficulty may arise partly from an ambiguity in how H_{c2} is determined from the measured resistive transition curves. From its definition, H_{c2} is the minimum magnetic field for a finite resistance to be observed. In real crystals, however, the transition curve is more or less rounded because of imperfections and the magnetoresistance in the normal state often observed even in low field. In real cases, therefore, H_{c2} is determined only with some arbitrariness.

The results of Meissner effect measurements [4.67] are also difficult to interpret. In superconductors the nearly full Meissner diamagnetism is usually observed from just below T_c, but in $TaSe_3$ the diamagnetic susceptibility is very small (even smaller than in $NbSe_3$) and is strongly dependent on temperature. It has been suggested that $TaSe_3$ is not a homogeneous superconductor but that the sample is composed of small fibres coupled weakly to each other by the Josephson effect. We must bear in mind that the results of such a measurement must be interpreted carefully; the sample is composed of number of fibrous crystals lossely packed together.

As explaied above, $TaSe_3$ should be regarded as a three dimensional (semi)metal; the phase transition related to CDW ordering is not observed but the compound becomes superconducting with anisotropic H_{c2}. It is probably possible to modify its structure, by e.g. intercalation, to synthesize more anisotropic metals, in which the superconductivity is replaced by the CDW state. It is well known that intercalating various ions or molecules changes conduction electron density and thereby the properties of layered materials such as MX_2 and graphite. Intercalants work as donors or acceptors depending on the species. Another effect of intercalation is to widen the interlayer spacing, just the opposite effect to that of externally applied pressure.

In $TaSe_3$, a strong one-dimensionality was expected from its crystal structure, but the expectation was not fulfilled. On the other hand, cubic conductors such as ZrV_2, HfV_2 [4.68] and CuV_2S_4 [4.69] were found to show incommensurate lattice distortion and conductivity anomalies – the characteristics of the CDW phase. Thus we are still unable to predict the properties of new materials, a fact which suggests that we can expect a quite fruitful future for this field.

5. Properties of KCP

In this chapter we shall study the properties of the 1d conductor $K_2Pt(CN)_4$-$Br_{0.3} \cdot 3.2H_2O$, which has 1d chains of platinum atoms. The name of this material is abbreviated to KCP or sometimes to KPC.

Until the end of Sect. 5.5, we shall regard KCP as a band conductor and summarize its properties from the view points of the Peierls transition and the charge-density waves. In Sect. 5.6 the discussion will focus on the behaviours of the water of crytallization and its effect on the 1d electron system. Section 5.7 will be devoted to a re-examination of experimental results from another view point. Here KCP will not be considered as band conductor but in terms of valence fluctuations. As we will find there, not all the properties of this material are consistently understood although for some reason investigations of KCP seem to have been set aside at the early stage of the studies of 1d conductors.

5.1 Mixed Valence Pt Compounds

The compound $K_2Pt(CN)_4X_{0.3} \cdot nH_2O$(X=Br,Cl) discovered by *Knop* [5.1] in 1842 is a material with a long history. It was quite recently, however, that this material was spot-lighted as one of the typical examples of a 1d conductor.

In the field of inorganic chemistry, platinum ions are known to have a stable valence of 2+ or 4+. In $K_2Pt(CN)_4X_{0.3} \cdot nH_2O$ [5.2] (hereafter abbreviated to KCP(X)), the Pt ion is oxidized by the halogen X and has an average valence of +2.3, so it is not expressed by single valence state. This state is called as a mixed-valence state.

In the crystal structure of KCP(X), the Pt ions form a 1d chain with an interatomic distance as small as 2.88 Å – comparable to the nearest neighbour distance in Pt metal of 2.78 Å. So the characteristic properties of a nearly 1d metal are expected.

Here, KCP(X) has been chosen as a typical example of the partial oxidation by halogen atoms. Other oxidizing materials such as H_2CO_3, H_2SO_4 etc. may also be used to attain partial oxidation [5.3, 5]. In this case of a typical oxidized complex salt $K_{2-x}Pt(CN)_4 \cdot nH_2O$, the partial oxidation is realized by a deficiency of alkali metal ions.

Table 5.1. Mixed valence Pt complexes and the Pt-Pt distance

Pt complexes	Degree of oxidation	Pt-Pt distance [Å]	Crystal structure	Ref.
$K_2Pt(CN)_4Br_{0.30} \cdot 3.2H_2O$	+2.30	2.88	tetragonal	[5.2]
$K_2Pt(CN)_4Cl_{0.32} \cdot 2.6H_2O$	+2.32	2.88	tetragonal	[5.2]
$Cs_2Pt(CN)_4Cl_{0.30}$	+2.30	2.86	tetragonal	[5.8]
$MgPt(CN)_4Cl_{0.38} \cdot 7H_2O$	+2.38	2.985	tetragonal	[5.4]
$(NH_4)_2Pt(CN)_4Cl_{0.33} \cdot 3H_2O$	+2.33	2.903	tetragonal	[5.12]
$K_{1.74}Pt(CN)_4 \cdot 1.8H_2O$	+2.26	2.96	triclinic	[5.5]
$H_{1.60}Pt(C_2O_4)_2 \cdot 3H_2O$	+2.40	2.82	orthorhombic	[5.6]
$Li_{1.64}Pt(C_2O_4)_2 \cdot 6H_2O$	+2.36	2.81	orthorhombic	[5.6]
$Na_{1.67}Pt(C_2O_4)_2 \cdot 2H_2O$	+2.33	2.88	triclinic	[5.6]
$K_{1.62}Pt(C_2O_4)_2 \cdot 2H_2O$	+2.38	2.85	triclinic	[5.6]
$Rb_{1.67}Pt(C_2O_4)_2 \cdot 1.5H_2O$	+2.33	2.83	triclinic	[5.11]
$(NH_4)_{1.64}Pt(C_2O_4)_2 \cdot H_2O$	+2.36	2.85	orthorhombic	[5.6]
$Zn_{0.81}Pt(C_2O_4)_2 \cdot 6H_2O$	+2.38	2.88	orthorhombic	[5.14]
$Mg_{0.82}Pt(C_2O_4)_2 \cdot 5.3H_2O$	+2.36	2.85	orthorhombic	[5.7]
$Co_{0.83}Pt(C_2O_4)_2 \cdot 6H_2O$	+2.34	2.84	orthorhombic	[5.13]
$Ca_{0.84}Pt(C_2O_4)_2 \cdot 4H_2O$	+2.32	2.85	orthorhombic	[5.6]
$Sr_{0.81}Pt(C_2O_4)_2 \cdot 4H_2O$	+2.38	2.85	orthorhombic	[5.6]
$Ba_{0.84}Pt(C_2O_4)_2 \cdot 4H_2O$	+2.32	2.85	orthorhombic	[5.6]
$Cs_2Pt(CN)_4(N_3)_{0.25} \cdot 0.5H_2O$	+2.25	2.88	tetragonal	[5.15]
$K_2Pt(CN)_4(FHF)_{0.3} \cdot 3H_2O$	+2.30	2.92	tetragonal	[5.9]
$Rb_2Pt(CN)_4(FHF)_{0.40}$	+2.40	2.80	tetragonal	[5.8]
$Rb_2Pt(CN)_4(FHF)_{0.36} \cdot 1.7H_2O$	+2.36	2.89	monoclinic	[5.9]
$Cs_2Pt(CN)_4(FHF)_{0.39}$	+2.39	2.83	tetragonal	[5.8]
$Cs_2Pt(CN)_4(FHF)_{0.23}$	+2.23	2.87	tetragonal	[5.10]
$Cs_2Pt(CN)_4F_{0.19}$	+2.19	2.89	tetragonal	[5.10]

In Table 5.1, the partially oxidized mixed-valence platinum complex compounds are listed. It should be noted that the interatomic distance between Pt atoms is about 3.5 Å in Pt compounds which consist of only Pt^{2+} ions. In the mixed-valend state however, the interatomic distance between Pt atom is about 10 ~ 20 % less. This fact suggests that there is a strong interaction between the Pt atoms within a 1d chain in the mixed-valence compounds.

In this chapter, we choose KCP(Br) as a typical example of the mixed-valence Pt complex compounds on which a great variety of different studies have been performed.

5.2 Crystal Structure and the Preparation of Single Crystals

5.2.1 Preparation of Single Crystals

By the addition of a halogen, such as Br, to the saturated aqueous soluton of $Pt(CN)_4$, a partially oxidized KCP(X) complex can be obtained as a needle crystal with metallic copper luster. The chemical reaction for this process is given as follows,

$$20K_2Pt(CN)_4 \cdot 3.2H_2O + 3X_2 \rightarrow 20K_2Pt(CN)_4X_{0.3} \cdot 3.2H_2O \quad . \qquad (5.1)$$

Large single crystals are obtained by the slow crystal growth of seeds within the saturated soluton: in this case the gradual evaporation of water contributes to compensate for the slow change of concentration of KCP(X). This is the usual crystal preparation method from the saturated water solution as used for CrKAlum ($K_2SO_4 \cdot Cr(SO_4)_3 \cdot 24H_2O$), CMN salts ($Ce_2Mg_3(NO_3)_{12} \cdot 24H_2O$) etc. [5.16]. But in the case of KCP(Br), there are several characteristics and we need to take special care. The first one is the pH-dependence of the crystal structure [5.17]. The pH of the saturated solution of KCP(Br) is 4.5, but we can only obtain the tetragonal crystal phase of KCP(Br) when the pH is above about 7 [5.18]. This condition can be realized by adding 0.5 mol KBr to the saturated solution. This addition also helps to prevent the nucleation of excess micro-crystals.

The second characteristic is the pH-dependence of the shape of the crystals. At around pH 7, the speed of crystal growth along the 1d axis and in the perpendicular plane are comparable, so we can obtain nearly cubic crystals. On the other hand, at around pH 11, the crystal grows much faster along the 1d axis than in the perpendicular plane, and needle crystals are obtained.

The best crystal of KCP(Br), which we have obtained for the electric conductivity and dielectric susceptibility measurements, was grown by the following methods [5.18]. We added the mixed aqueous solution of 0.1 mol KBr and 1 mol urea to the saturated aqueous solution of KCP(Br). After filtration by 0.5 μm millipores or a fine sintered glass funnel into a plastic beaker, the solution was warmed to dissolve any micro-crystals and kept $23 \pm 2°$ C in a desiccator. The slow evaporation of the soluent at $23 \pm 2°$ C produced crystals within 24 hours. These crystals were either used directly for physical measurements, such as electrical conductivity, or used as seeds to grow larger crystals with centimetre size over a period of weeks. The crystal growth media should be tended daily through the addition of the mother liquor to keep the salt and urea concentrations nearly constant and to take off the micro KCP(Br) crystals floating on the surface of the solutions.

Single crystals of KCP(Br) can also be obtained by the diffusion method. The saturated solutions of $K_2Pt(CN)_4 \cdot 3H_2O$ and $K_2Pt(CN)_4Br_2$ are separated by a special Teflon membrane composed of tetrafluoroethylene (TFE) fibres in thin-sheet form with a relatively uniform pore size of $2 \sim 5\,\mu$m. The permeated $K_2Pt(CN)_4Br_2$ ions crystallize on the $K_2Pt(CN)_4 \cdot 3H_2O$ side as KCP(Br) on the surface of the membrane [5.17]. It should be mentioned that the crystals obtained are dried by contact with a Kimwipe or filter paper and then stored at constant humidity to prevent the water from evaporating.

Many of the physical properties of KCP(Br) are sensitive to the amount of water of crystallization [5.19], as we discuss later in Sect. 5.6, so the crystals should be kept at a humidity of 72 % at 23° C to prevent water evaporation. When we keep the crystal at a low temperature, such as 77 K, the vapour pressure of the water of crystallization of KCP(Br) becomes low and the evaporation is negligible.

The problem of the amount of water of crystallization in KCP(Br) is important not only for the preservation of the quality of crystals but also particularly important for the measurement of optical properties [5.20], which are sensitive to the surface of crystals. So we must pay special attention to the treatment of crystals in these measurements.

With respect to the concentration of Br ion in KCP(Br), the most reliable determination of the Br concentration gave a value of 0.300 ± 0.006 from x-ray fluorescence analysis and neutron-activated radioactivity. Until now no other Br concentration has been found in KCP(Br) [5.18].

5.2.2 Crystal Structure

The detailed crystal structure analysis by *Krogmann* and *Hausen* [5.2] in 1968 on $K_2Pt(CN)_4Cl_{0.3} \cdot 2.6H_2O$ and $K_2Pt(CN)_4Br_{0.3} \cdot 2.6H_2O$ using x-ray diffraction stimulated much research into quasi-1d conductors, and KCP(X) was spot-lighted thereafter.

The crystal structure of KCP(X) is tetragonal and the platinum ions are situated at the corner of the unit cell as shown in Fig. 5.1, and they stack along the *c* axis. The interatomic distance between Pt ions is 2.88 Å and comparable with the 2.78 Å of metallic platinum, so we can expect metallic behaviour along the *c* axis in KCP(X).

On the other hand, in the perpendicular directions, the platinum ions are surrounded by large CN bases, so the interatomic distance between Pt ions in *a*-*b* plane is as large as 9.87 Å and each Pt chain along the *c* axis is well separated from neighbouring chains.

On the basis of these structural characteristics, KCP(X) is expected to be a candidate for a 1d metal. Next, we look at the more detailed crystal structure of KCP(X); the structural unit $[Pt(CN)_4]$ is shown in Fig. 5.2a. The *d*-electron orbital of the Pt ions stretches to the *c* axis where the interatomic distance between Pt atoms is comparable with that in metallic platinum as shown in Fig. 5.2b. So the d_{z^2}-electrons form strong bonds along the *c* direction as seen in Fig. 5.2b. Due to the partial oxidation by the halogen X, these $5d_{z^2}$ orbitals have a hole in the corresponding bands. This results in a metallic $5d_{z^2}$ band in KCP(X) as shown in Fig. 5.3.

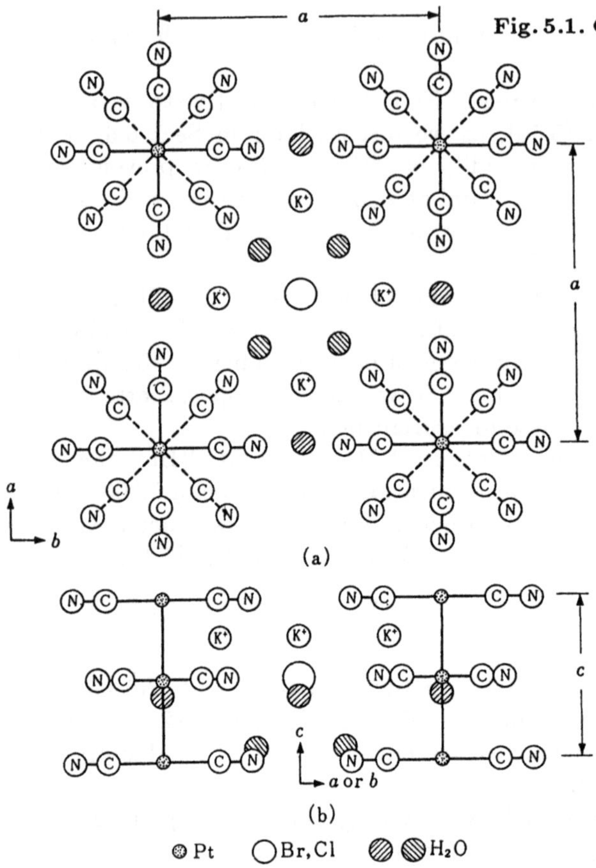

Fig. 5.1. Crystal structure of KCP(Br)

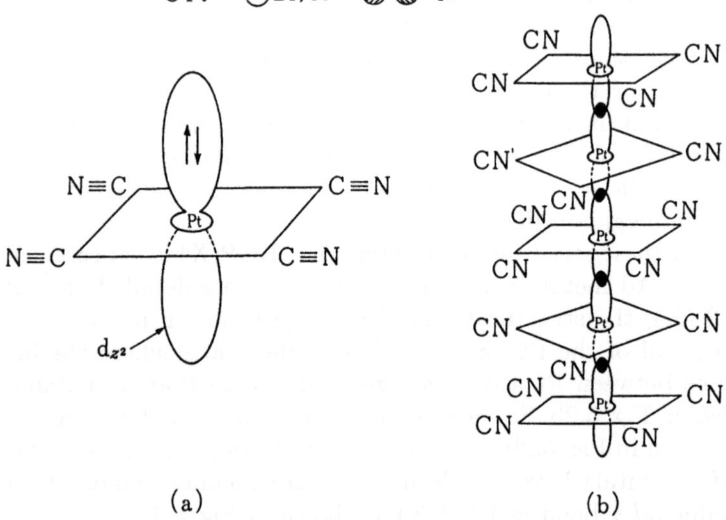

Fig. 5.2. (a) d_{z^2} orbital of $Pt(CN)_4$; (b) stacking of d_{z^2} orbitals in KCP(Br)

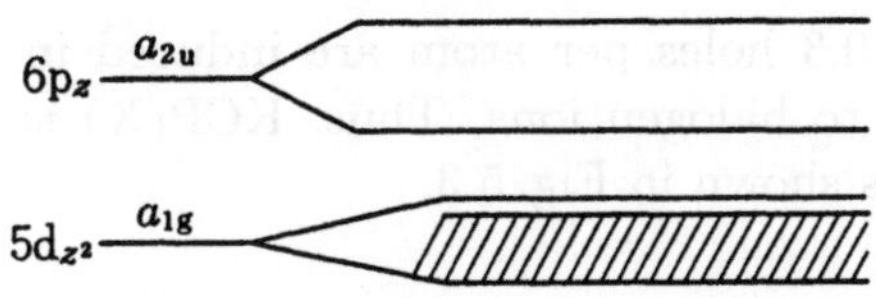

Fig. 5.3. Hole in the $5d_{z^2}$ orbital of the Pt ion due to partial oxidation

It is interesting to note that the interatomic distance between Pt ions is reduced by about 20 % in a partially oxidized Pt complex salt with respect to non-oxidized ones (cf. Table 5.1). Due to the strong metallic bonding between Pt ions, the interplanar distance is reduced and the strong repulsive force between CN bases makes the direction of CN bonds 45° between adjacent c planes as shown in Fig. 5.2b.

Halogen ions, which contribute to the partial oxidation, occupy the centre of the unit cell with a probability of 0.60. There has been no report of any periodicity in the Br ion occupation, so Br ions are believed to be distributed random. (The periodic occupation by Br ions can in any event hardly be observed by diffraction experiments, so there remains some doubt as to whether there is a periodic occupation of Br ions or not.) The halogen ion sites remain vacant with a probability of 0.40, on the basis of x-ray [5.21] and neutron diffraction [5.22] studies. *Peter* et al. found that these sites are fully occupied by molecules of water of crystallization as shown in Fig. 5.1 as the open circle. These results are confirmed by ^{1}H NMR studies [5.23] of the water of hydration in KCP(Br). In KCP(X), the water molecule exists at the other positions on the hatched circle, ◍ and ▨. One water molecule occupies the site I ◍ which is on the middle of the CN base in the (100) direction and two molecules are located on the site II ▨ between a platinum atom and the halogen ions. There exist 3.2 moles of water in 1 mole KCP(Br) such as $K_2Pt(CN)_4Br_{0.3} \cdot 3.2H_2O$.

Lastly we mention the K^+ ions shown in Fig. 5.1. The position has no inversion symmetry with respect to the c axis [5.24]. The crystal is thus likely to grow as twin crystals parallel to the c axis, depending on the conditions of the crystal growth. But the existence of twin crystals of KCP(X) has not been found by x-ray or neutron diffraction studies. This may be understood by assuming that the whole crystal grows from a single seed.

5.2.3 Crystal Structure and Electronic States

Interrante and *Messmer* [5.25] have calculated the electronic states of platinum ions by the molecular orbital method. Although their discussion was based on a rough approximation which neglects the spin-orbit interaction, they suggested a qualitatively reasonable model of KCP(X). According to this the valence state is given as a_{1g}, consisting mainly of d_{z^2} orbitals, and the lowest excited state as a_{2u}, a covalent admixture of $2p$ electrons of the carbon atom and d_{xy} electrons. Here, 0.3 holes per atom are induced in the d_{z^2} band by partial oxidation due to halogen ions. Thus, KCP(X) is expected to have a metallic character as shown in Fig. 5.3.

5.3 Electrical Properties

5.3.1 Electrical Conductivity

On the basis of structural and electronic considerations, metallic behaviour is expected for KCP(Br). The dc conductivity has been measured and it reveals that the conductivity is sensitive to any impurities or defects in the crystals, as is found in other quasi 1d conductors. In particular, the anisotropy of the conductivity, that is the ratio $\sigma_{\parallel}/\sigma_{\perp}$, is found to be a measure of the quality of crystals. It is believed that the larger $\sigma_{\parallel}/\sigma_{\perp}$ is, the better the quality of the crystal.

The most reliable experimental results for electrical conductivity in KCP(Br) are shown in Fig. 5.4 as a function of the temperature. From this figure, the electrical conductivity along the 1d Pt chain is about (3–4) $\times$ $10^4\,\Omega^{-1}\mathrm{m}^{-1}$ at room temperature and is nearly comparable to that of

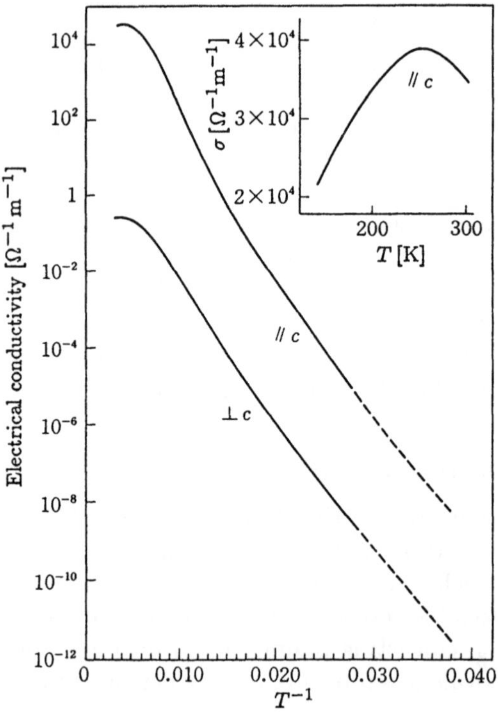

Fig. 5.4. Temperature dependence of the electrical conductivity of KCP (Br) [5.16]

typical semimetals. The conductivity near 250 K shows a broad peak [5.26]. At temperatures above 250 K, the resistivity shows metallic behaviour i.e. it increases with temprature. The resistivity is semiconductor-like below 250 K, and it increases here with decreasing temperature. The ratio of the conductivity along the chain to that in the perpendicular direction, $\sigma_{\parallel}/\sigma_{\perp}$, reaches more than 10^4 at room temperature. This allows KCP(Br) to be considered as one of the typical 1d metals. In Fig. 5.4 one can see that the temperature dependences of $\sigma_{\parallel}$ and $\sigma_{\perp}$ are nearly parallel below 100 K and seem to have a similar activation energy. This can be explained as follows: Below 100 K the conductivity $\sigma_{\parallel}$ is dominated by a hopping motion of electrons across defects or impurities but not by an activation across the intrinsic band gap. At the defect or impurity sites, electrons will be able to hop also to the neighbouring chains of Pt. Hence both $\sigma_{\parallel}$ and $\sigma_{\perp}$ are dominated by the same origin: the hopping at the defect or impurity sites. The hopping probability itself will be different for $\sigma_{\parallel}$ and $\sigma_{\perp}$ but its temperature dependence can be the same.

Bloch et al. [5.27] have already examined this phenomenon from a general view point. They proposed the one-dimensional variable range hopping model. In the interrupted stands, the electron hopping motion between 1d chains is important at low temperature. They found that the observed conductivity of the various types of quasi-1d conductors is proportional to $\exp(-A/\sqrt{T})$ at low temperature. In KCP(Br), the observed conductivity was found to have the same temperature dependence, $\exp(-A/\sqrt{T})$, below 100 K [5.27]. The previous discussion suggests that the dc electrical conductivity of KCP(Br) is not intrinsic below 100 K. Evidence for this is clearly given by the experimental results of microwave conductivity measurements. In the measurement of dc conductivity along the chain, electrons must move over macroscopic distances, but it is impossible to obtain perfect crystals without any defects between the two electrodes. Therefore, the dc conductivity measurements usually suffer from crystal imperfections and impurities in the case of 1d conductors. On the other hand, the microwave conductivity measurements [5.26] give us more reliable results. If there are no imperfections or impurities in a 1d chain within the length $l<10^{-2}$ cm, we can expect that the electron does not encounter any defects (the electron moves with a period of 10^{-10} s (10 GHz), at the Fermi velocity of $\sim 10^8$ cm/s) and the observed electrical conductivity will be an intrinsic one.

As we see above, the conductivity measurement using microwaves is less sensitive to imperfections in the sample crystal than that at dc. Therefore the conductivity of quasi-1d conductors and its frequency dependence can be used to estimate the degree of crystal perfection.

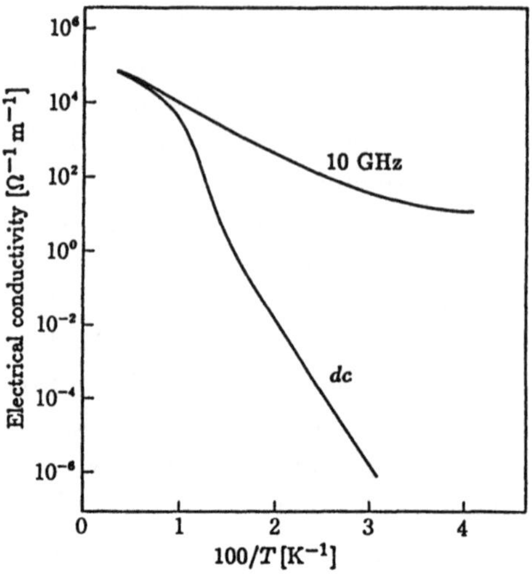

Fig. 5.5. Temperature dependence of electrical conductivity at microwave (10 GHz) and dc [5.26]

In Fig. 5.5, the experimental results for the microwave conductivity [5.26] are shown; one can see that below 150 K it is larger than that obtained by the dc method. The dc conductivity shown in Fig. 5.4 in this temperature range may not be an intrinsic one. As we have seen above, the dc conductivity of a 1d conductor is sensitive to the crystal condition. Thus we need special care in the method of sample fixing and cooling, so as not to induce imperfections by stress during dc conductivity measurements, in particular since KCP(Br) has a large thermal linear expansion coefficient. But the other measurements, such as the microwave conductivity, the optical and the magnetic properties, are less sensitive to defects, since the effect of 1d chain breaking is of the order of $1/N$.

In the case of the organic 1d conductor TTF-TCNQ, the dc conductivity results nearly agree with those of the microwave method. So the crystal condition is better in TTF-TCNQ than in KCP(Br). KCP(Br) shows metallic behaviour above 250 K and at temperatures below 250 K it becomes gradually non-metallic. This gradual metal to non-metal transition, with no definite transition temperature, is the most remarkable characteristic of KCP. This phenomenon is rarely found in other typical 1d conductors.

5.3.2 Optical Properties

As mentioned previously, in KCP(Br) the conductivity along the c axis is as high as that of a typical semimetal and the conductivity in perpendicular directions is smaller by more than four orders of magnitude at room temperature. The anisotropy of the conductivity is also remarkable in the measurements of optical reflectivity [5.28], where the electromagnetic waves have a much shorter wavelength than microwaves. The wave-number de-

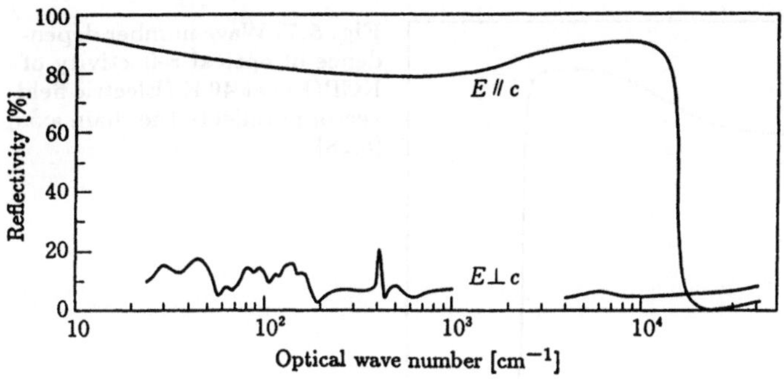

Fig. 5.6. Wave number dependence of optical reflectivity of KCP(Br) at room temperature [5.28]

pendence of the optical reflectivity is shown in Fig. 5.6. When the electric field vector lies parallel to the platinum chains, KCP(Br) shows high optical reflectivity due to the shielding of the electric field by the conduction electrons, but becomes transparent for light polarized perpendicular to the chains. The plasma absorption edge is observed as an abrupt decrease in reflectivity at about $2 \times 10^4\,\mathrm{cm}^{-1}$ (4×10^{-19} J). The experimental results can be analysed using Drude's equation

$$\sigma(\omega) = \frac{\varepsilon\omega_\mathrm{p}^2\tau}{4\pi(1 - \mathrm{i}\omega\tau)} \quad ; \quad \omega_\mathrm{p}^2 = \frac{ne^2}{\varepsilon_0 m} \tag{5.2}$$

where ω_p is the plasma frequency, τ the relaxation time of the conduction electron scattering, n the number of conduction electrons which contribute to the conductivity, ε the dielectric constant and m the mass of a conduction electron. The values $\omega_\mathrm{p} = 1.88\,\mathrm{eV} = 15\,200\,\mathrm{cm}^{-1}$ (3×10^{-19} J), $\varepsilon = 2.10$ and $\tau = 3.2 \times 10^{-15}$ s are obtained experimentally [5.28, 29]. The experimental results for the optical reflectivity at 40 K as a function of the wave number of light are shown in Fig. 5.7. These results are analysed using the Kramers-Kronig relation and the wave-number dependence of the optical conductivity of KCP(Br) in a low energy range is obtained as shown in Fig. 5.8. The broad peak at about $1.5 \times 10^3\,\mathrm{cm}^{-1}$ is believed to originate from the one electron excitation term across the energy gap. The small peak at around $20\,\mathrm{cm}^{-1}$ (4×10^{-22} J) has been attributed by *Brüesch* et al. [5.20] to the excitation mode of the pinned phason. They found that this peak is sensitive to temperature variation as shown in Fig. 5.9.

Up to now, definite evidence for the pinned phason mode has not been obtained. But if this mode exists in a good crystal of KCP(Br), we may conclude that conductivity of the Peierls-Fröhlich type is realized in KCP(Br). This is an interesting problem for the future.

In order to obtain information about molecular motion and lattice oscillations, measurements of the Raman scattering [5.30–32], and far infrared

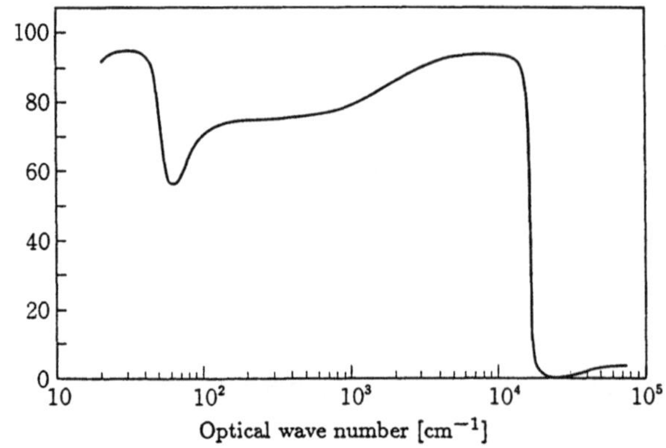

Fig. 5.7. Wave number dependence of optical reflectivity of KCP(Br) at 40 K (Electric field vector parallel to the chain axis) [5.28]

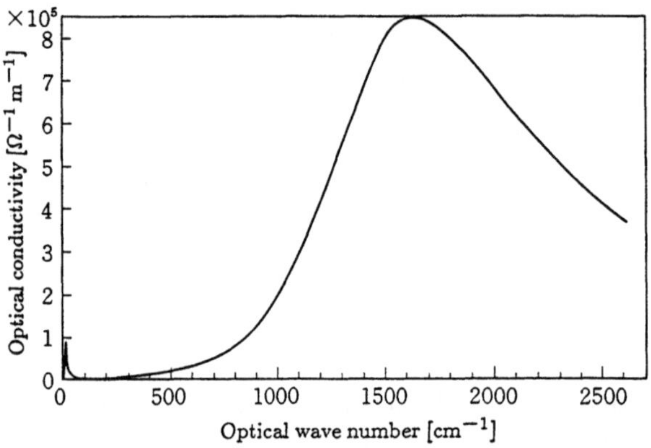

Fig. 5.8. Optical conductivity of KCP(Br) at 40 K [5.28]

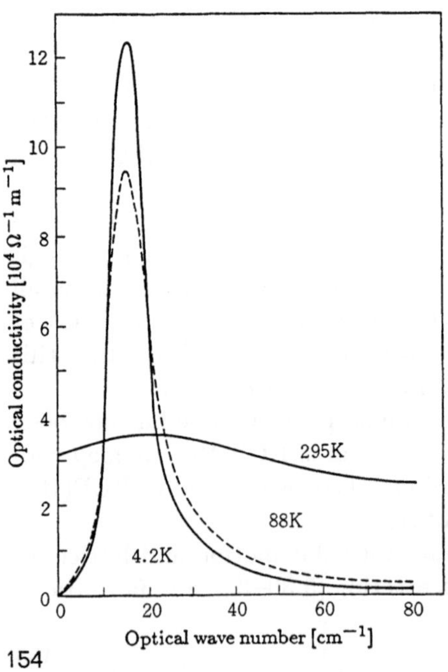

Fig. 5.9. Temperature variation of the optical conductivity peak around 20 cm^{-1} [5.20]

Table 5.2. Raman active modes in KCP complexes

	Symmetry	KCP(Br)	$K_2Pt(CN)_4 \cdot 3H_2O$	$K_2Pt(CN)_4Br_2$
C-N stretching	A_{1g}	2180 cm^{-1}	2168 cm^{-1}	2193 cm^{-1}
C-N stretching	B_{2g}	2172	2150	2178
		2166		
C-N stretching	E_u	2156	2132	2172
		2148		
H_2O stretching	–	3500	–	–
Pt-C stretching		481, 476, 467		
	A_{1g}, B_{2g}	470, 450	465, 455	

absorption [5.30, 31] have been carried out. The back scattering Raman spectra under conditions of parallel and perpendicular laser beam incidence to the 1d chain are studied by varying the polarization direction of light. In these measurements, some damage to the crystal due to heating by the laser beam is expected, so special care is taken by defocusing the laser beam by a cylindrical lens and in cooling the sample. The observed spectra of KCP(Br) are given in Table 5.2 which also includes two other compounds for comparison. It is interesting to note in this table that the C-N stretching modes having symmetry B_{2g} and E_u have a doublet fine structures. This phenomenon was only seen in partially oxidized KCP(X). The crystal symmetry of KCP(Br) is D_{4h} and these modes are expected to be degenerate. So these fine structures are considered to be a clue to the fact that there are two nonequivalent local atomic coordinations surrounding the C-N molecule [5.16]. But according to this model, the observed singlet spectrum of the

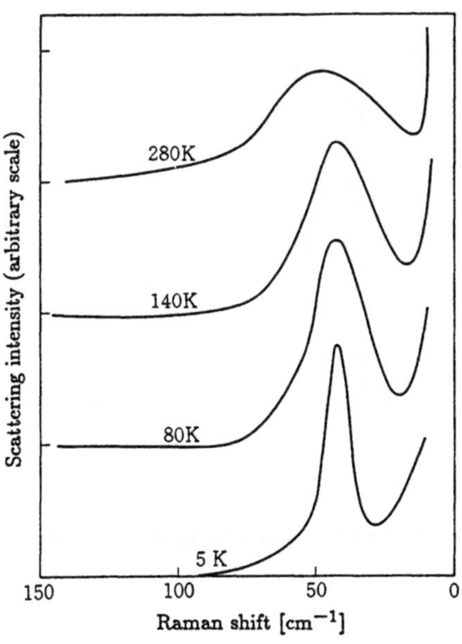

Fig. 5.10. Temperature variation of the Raman shift of polarized light along Pt chain in KCP(Br) [5.32]

non-oxidized KCP cannot be understood because this crystal has a more complex crystal structure containing inequivalent local coordination of C-N molecules.

Another problem is the Raman scattering peak [5.32] around $\sim 50\,\mathrm{cm}^{-1}$ (1×10^{-21} J) as shown in Fig. 5.10. There is not yet any decisive evidence on the origin of this peak. *Steigmeier* et al. [5.32] postulated that this scattering peak may be due to the amplitude mode of the CDW (A_+ mode in Sect. 2.2.2), on the basis that the observed peak position and width are variable with respect to temperature. This problem also remains to be solved.

5.4 Magnetic Properties

5.4.1 Magnetic Susceptibility

The magnetic susceptibility of KCP(Br) has been measured by several groups [5.33, 34]. The diamagnetic susceptibility due to closed-shell ions exceeds the paramagnetic susceptibility of the 5*d* electrons of the Pt ions and it is not easy to separate them. So a reliable temperature dependence of the paramagnetic susceptibility of KCP(X) cannot be obtained by static methods. A steep increase of the Curie-like susceptibility was observed in some cases [5.33] at low temperatures. But this may originate from defects in the crystal or from impurities, and is believed not to be intrinsic. In order to understand this circumstance, the temperature dependence of the magnetic susceptibility of powdered KCP(Br) [5.35] is shown in Fig. 5.11 as a function of concentration of water of crystallization. One can easily decrease

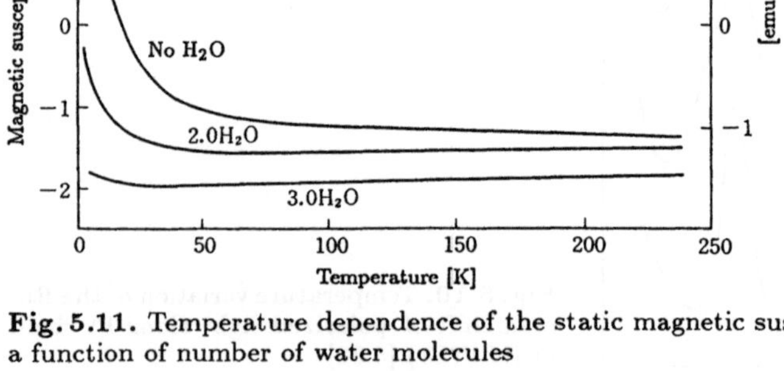

Fig. 5.11. Temperature dependence of the static magnetic susceptibility of KCP(Br) as a function of number of water molecules

the amount of water in KCP(Br) by pumping on the powdered sample at room temperature. The amount of evaporated water was measured by using a magnetic balance to detect the weight change of the sample. The magnetic susceptibility was then measured by applying a magnetic field. The magnetic susceptibility due to localized moments was found to increase with decreasing water concentration as shown in Fig. 5.11.

5.4.2 Electron Spin Resonance

As discussed in Sect. 5.4.1, from the static magnetic susceptibility measurements one cannot obtain the susceptibility of a *d*-electron of a Pt ion independent of the large diamagnetic susceptibility of the inner closed shells and the other nonmagnetic ions. Thus the electron spin resonance (ESR) method is useful to obtain the paramagnetic spin susceptibility of KCP(Br) [5.36, 37].

The ground state of the $5d$ orbital has been determined to be mainly a $5d_{z^2}$ orbital and the slight admixture of $5d_{x^2-y^2}$ due to spin-orbit interaction was estimated on the basis of the observed g value of $g_{\parallel} = 1.946(5)$, $g_{\perp} = 2.336(5)$. The signal intensity of ESR is proportional to the spin susceptibility [5.39] and is shown in Fig. 5.12 between 60 and 200 K in KCP(Br). This temperature dependence is believed to be an important experimental result which directly reflects the electronic states at the metal-insulator transition of KCP(Br). It is interesting to note that this temperature dependence roughly corresponds to that of the Pauli paramagnetic susceptibility calculated by Lee-Rice-Anderson theory [5.40] in Sect. 2.3.2 as shown in Fig. 2.20. The temperature dependence of the magnetic susceptibility of KCP(Br) will be discussed later in the section on NMR of ^{195}Pt nuclei.

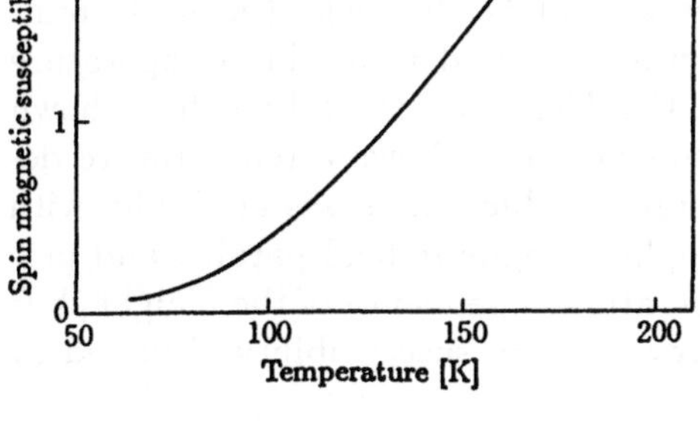

Fig. 5.12. Temperature dependence of the spin susceptibility of KCP(Br)

In the determination of the absolute value of spin susceptibility by ESR, we need an appropriate standard ESR sample for which we can estimate the magnitude of susceptibility by the relative signal intensity.

KCP(Br) without water of crystallization is used as the standard sample for ESR. We observed the ESR signal of a KCP(Br) crystal with water of crystallization and again after the evacuation of the water. The latter signal is used as a standard value of susceptibility and the static magnetic susceptibility [5.35] has already been measured as shown in Fig. 5.11. The spin susceptibility obtained below 180 K is found to be described by the following equation [5.41]

$$\chi_{\text{spin}}[J \cdot T^2 \,\text{mol}^{-1}] = \frac{0.983}{T} \exp\left(-\frac{700}{T}\right) \quad . \tag{5.3}$$

The observed resonance line shape fits well with the calculated Lorentzian form as $1/[1 + (H - H_0)^2/\Delta H^2]$ in a wide magnetic field range of $\pm 5\Delta H$. This result suggests that the ESR line is motionally narrowed. On the other hand, the observed resonance linewidth is found to increase as T^3 with temperature [5.37].

The resonance linewidth shows a characteristic anisotropy according to the angle between the crystal axis and the applied magnetic field. That is to say, the linewidth is maximum when the magnetic field is applied parallel to the 1d c axis and minimum at about 50° from the c axis. The temperature dependence and the angular dependence of the ESR linewidth gives us important information about the motion of electrons in KCP(Br). For further discussions we need a more concrete model which will be described in Sect. 5.7.

5.4.3 Nuclear Magnetic Resonance

It becomes clear from ESR that the 5d electrons of the Pt ions play an important role in the metal-insulator transition of KCP(Br). Another useful method for probing the 5d electron states is nuclear magnetic resonance (NMR) of ^{195}Pt nuclei. In metals, the NMR resonance shift is known as the Knight shift (Sect. 3.3.2) and it directly reflects the local susceptibility of a metal atom. The resonance line shift of ^{195}Pt nuclei (natural abundance of 33 %, $I = \frac{1}{2}$) in KCP(Br) was measured in a wide temperature range between 4.2 K and 350 K [5.42, 43]. The experimental result is shown in Fig. 5.13; one can see that the resonance shift shows a temperature dependent negative value at low temperature which increases gradually with temperature for both directions of applied magnetic field parallel and perpendicular to the 1d chain. This temperature dependence of the Knight shift in ^{195}Pt nuclei corresponds to the previous spin susceptibility obtained by

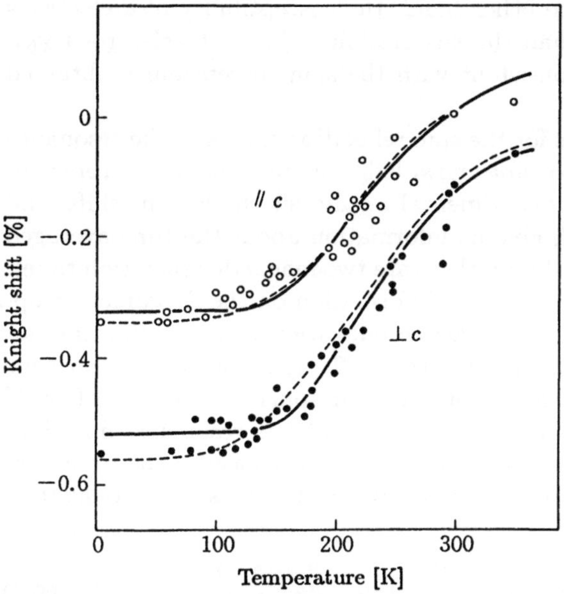

Fig. 5.13. Temperature dependence of the Knight shift of ^{195}Pt nuclei in KCP(Br). [The solid line in the figure is calculated on the basis of Lee-Rice-Anderson model. The dotted line is calculated by the mixed-valence model discussed in Sect. 5.7.2]

ESR. The solid lines in Fig. 5.13 are the calculated Pauli paramagnetic susceptibility on the basis of the Lee-Rice-Anderson theory [5.40]. The numerically calculated results are scaled by the appropriate magnitude of shift and temperature in order to fit the observed results. The agreement is satisfactory, so the value used in scaling of temperature 480 K may correspond to the Peierls transition temperature T_P in the Lee-Rice-Anderson theory of the mean-field approximation.

^{195}Pt NMR studies were carried out on other tetra-cyano Pt salts. It was known that the ^{195}Pt Knight shift of KCP(Br) at low temperatures agrees with the observed resonance shift of tetra-cyano Pt complex compounds having only divalent Pt ions [5.38].

The temperature dependence of the ^{195}Pt Knight shift in KCP(Br) shows that the positive component of the resonance shift increases with increasing temperature, according to the increase of spin susceptibility. Usually the Knight shift due to d-electrons is given by $K_d = H_{\text{hf}}^{5d}\chi_d/N\mu_B$. Here H_{hf}^{5d} is the internal field of the nuclei caused by the polarization of inner shell electrons and known to be negative in Pt compounds. The experimental results for KCP(Br) cannot be understood by the simple picture that the valence electrons of the Pt ions consist of only $5d$ electrons, since Pt ions also have strong covalent bonding with sp-electrons of neighbouring C-N bases. So there exists appreciable $6s$-electron density at the Pt nuclei as expected from band calculations. In this case the internal magnetic field of Pt nuclei due to $6s$-electrons is positive and its contribution is much larger than that from $5d$-electrons. So the resonance shift from s-electrons exceeds that from d-electrons. On the other hand, the susceptibility of d-electrons χ_{5d} is always much larger than the susceptibility due to $6s$-electrons χ_{6s}, and thus these results are consistent with the spin susceptibility obtained by ESR.

The experimental results for the spin relaxation time and the resonance line width [5.43] give us important knowdledge about dynamical properties of the electronic states. The experimental results for the Knight shift, and the resonance magnetic field, give us information about the time averaged electronic state. On the other hand, there are two magnetic relaxation times, defined as follows which give us direct information on the dynamical properties of electronic states: (1) Spin-lattice relaxation time T_1; the time for energy transfer from spin system to lattice. (2) Spin-spin relaxation time T_2; the time for energy transfer among nuclear spins. In the case of rapid fluctuations, there is a relation between T_2 and the resonance linewidth $2/T_2 \cong \gamma_N \Delta H$, therefore one can estimate T_2 on the basis of the resonance linewidth. In the extreme limit of rapid narrowing, these spin relaxation times are given by [5.39]

$$\frac{1}{T_1} \sim \frac{1}{T_2} \sim (\gamma_N \Delta H_{loc})^2 \tau_c \tag{5.4}$$

as a function of the magnitude of the fluctuating local field ΔH_{loc} and the characteristic time of fluctuation τ_c. If we know the fluctuating local field ΔH_{loc}, we can obtain a knowledge of the time dependent fluctuation, which is important in a wide range of branches of solid state physics. Figures 5.14 and 5.15 show the temperature dependence of $1/T_1$ and ΔH for KCP(Br) [5.43]. Here, we discuss the spin-lattice relaxation time T_1 of ^{195}Pt nuclei.

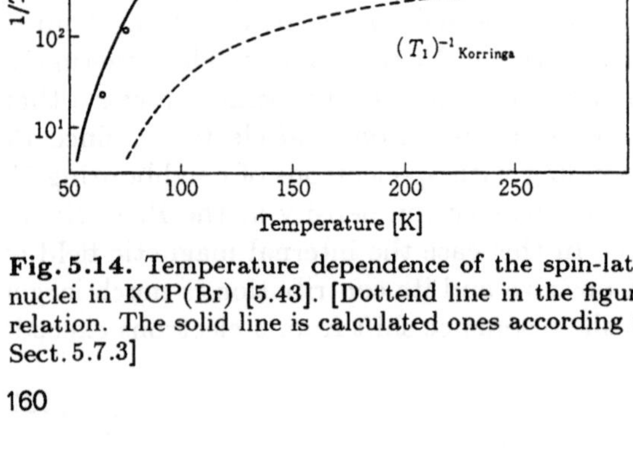

Fig. 5.14. Temperature dependence of the spin-lattice relaxation rate $(1/T_1)$ of ^{195}Pt nuclei in KCP(Br) [5.43]. [Dottend line in the figure is the calculated one by Korringa relation. The solid line is calculated ones according to the mixed valence model [5.64] in Sect. 5.7.3]

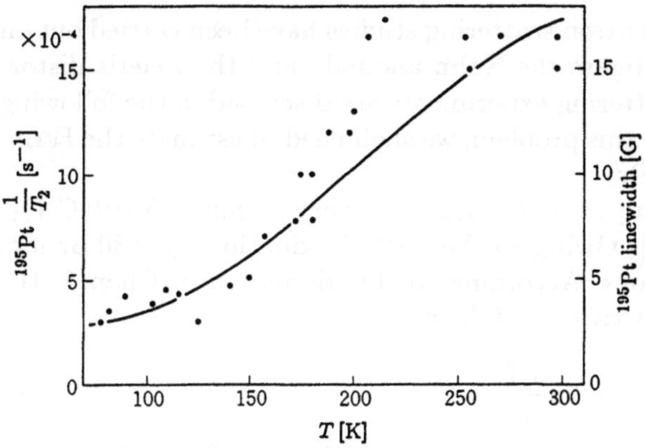

Fig. 5.15. Temperature dependence of the resonance linewidth of ^{195}Pt nuclei in KCP(Br). The solid line in the figure is calculated according to the mixed valence model in Sect. 5.7.3 [5.64]

In a normal metal, there is a simple relation called the "Korringa relation" between the Knight shift K (the time averaged quantity $\langle K \rangle$) and T_1 (the fluctuation term) as described in Sect. 3.3.2 [5.39]

$$4\pi k_B \langle K \rangle^2 T_1 T = \alpha \left(\frac{\gamma_e}{\gamma_N} \right)^2 \quad . \tag{5.5}$$

This relation is based on the following assumptions: 1) The electron near the Fermi level is moving with the Fermi energy in a metal and the conduction electron density is estimated from the Pauli paramagnetic χ and the Knight shift. 2) The number of electrons which contributes the nuclear spin energy relaxation to the lattice is proportional to $k_B T$. Here γ_e and γ_N are the gyromagnetic ratios of electron and nuclear spin respectively and α is a numerical constant of the order of one. In the case of KCP(Br) the experimental results, as shown in Fig. 5.14, give a value for T_1 that is about 1/60 of that of usual metals and this disagreement leads us to question the frequent assumption that KCP(Br) is a normal metal.[1]

5.5 Charge Density Waves and the Kohn Anomaly

In 1d metals, the softening of $2k_F$ phonons in the dispersion relation between energy and wave number is observed as predicted by *Peierls* [5.46]. We call this phenomenon a Kohn anomaly. In cases where the softening becomes perfect so that $\omega_{2k_F} = 0$, the static lattice distortion with wave number $2k_F$ is realized as the Peierls distortion and a charge density wave is present (Chap. 2).

[1] It has been suggested [5.45] that an enhancement of the spin-lattice relaxation rate is expected in the case of 1d systems due to the susceptibility enhancement of the $\chi(2k_F)$ term.

Extensive x-ray and neutron scattering studies have been carried out on KCP(Br) to directly investigate the Kohn anomaly and the Peierls distortion. The details of the scattering experiments are discussed in the following sections. Before we address this problem, we shall need to estimate the Fermi wave number k_F in KCP(Br).

Two electrons can occupy the d_{z^2} band for every Pt ion in $K_2Pt(CN)_4$-$Br_{0.30} \cdot 3.2H_2O$ [KCP(Br)]. Owing to the partial oxidation by 0.30 Br ion, each Pt ion has 1.7 electrons. According to the discussion in Chap. 2, the Fermi wave number k_F is written as follows

$$2k_F[\mathrm{KCP(Br)}] = 2\pi\frac{1.7}{2c'} = 2\pi\frac{1.7}{c} = 1.7c^* \tag{5.6}$$

using the interatomic distance between Pt atoms $c' = 2.88$ Å. Here it should be noted that the unit cell length c of c axis in KCP(Br) corresponds to $2c'$. Therefore we can expect the Kohn anomaly and the Peierls distortion to have a wave number of $2k_F = 1.70c^*$.

5.5.1 X-ray Scattering

It began with an x-ray film of KCP(Br) by *Comes* et al. [5.47] at the end of 1972. This clearly showed the existence of the Peierls distortion and caused 1d conductors to enter the limelight in solid state physics. The historic photograph is shown in Fig. 5.16a. One can see the diffuse scattering lines between the layer lines of Bragg spots separated by 1/6 of the inter layer line distance. This scattering line, is the diffuse streak peculiar to 1d chains. From these photographs, the following conclusions may be drawn:

i) the scattering vector $\boldsymbol{K}$ in the reciprocal plane is given by

$$K_{\|} = (l \pm 0.30)c^* \quad (l:\text{ even}) \quad , \tag{5.7}$$

ii) the scattering intensity $I(K_{\|})$ along the 1d chain is proportional to $f_0^2 K_{\|}^2$,

iii) the scattering intensity is constant in every scattering plane and it decreases as the vertical component of the diffuse scattering wave number $K_{\perp}$ increases.

These observations suggest that the interatomic distance along the 1d chain of KCP(Br) is modulated sinusoidally with the wave number $q = 0.3c^*$. The intensity $I(K)$ of diffuse scattering sheets of KCP(Br) at room temperature is given by the correlation function $G(r)$ as shown in (3.17)

$$I(K) \propto f_0(\boldsymbol{K} \cdot \Delta_c)^2 \frac{1}{\xi^{-2} + [K - (l \pm 0.30)c^*]^2} \quad (l:\text{ even}) \quad . \tag{5.8}$$

The coherence length ξ of KCP(Br) could not be determined by x-ray diffraction due to the resolution of the instruments. We will return later to estimates of the coherence length on the basis of neutron diffraction studies. It should be noted that the diffuse scattering vector due to the lattice distortion $K_{\|} = \pm 0.30c^*$ corresponds to a wave vector $2k_F[\mathrm{KCP(Br)}] = 1.70c^* = (2 - 0.30)c^*$. As the scattering factor of x-rays becomes larger for a heavier atom, we can conclude that this diffuse x-ray scattering may originate from a phenomenon associated mainly with the heavy platinum chains.

In (5.8), l is limited to be an even integer as a consequence of the extinction law, because there are two Pt atoms within a lattice parameter $c = 2c'$. Since the x-ray scattering factor depends on the time averaged electron density distribution, by the x-ray diffuse scattering peak shown in Fig. 5.16b, one could not determine whether it originates from the static modulation of the interatomic distance or from a dynamical phonon anomaly.

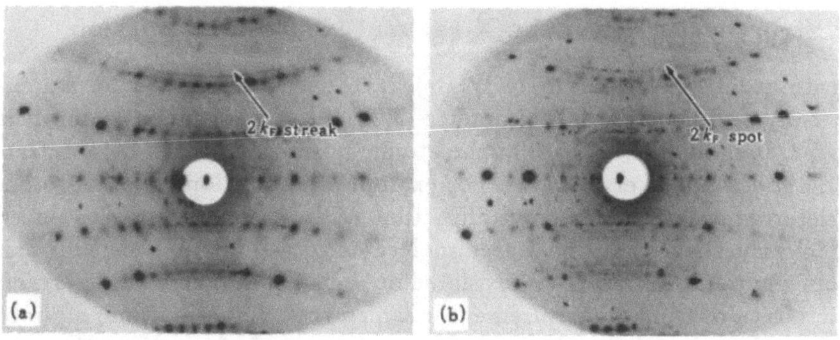

Fig. 5.16. Photograph of x-ray diffuse scattering patterns of KCP(Br) **(a)** at room temperature **(b)** at 77 K [5.47]

Next, we consider the difference between the two x-ray photographs [5.47] of KCP at 77 K and at room temperature. The continuous diffuse streak at room temperature changes to discrete spots at 77 K as shown by the arrow in Fig. 5.16. This change means that interchain interactions have induced a periodicity of twice the unit cell length in the c plane as determined by the position of spots along a and b axes. This periodicity perpendicular to the 1d axis can be understood as the alternating arrangement of charge density waves with phase factors 0 and π. In this case, the correlation length between chains is rather short range, so the width of the satellite reflection spots is found to be much greater than that of usual Bragg reflections in KCP(Br).

Table 5.3. Atomic positions in KCP(Br) at 77 K. $a = 9.841$ Å; $c = 5.714$ Å

	10^4x	10^4y	10^4z
Pt(1)	0	0	0
Pt(2)	0	0	5022
C(1)	2025	0	−9
C(2)	1438	1438	5055
N(1)	3206	0	−49
N(2)	2270	2270	5133
Br(1)	5000	5000	4990
O(1)	0	5000	4073
O(2)	3379	3379	678
K(1)	1940	5000	7475
H(1)	4234	0	3048
H(2)	2819	2819	1508
H(3)	3815	3815	1837
O(3)	5000	5000	6565

5.5.2 Structural Analysis by X-ray and Neutron Diffraction

Accurate atomic positions in KCP(Br) have been determined by a four-circle x-ray and neutron diffractometer [5.22, 48]. The results are shown in Table 5.3. It should be noted that the atomic position obtained by a diffraction method is the averaged one. For example the positions of the protons were determined by the neutron diffraction study, but NMR studies on ^{1}H nuclei clearly show that the water molecules rotate about a certain axis at room temperature, as discussed later in Sect. 5.6.4. When we determine atomic positions from a measurement in which the interaction time of probe particle and nucleus is much shorter than the period of rotation, then the nucleus appears stationary. The measurement being the average over very many scattering events leads however to the derivation of an average atomic position.

5.5.3 Elastic Neutron Scattering

From the x-ray diffuse scattering study it can be concluded that a correlation between platinum chains arises at low temperature and that this is due to a 3d ordering of the lattice distortions. Further more detailed neutron diffraction studies have been carried out on this system [5.49, 50]. In Fig. 5.17, the intensity of elastic neutron scattering [5.49] is shown as a function of reciprocal lattice vector around $(\pi/a, \pi/a, 2k_F)$. From this figure which shows the direction $(\xi, \xi, 2k_F)$, one can see the peak of scattering intensity at $\xi = 0.5a^*$. This means that the unit cell of KCP(Br) is doubled in the direction perpendicular to the chains. In other words, neighbouring 1d Pt chains are displaced in opposite directions.

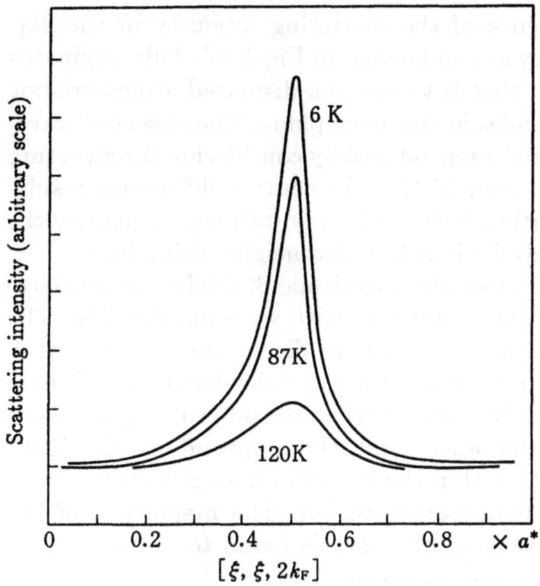

Fig. 5.17. Wave number dependence of scattering intensity in the $2k_F$ plane at 6 K, 87 K and 120 K [5.49]

The temperature dependence of the scattering intensity of the $(\pi/a, \pi/a, 2k_F)$ satellite diffraction is shown as Fig. 5.18 and the abrupt growth of the 3d correlation between 80 K and 120 K is clearly seen. But the correlation length perpendicular to the 1d chain is limited to at most 50 Å, and no development of true long-range order is found. However the correlation length along the 1d chains in KCP(Br) is more than 100 atomic distances (~ 300 Å). Therefore, we expect the existence of micro-domains of $\sim 300 \times 50 \times 50\ \text{Å}^3$ below about 100 K.

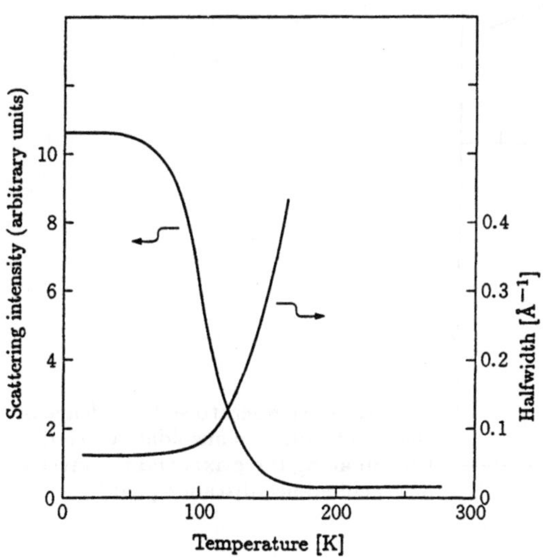

Fig. 5.18. Temperature dependence of neutron scattering intensity and the half width of the scattering peak at $(\pi/a, \pi/a, 2k_F)$ [5.49]

The wave-number dependence of the scattering intensity in the $2k_F$ plane shows the clear asymmetry as can be seen in Fig. 5.17. This originates from the structure factor effect, that is to say, the displaced atoms are not only Pt ions but also C-N molecules in the same plane. The observed wave-number dependence was found to be reproduced by considering the structure factor of the coupled $Pt(CN)_4$ system [5.49]. The neutron diffraction results give us complementary information to that of x-ray diffraction, namely the atomic displacements of not only Pt ions but also neighbouring ions.

Lynn et al. [5.49] have estimated the magnitude of the lattice displacement assuming a simple sinusoidal modulation with wave number $2k_F$. The estimated temperature dependence of the lattice displacement is shown in Fig. 5.19. The inverse of the correlation length perpendicular to the 1d chain is also shown in the same figure. One can see that the 3d ordering increases significantly below 100 K. On the other hand, the amplitude of the CDW (sinusoidal lattice displacement) within the 1d chain does not show an appreciable change in the same temperature region. The magnitude of the lattice displacement with the wave number $2k_F$ is found to be as small as about 1 % of the interatomic distance between Pt ions.

Here we summarize the experimental results of x-ray and neutron diffraction studies as follows.

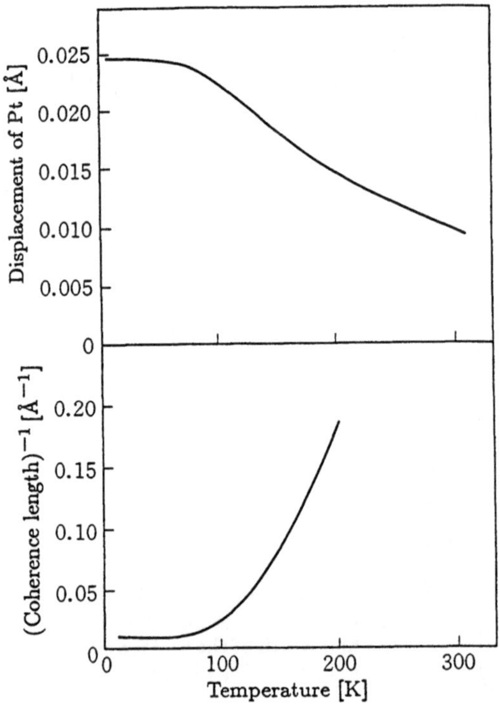

Fig. 5.19. Temperature dependence of the amplitude of sinusoidal lattice distortion along the c axis and the inverse of its coherence distance [5.49]

1) There exists a sinusoidal displacement of $Pt(CN)_4$ molecules along the 1d chain with the wave number $2k_F$. Higher harmonics have not been found. The coherence length of this displacement is greater than 100 Pt-Pt atomic distances (this is the limit of resolution of experimental equipment). Here we note that it cannot be determined whether these displacements are static or dynamic in nature.

2) The coherence length between the 1d chains increases rapidly below 100 K. But it cannot exceed the micro-domain size of 50 Å found in the limit of low temperature. Here, the question naturally arises as to why the 3d ordering cannot exist on a macroscopic scale. At first it was believed that the random distribution of bromine atoms between platinum chains might be the cause of the micro-domains. But recent theoretical work [5.44] demonstrates that the large anisotropy in the coherence length (larger than 100 interatomic distances at room temperature along the 1d chains and $\sim$50 Å perpendicular to this direction) cannot be understood by the usual CDW theory. Therefore, it was suggested that the Br atoms may actually be ordered within the size of micro-domains and not in a random distribution [5.44]. Up to now, experimental difficulties have prevented a decisive determination of the distribution of Br atoms.

To explain the origin of the micro-domains of KCP(Br), an inhomogeneous model for the 1d chains (i.e. the coexistence of two phases, the distorted insulating state and the metallic state, as separate domains) has been proposed [5.51] in order to reconcile the experimental results for the electrical conductivity with the neutron diffraction. But the observed experimental results of ^{195}Pt nuclear magnetic resonance clearly show the existence of the homogeneous rapid narrowing mechanism. In order to fit this result the micro-domain would have to move very fast within the crystal. But evidence for such rapid motion of micro-domains has not been found. At present this inhomogeneous model lacks any real confirmation.

5.5.4 Inelastic Neutron Scattering

In the 1d metal, the softening of the $2k_F$ phonon is known as the Kohn anomaly. To observe this phenomenon, we can directly measure the phonon spectrum using inelastic neutron scattering. The Kohn anomaly in KCP(Br) was first discovered by *Renker* et al. [5.50]. Recently *Carneiro* et al. [5.52] carried out the detailed studies on this problem.

Scattering data were collected by employing the technique of constant-E scans (fixing the energy and varying the momentum transfer); as the scattering near $2k_F$ is very steep in energy, the usual constant-Q method of scanning (fixing the momentum transfer and varying the energy transfer) is not useful. In Fig. 5.20b, the normalized intensity contours of the neutron

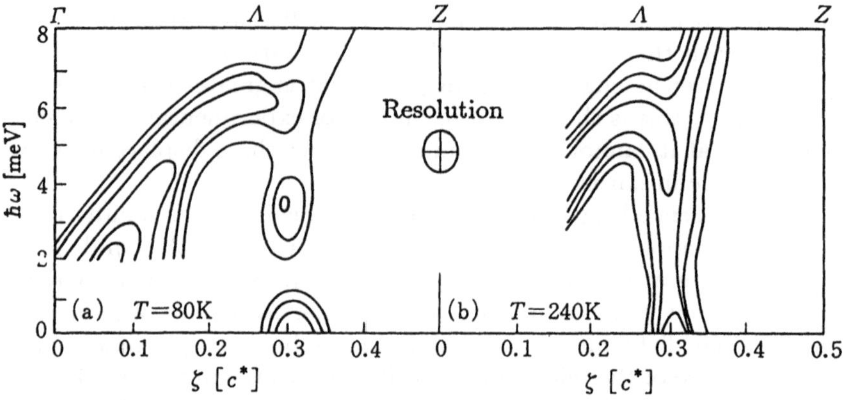

Fig. 5.20. Neutron elastic scattering intensity contours from the excitations of wave vector $q = \xi c^*$, measured at 80 K and 240 K with an average resolution as indicated [5.52]

scattering of KCP(Br) are shown. The phonon spectrum at 240 K, where the metallic characteristic is expected, shows a softening of the phonon of wave number $\zeta = 0.3c^*$. This corresponds to the Kohn anomaly. At the same time, rather surprisingly, a broad scattering peak around $\omega = 0$ was observed in the same direction at wave number $\zeta = 0.3c^*$. This result implies the existence of the soft mode with maximum at $\omega = 0$. At 80 K two scattering peaks at $\sim$5.5 meV (8.8×10^{-22} J) and 3.5 meV (4.0×10^{-22} J) in the direction $\zeta = 0.3c^*$ were found as shown in Fig. 5.20a. Carneiro et al. tried to understand these scattering peaks in terms of a damped harmonic oscillator with characteristic oscillation energies of 5.5 meV and 3.5 meV. It is also interesting to note that these two energies correspond to the excitation energies of the amplitude and the phase modes of the CDW on the basis of the Raman scattering and the infrared absorption measurements of *Steigmeier* et al. [5.32] and *Brüesch* et al. [5.53] respectively. The latter interpretation contradicts the harmonic oscillator model. In order to settle this problem, further more careful and detailed studies will be necessary. In both models the damping effect may become significant at higher temperatures so that the scattering energy has no clear peak and white spectra are expected.

The unsolved problem of inelastic neutron scattering is the large scattering peak at $\omega = 0$. If we adopt the most simple-minded model, we are forced to assume that a perfect phonon softening and Peierls distortion occur in some parts of the crystal but that in the remaining parts, the softening of the $2k_F$ phonon is imperfect.

In fact, if there were an inhomogeneity in the KCP crystals showing up in the neutron diffraction experiment, we would need to make a correction to the discussion of the displacement of Pt ions in the 1d chain as mentioned in

Sect. 5.5.3. However no definite conclusions concerning the observed central peak can be made at present.

It is found by x-ray and neutron scattering experiments that 3d ordering in KCP(Br) occurs only in a small region and cannot develop to a macroscopic size even at low temperature. In other words, it would seem reasonable to suppose that there is an intrinsic mechanism that disturbs the inter-chain interactions. As we have seen above, the observed phenomena are not so simple as the Peierls theory predicts. In this sense, one might question whether KCP(Br) is really a typical example of a 1d conductor [5.54].

5.6 Behaviour of the Water of Crystallization

5.6.1 Position of the Water of Crystallization

Detailed neutron diffraction studies [5.22] have determined the position of the water molecules. It is found that 3.2 molecules occupy the three positions as listed in Table 5.4. In Fig. 5.1 the three positions of water molecules are shown. There has been some discussion [5.22, 24] on the existence of the site III position of water molecules and this is not completely settled yet. But the recent NMR results [5.53, 55] support the existence of water molecules at site III.

Table 5.4. Water of crystallization in KCP(Br)

	Number of water molecules	Position
Site I	1.0	Between CN bases in (a, c) plane
Site II	2.0	Between Br and Pt ions in (110) plane
Site III	0.2	Vacant Br positions

5.6.2 The Amount and Physical Properties of the Water of Crystallization

From the very beginnings of studies on KCP(X), it has been known that the crystal is very sensitive to its environment [5.29]. Now it becomes clear that this phenomenon originates from the fact that the water in KCP(Br) easily evaporates out from the surface and that various physical properties change depending on the concentration of water. First, we consider the electrical conductivity. The dependence of the electrical conductivity at room temperature on the concentration of water is reported [5.56] as Fig. 5.21. The electrical conductivity of KCP(Br) decreases by more than four orders

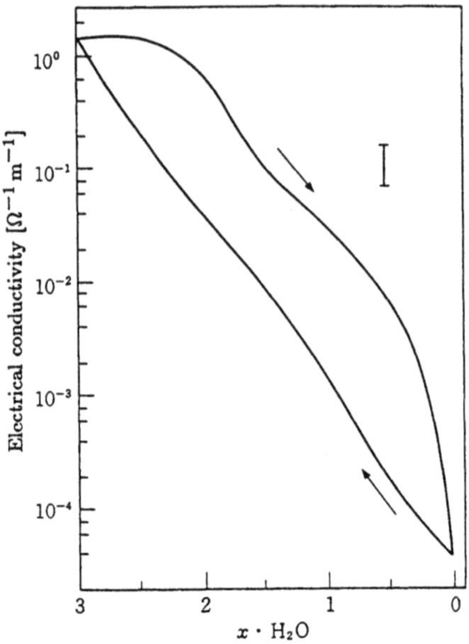

Fig. 5.21. The electrical conductivity of KCP(Br) along the c axis at room temperature on dehydration and hydration cycle [5.56]

of magnitude as a result of dehydration. If we then add water gradually to dehydrated KCP(Br) until the full water content is reached, the electrical conductivity returns to the original value as seen in Fig. 5.21. With respect to the crystal appearance, a KCP(Br) crystal with its full water complement has a copper-bronze like luster, but after dehydration by pumping it loses this luster and turns to a brittle grey crystal. However it regains a metallic luster with the addition of water.

The effect of dehydration in KCP(Br) can be seen, not only in the electrical conductivity, but also in the magnetic properties. An increase of the number of localized moments was observed in the static susceptibility measurements [5.37] upon dehydration as shown already in Fig. 5.11.

These experimental results can be qualitatively understood as follows: The water molecules (site II) in KCP(Br) are situated between the halogen atom Br and the platinum ions and play an important role in screening the electric field at the Pt site from the Br ions. If there were no water molecules between the halogen and platinum ions, the large bare crystalline electric field may localize d-electrons within a Pt chain. Thus this explains the decrease of electrical conductivity and the increase of the number of Pt ions with a localized magnetic moment upon dehydration.

5.6.3 Motion of Water Molecules and Ultrasonic Anomalies

As we have seen previously, the "gradual metal-nonmetal transition" at around 100 K is the most important characteristic of KCP(X), and is reflected also in the magnetic properties. Therefore, it seems unlikely that phenomena which occur below 100 K, can originate from the phenomenon of the one-dimensionality of the electronic state.

In KCP(Br), the elastic constants were determined from ultrasonics measurements [5.57] far below the metal-nonmetal transition temperature. The elastic constant C_{33}, which relates to the longitudinal phonon along the c axis, shows a monotonic temperature dependence. This result is consistent with the results of inelastic neutron scattering. The softening was observed for the phonon of wave number $q = 0.3c^*$, but no anomaly was found for the long wavelength $q \cong 0$ phonons. Next, the temperature dependence of the elastic constant C_{66}, which relates to the transverse acoustic phonon propagating along the c axis is shown in Fig. 5.22. Although the temperature range of the measurements is limited to between 110 K and 190 K, a remarkable softening is observed for C_{66} phonon. On the other hand, this softening of the transverse acoustic phonon at long wavelength cannot be observed by inelastic neutron diffraction [5.52].

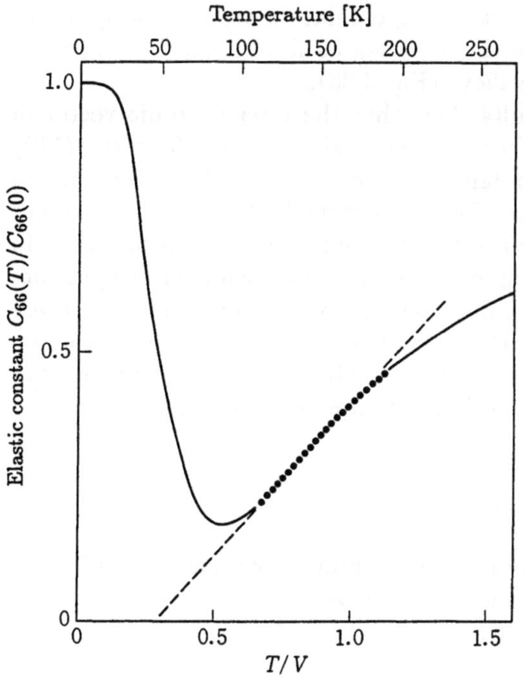

Fig. 5.22. Temperature dependence of the elastic constant C_{66} obtained by the transverse acoustical phonon [5.57] propagating along the a axis. Solid line is calculated [5.58] according to (5.9)

This discrepancy has been removed by the spin model of *Kurihara* et al. [5.58]; they considered the vibrational motion of the hydrogen atoms of water molecules in KCP(Br). The existence of the rotational and vibrational motion of water molecules has been demonstrated by NMR of ^1H nuclei [5.23] and by neutron diffraction experiments [5.22]. *Kurihara* et al. [5.58] paid particular attention to the water molecules at site II, where the H-H

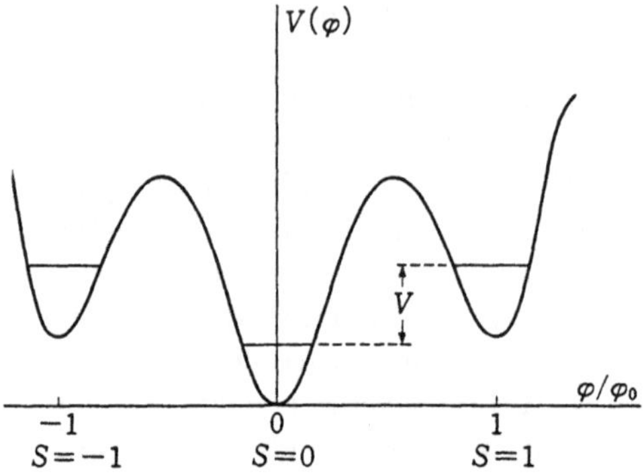

Fig. 5.23. Spin analogue model [5.58] of the three stable directions of H-H vector of the water molecule at site II in KCP(Br)

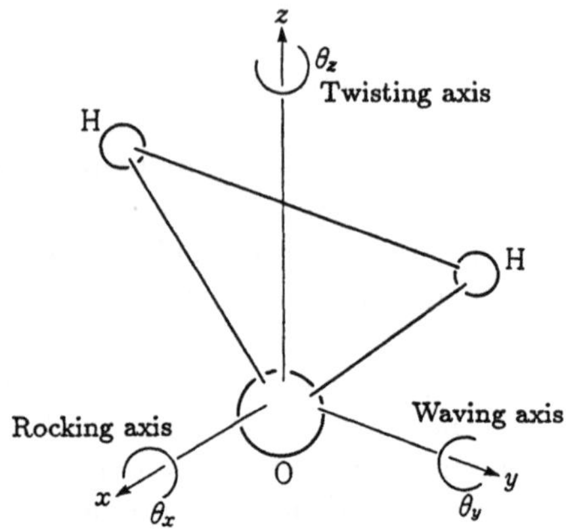

Fig. 5.24. Structure of H_2O molecule in principle axes of its twisting or vibrational motion

vector of the water is almost parallel to the direction (±1, ±1, 0) in the (ab) plane and for which there exists a twisting oscillation around the z axis as shown in Fig. 5.24. They analyze this twisting oscillation as the motion of particles within three potential valleys (Fig. 5.23).

The neutron diffraction results show that the inter-protonic vector of site II water molecules oscillates between the angles $\phi\leq\pm16°$ from the $\langle110\rangle$ direction. Thus the most stable potential minimum is in $\langle110\rangle$ direction and the excited potential minimum positions correspond to $\phi_0 = \pm16°$. *Kurihara* et al. proposed a spin model for the three potential minimum problem. In this, the most stable state corresponds to an $S_z = 0$ state and the degenerate excited states to $S_z = \pm1$ states, and the energy difference between these states is defined as V as seen in Fig. 5.23. They considered the interaction between these spins and lattice systems, and obtained the elastic constant C by a simple calculation as a function of temperatur as

$$\frac{C(T)}{C_0} = 1 - \frac{8\lambda V}{T}\frac{e^{-V/T}}{(1+2e^{-V/T})^2} \quad . \tag{5.9}$$

By using $V = 170\,\mathrm{K}$ and the spin-lattice coupling constant $\lambda = 0.61$, the good agreement between (5.9) and the observed results was obtained as shown in Fig. 5.22 by the solid line.

5.6.4 Motion of the Water of Crystallization Studied by ^{1}H NMR

The ultrasonic anomaly in KCP(Br) was found to originate from the motion of water molecules and we need therefore to know the details of these motions. The most appropriate experiment is the nuclear magnetic resonance of ^{1}H nuclei. The two protons in a water molecule are separated by a distance $r_{\text{H-H}}$ as shown in Fig. 5.24 and *there is* a nuclear dipole interaction between them.

As ^{1}H nuclei have a spin $I = \frac{1}{2}$, the total spin of water molecule $I = I_1 + I_2$ becomes a good quantum number. There are three energy levels due to the Zeeman splitting of the total spin 1 system. So the nuclear resonance frequency, which corresponds to the transition $\Delta I = \pm 1$, has a fine structure called the Pake doublet [5.59]

$$\hbar\omega_{\text{NMR}} = \gamma_{\text{H}} H_0 \pm \frac{3\gamma_{\text{H}}\hbar}{2r_{\text{H-H}}^3}(3\cos^2\theta - 1) \quad . \tag{5.10}$$

Here θ is the angle between external field and the interatomic vector $r_{\text{H-H}}$ between the two protons in a single water molecule. By measurements of the Pake splitting and its angular dependence, one can determine the interprotonic distance and the direction of the $r_{\text{H-H}}$ vector. When the water molecules are in motion, e.g. rotational motion, the Pake doublet varies as a function of time and the term of $(3\cos^2\theta - 1)$ is averaged over time. The Pake doublet splitting is given by

$$\Delta H_{\text{P.D.}} = \frac{3\gamma_{\text{H}}\hbar}{r_{\text{H-H}}^3}\langle 3\cos^2\theta - 1\rangle \quad . \tag{5.11}$$

When the external magnetic field is applied parallel to $r_{\text{H-H}}$, that is $\langle\theta\rangle = 0$, the observed maximum splitting is determined by the time-averaged value of $\langle\cos^2\theta\rangle$.

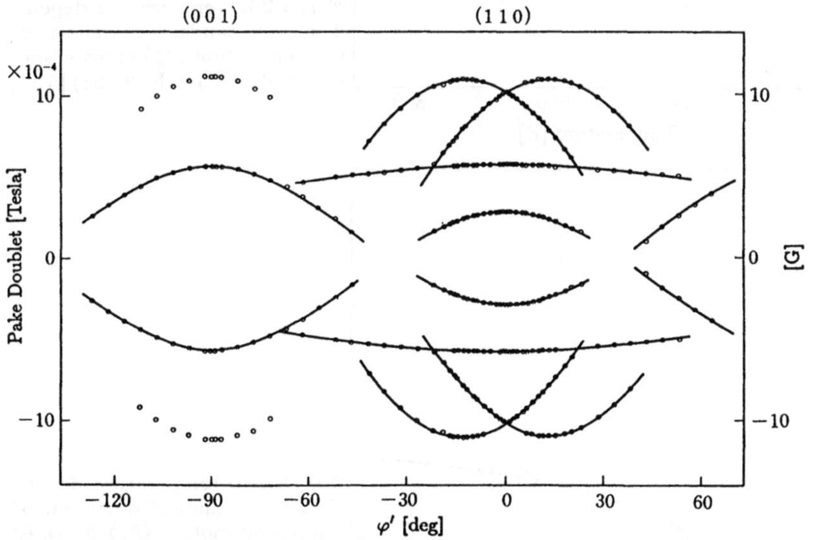

Fig. 5.25. Angular dependence of ^{1}H NMR spectra in (1, 1, 0) plane in KCP(Br) [5.55]. The maximum splitting along $\phi = \pm 13°$ shows that the inter-proton vector deviates 13° from (a, b) plane

As a typical example, the angular dependence of the Pake doublet at 77 K in the (110) plane is shown in Fig. 5.25 [5.55]. The normal modes of vibration of a water molecule are shown in Fig. 5.24; (a) the twisting oscillation about the two fold symmetric z axis (b) the rocking oscillation about an axis normal to the plane of the water molecule (x axis), (c) the waving oscillation about an axis parallel to the p-p vector of water molecule (y axis). Of these three oscillations, only the waving oscillation maintains the direction of the inter-protonic vector and in this case one cannot obtain any knowledge on the motion of water molecules by NMR. For the motion around the x and z axes, the mean square deviation of the oscillating angles $\langle\theta_x^2\rangle$ and $\langle\theta_z^2\rangle$ are shown as a function of temperature in Figs. 5.26 and 5.27, assuming that the inter-protonic distance $r_{\text{H-H}}$ of 1.51 Å is temperature independent [5.55].

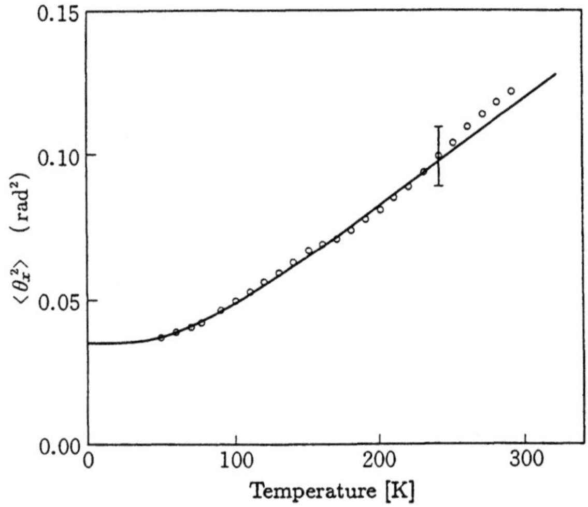

Fig. 5.26. Temperature dependence of standard deviation of rocking motion $\langle\theta_x^2\rangle$ of water molecules at site II in KCP(Br) [5.55]

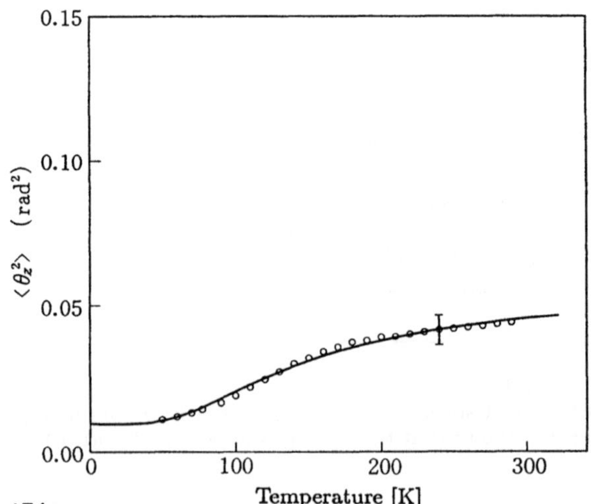

Fig. 5.27. Temperature dependence of standard deviation of rotational motion $\langle\theta_z^2\rangle$ of water molecules at site II in KCP(Br) [5.55]

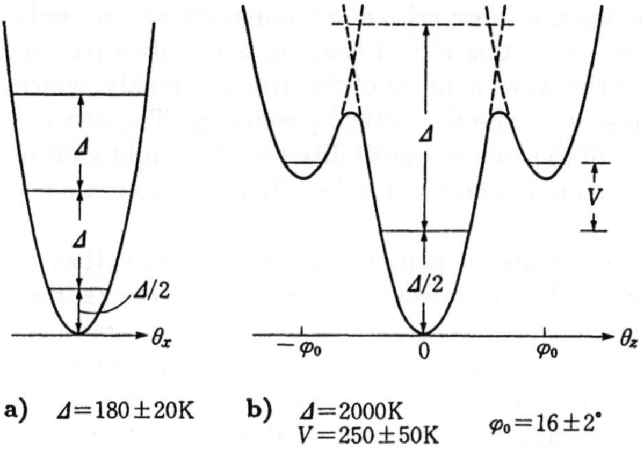

Fig. 5.28. Vibrational energy levels of water molecules at site II in KCP(Br) determined by ^{1}H NMR [5.55]

From Fig. 5.26, the rocking motion of the water molecules at site II can be understood in terms of a simple harmonic oscillator model (Fig. 5.28a), where $\langle\theta_x^2\rangle$ is calculated as

$$A\left[\frac{1}{2}+\frac{1}{\exp\left(\Delta/T\right)-1}\right]$$

in the solid line in the figure. On the other hand, the twisting motion about the z axis $\langle\theta_z^2\rangle$ cannot be simulated by the simple harmonic oscillator model because the observed $\langle\theta_z^2\rangle$ shows a tendency to saturate as seen in Fig. 5.27. But a good agreement with the experimental results was obtained by the three potential valley model (Fig. 5.28b), which is used in ultrasonic anomalies and is shown as the solid line in Fig. 5.27. In the previous discussion of the acoustic attenuation, V was determined to be 170 K; here by the NMR experiment V is estimated to be 250 ± 50 K – the agreement is satisfactory. As we have seen above, it has been confirmed both theoretically and experimentally that the acoustic anomalies originate from the motion of the water molecules.

5.7 Band Model and Mixed Valence Model

The characteristic properties of KCP(Br) have been described in the previous section. In Sects. 5.3–5.5, we have discussed the experimental results on the basis of a 1d electron-gas model. We have seen that almost all experimental results, except for a few, can be understood qualitatively in terms of this model by considering the small corrections, such as the effects of the inter-chain interaction. But upon a more critical examination we find only two experimental results for which this model leads to a quantitative understanding. One of these is the wave number of the Kohn anomaly, which at $Q = 0.3c^* = 2k_F$, coincides with the theoretical prediction. The other is the temperature dependence of the spin susceptibility (NMR Knight shift of the ^{195}Pt nuclei) which can be understood by the Lee-Rice-Anderson theory [5.27] (Fig 5.13).

On the other hand, the results of neutron diffraction show that in KCP(Br) the Peierls distortion does not occur as the theory predicts, that is to say, the softening of the Kohn anomaly does not develop gradually into the static lattice distortion. In this sense, the physical properties of KCP(Br) are not yet well understood. In the following sections, we reconsider experimental results relating to the gradual metal-insulator transition of KCP(Br). These are now interpreted on the basis of a valence fluctuation of Pt ions, which is the opposite limit to the 1d band picture.

5.7.1 X-ray Photoelectron Spectroscopy (XPS)

X-ray Photoelectron Spectroscopy is one of the most useful methods for investigating the electronic states of atoms in materials. By irradiation of the x-rays, electrons of surface atoms in the materials can be excited and emitted from the surface with a kinetic energy that is the energy difference between the x-ray energy and the excitation energy of electron to the Fermi level plus the work function. The merit of this method is that one can deduce the binding energy of a specified electron orbital of a particular atomic species, and hence it has found application in a wide range of fields in chemistry and physics. An x-ray photoelectron spectrum of KCP(Br) [5.60] is shown in Fig. 5.29 (II). In order to determine the valency of the Pt ions, the spectra of the inner $4f$-shell electrons with its narrow energy width is usually used. As examples of XPS spectra of the Pt^{2+} and Pt^{4+} valence states, the spectra of $K_2Pt(CN)_4 \cdot 3H_2O$ and $K_2Pt(CN)_4Br_2$ are shown as I and III respectively in the same figure. The binding energies and their energy widths are collected in Table 5.5. The XPS spectra of KCP(Br) were obtained at 153 K because the sample was evacuated at low temperature in order to avoid loss of water from the surface.

The XPS spectra of the $4f$ electrons of the Pt ion in KCP(Br) were found to be reproduced by the simple addition of Pt^{2+} spectra and Pt^{4+} spectra in the ratio of 5 : 1 (average valence corresponds to 2.3+). On the basis of XPS results, the valence state of a Pt ion in KCP(Br) is concluded to be Pt^{2+} or Pt^{4+} state, changing on a short time scale of $\sim 10^{-15}$ s. If a metallic band exists, the averaged valence state is $Pt^{2.3+}$ and the coexistence of Pt^{2+} and Pt^{3+} would be expected. Thus the absence of a Pt^{3+} state in the XPS spectra is a result which casts doubt on the band model of KCP(Br).

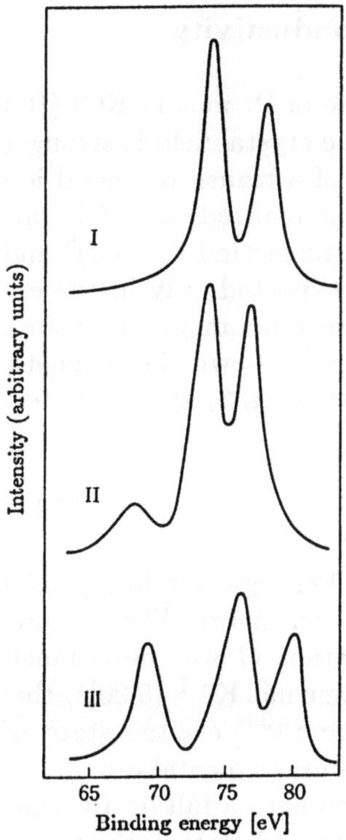

Fig. 5.29. X-ray photoelectron spectra of KCP salts [5.60]. (*I*) $K_2Pt(CN)_4 \cdot 3H_2O$, (*II*) $K_2Pt(CN)_4Br_{0.30} \cdot 3.2H_2O$, (*III*) K_2 Pt $(CN)_4$ Br_2

Table 5.5. Binding energy of electrons in KCP complexes measured by XPS

Complex	Binding energy [eV]							
Complex	Pt^{+2}	Pt^{+2}	Pt^{+4}	Pt^{+4}	Br	Halfwidth [eV]		
	$(4f^{5/2})$	$(4f^{7/2})$	$(4f^{5/2})$	$(4f^{7/2})$	$(3d^{5/2})$	Pt^{+2}	Pt^{+4}	Br^-
$K_2Pt(CN)_4 \cdot 3H_2O$	76.95 ±0.01	73.59 ±0.01				1.53 ±0.01		
$K_2Pt(CN)_4Br_2$			79.18 ±0.01	75.80 ±0.01	69.23 ±0.01		1.59 ±0.01	2.48 ±0.01
$K_2Pt(CN)_4Br_{0.30} \cdot 3.2H_2O$ (153 K)	76.07 ±0.01	72.73 ±0.01	78.50 ±0.01	75.45 ±0.08	67.93 ±0.04	1.96 ±0.02	2.44 ±0.17	3.23 ±0.14

The stable valence states of Pt in inorganic chemistry are believed to be Pt^{2+} and Pt^{4+}. No Pt^{3+} state has been found except as a dilute impurity ion. Thus one may naturally understand the existence of the Pt^{2+} and Pt^{4+} states found in KCP(Br).

5.7.2 Magnetic Properties and Electrical Conductivity

The XPS results show that the stable valence state of Pt ions in KCP(Br) at low temperatures is divalent or tetravalent. If the crystal field is stronger than the Coulomob interaction, the electron spin of a transition metal ion does not obey Hund's rules and a low spin state is realized. In a CN salt like KCP(X) a strong crystal field induces the nonmagnetic Pt^{2+} $(5d)^8$ and Pt^{4+} $(5d)^6$ ground states. So a magnetic state is expected only in the excited state, and assuming an energy difference between nonmagnetic ground state and the excited magnetic state Δ, one can write down the magnetic susceptibility χ at low temperatures (much lower than Δ/k) approximately as follows

$$\chi = \frac{C}{T} \exp\left(-\frac{\Delta}{T}\right) \quad . \tag{5.12}$$

The spin susceptibility obtained by ESR (cf. Fig. 5.12) is plotted in Fig. 5.30 and found to be well fitted by (5.12) and the activation energy Δ has a value between 500 K and 700 K [5.41]. The Curie constant C was determined to be $0.98 \pm 0.20\,\mathrm{J}T^{-2}\,\mathrm{mol}^{-1}\,\mathrm{K}^{-1}$ (0.098 ± 0.020 emu mol K^{-1} [5.20]), this value nearly equals to the value expected for the Pt^{3+} excited state of $1.12\,\mathrm{J}T^{-2}\,\mathrm{mol}^{-1}\,\mathrm{K}^{-1}$ (0.112 emu mol K^{-1}) within experimental error.

The electronic state of Pt^{2+} or Pt^{4+} corresponds to a full or a vacant $5d_{z^2}$ orbital respectively, so the electrons in the ground state cannot contribute to the electrical conductivity. The electrical conduction mechanism

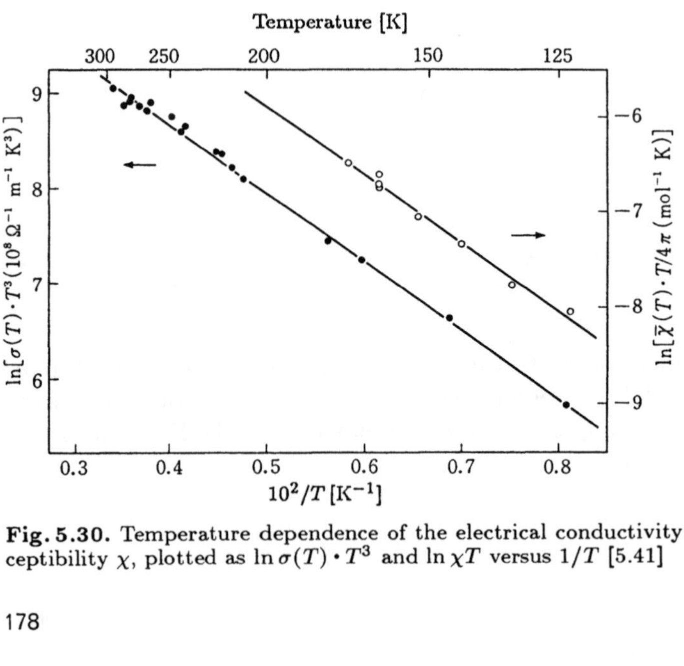

Fig. 5.30. Temperature dependence of the electrical conductivity σ, and magnetic susceptibility χ, plotted as $\ln \sigma(T) \cdot T^3$ and $\ln \chi T$ versus $1/T$ [5.41]

is expected through the thermally excited Pt^{3+} state. The temperature dependence of conductivity shown in Fig. 5.4, is plotted as $\ln[\sigma(T)T^3]$ versus $1/T$ in Fig. 5.30. One can see that these results show nearly the same temperature dependence so they have the same activation energy. The conductivity can be written as [5.41]

$$\sigma(T) = \frac{A}{T^3} \exp\left(-\frac{\Delta}{T}\right) \quad . \tag{5.13}$$

Equation (5.13) means that the term $\exp(-\Delta/T)$ corresponds to the term of the carrier density and the A/T^3 term reflects the temperature dependence of the mobility. The mobility of KCP(Br) at room temperature is estimated to be order of $10^{-4}\,m^2V^{-1}\,s^{-1}$ and in this sense, the transport properties of KCP(Br) are intermediate between the hopping model and the metallic band model. On the basis of the mixed valence model of Pt^{2+} and Pt^{4+}, a schematic energy level diagram is shown in Fig. 5.31.

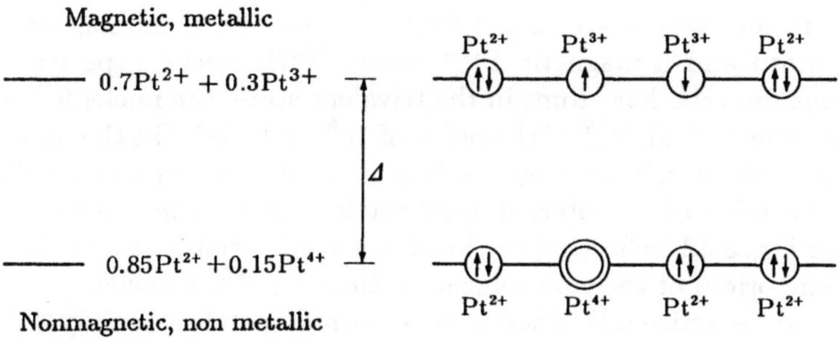

Fig. 5.31. Energy levels of mixed valence model in KCP(Br) [5.64]

In Sect. 5.4.3, the Knight shift of ^{195}Pt nuclei is shown to be a quantity that reflects the spin susceptibility of KCP(Br); it is found that the resonance shift is given as

$$K(T) = K_0 + \frac{C}{T}\frac{e^{-\Delta/T}}{1+e^{-\Delta/T}} \quad . \tag{5.14}$$

In this equation, $\Delta = 700\,K$ and K_0 and C are temperature independent parameters. The observed experimental results can be well reproduced as the dotted line in Fig. 5.13.

At low temperature, far below Δ, the electrons is rarely excited to Pt^{3+} state and the valence state of Pt^{2+} or Pt^{4+} ion are nearly stable. Therefore, in the case of no valence fluctuation, we can expect that the different valence states of Pt^{2+} and Pt^{4+} should be observable as a difference of the chemical shift of ^{195}Pt NMR. Unfortunately the spin-lattice relaxation time of ^{195}Pt

NMR in KCP(Br) at 4.2 K becomes as long as several minutes [5.42], so it is not possible to observe the ^{195}Pt NMR signal with a good signal to noise ratio. Using the Mössbauer effect an attempt has been made [5.61] to detect the different valence states of Pt ions from the measurement of the isomer shift. But the halfwidth of resonant γ-ray absorption was found to be much larger than the difference of isomer shift, so this experiment also failed to produce useful results.

5.7.3 Nuclear Spin Relaxation of ^{195}Pt Nuclei

According to the mixed-valence model, of which the schematic energy levels are shown in Fig. 5.31, one can see that an electron transfer from Pt^{2+} ion to Pt^{4+} ion causes the presence of two Pt^{3+} ions but that this process requires an excitation energy. For the thermal excitation of the Pt^{2+} and Pt^{4+} into the Pt^{3+} state, the thermal attempt frequency of the electronic state of Pt ions is given as $\nu_{\text{thermal}} \sim kT/h$ and the probability c of an electron producing the excited Pt^{3+} ions is found to be $c/(1-c) = \exp(-\Delta/T)$. If we watch a certain Pt ion, we can see that it fluctuates between a nonmagnetic Pt^{2+} or Pt^{4+} state and a magnetic Pt^{3+} state. ^{195}Pt nuclei experience these fluctuations directly. For atoms in the trivalent state, the nuclei feel a large internal magnetic field H_{hf}^{5d} of the order of 10^6 Oe [5.62]. On the other hand, the Pt nuclei see nearly no magnetic field in divalent or tetravalent Pt ions. The time variation of the internal magnetic field at a nucleus is shown schematically in Fig. 5.32, reflecting the local valence fluctuation; this time dependence is the origin of the spin-relaxation time for ^{195}Pt nuclei.

The nuclear spin-lattice relaxation time of normal metals is mainly due to the Korringa mechanism. The observed $1/T_1$ of ^{195}Pt nuclei in KCP(Br) is about a factor of two larger than that calculated using the Korringa relation. But according to the valence fluctuation model, the nuclear spin relaxation time T_1 or T_2 is given by Eq. (5.15) [5.63, 64] using the thermal attempt frequency $1/\tau_{\text{thermal}}$ and the random time variation of local hyperfine field (Fig. 5.32).

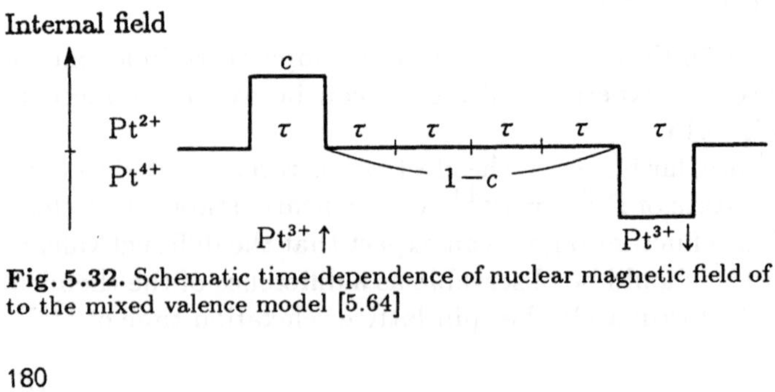

Fig. 5.32. Schematic time dependence of nuclear magnetic field of ^{195}Pt nuclei according to the mixed valence model [5.64]

$$\frac{1}{T_1} \sim \frac{1}{T_2} = 0.3(\gamma_N H_{hf}^{5d})^2 \frac{\hbar}{k_B T} \frac{e^{-\Delta/T}}{1+e^{-\Delta/T}} \quad . \tag{5.15}$$

Then, using (5.15) and $H_{hf}^{5d} = -1.2 \times 10^6\,\mathrm{Oe}\,(-9.5 \times 10^7\,\mathrm{Am^{-1}})$ which was obtained for Pt metal, the spin-lattice relaxation rate $1/T_1$ shown as the solid line in Fig. 5.14 can be calculated without any adjustable parameters [5.64]. On the other hand, two experimental results, (1) the Lorentzian resonance line shape and (2) the temperature dependence of the resonance linewidth, defined as $2/T_2$, are understood in the temperature range of the metal-nonmetal transition above 120 K. We show the calculated temperature dependence of ^{195}Pt NMR linewidth according to (5.15) as the solid line in Fig. 5.15 [5.64]; the agreement is satisfactory.

5.7.4 Electron Spin Relaxation

We have seen in Sect. 5.2, that the ESR linewidth of KCP(Br) shows a T^3 temperature dependence and a characteristic angular dependence [5.37]. Here on the basis of the mixed valence model, we shall discuss the problem of the ESR resonance linewidth. The resonance linewidth is given as two terms, one comes from the spin-lattice relaxation $1/T_2^{\text{s-p}}$ and the other from the spin-spin interaction $1/T_2^{\text{s-s}}$ as shown in the following

$$\frac{1}{T_2} = \frac{1}{T_2^{\text{s-p}}} + \frac{1}{T_2^{\text{s-s}}} \quad . \tag{5.16}$$

A conduction electron in the metal is scattered by a lattice vibration, and at the same time spin flip events occur with a certain probability. This is the so-called Elliot mechanism [5.65] for $1/T_2^{\text{s-p}}$ term. In 3-dimensional metals, this contribution is usually dominant, but in KCP(Br) we should take into consideration the one-dimensionality of system. In the other contribution to the spin-spin relaxation, this dimensionality effect also plays an important role.

a) Spin-lattice Relaxation

The observed g value of the electron spin resonance in KCP(Br) is different from 2, and has a large anisotropy. This result indicates that the spin-orbit interaction is large in this mixed valence system. Thus the Elliot mechanism [5.65] which is known as the most important spin-lattice relaxation mechanism for conduction electron spin resonance in metals, is expected as follows

$$\frac{1}{T_2^{\text{s-p}}} = a\frac{(\lambda/\Delta E)^2}{\tau_{tr}} \quad . \tag{5.17}$$

On the basis of the experimental results of the g value, $\lambda/\Delta E$ is determined as 0.068. Here τ_{tr} is the scattering time of a conduction electrion, a is a numerical constant – whose value is about 30 in the case of 3d alkali metals.

In an ideal 1d system, the electron motion is limited to motion along the chain axis, so electron scattering can occur only in backward or forward directions. Scattering accompanied by spin flipping cannot be realized, since it breaks the time reversal symmetry. Therefore, the relaxation from the Elliot mechanism can contribute only when the electron can move in 3d.

In 3d the scattering time $\tau_{\mathrm{tr}}^{3\mathrm{d}}$ is defined as follows

$$\frac{1}{\tau_{\mathrm{tr}}^{3\text{-}\mathrm{d}}} = \frac{\tau_{\parallel}}{\tau_{\perp}} \frac{1}{\tau_{\mathrm{tr}}^{1\mathrm{d}}} \quad . \tag{5.18}$$

The quantities $\tau_{\parallel}$ and $\tau_{\perp}$ are the hopping times of electrons parallel and perpendicular to the chain respectively, and by using the transfer integral $t_{\parallel}$ or $t_{\perp}$ and the electron density of states $\varrho(E_{\mathrm{F}})$, they are given by

$$\frac{1}{\tau_{\parallel}} = \frac{2\pi}{\hbar} |t_{\parallel}|^2 \varrho(E_{\mathrm{F}}) \quad ; \quad \frac{1}{\tau_{\perp}} = \frac{2\pi}{\hbar} |t_{\perp}|^2 \varrho(E_{\mathrm{F}}) \quad . \tag{5.19}$$

On the other hand $\tau_{\mathrm{tr}}^{1\mathrm{d}}$ has the following relation with the conductivity along the chain,

$$\sigma_{\parallel} = \frac{ne^2}{m^*} \tau_{\mathrm{tr}}^{1\mathrm{d}} \tag{5.20}$$

where m^* is the effective mass of the electrons. According to the discussion in Sect. 5.7.2, the experimental conductivity along the 1d axis is found to obey the relation $\sigma_{\parallel} \propto \exp(-\Delta/T)/T^3$ [5.3, 64]. The number of electrons, n, participating in the electrical conduction is usually given by $n \sim n_0 \exp(-\Delta/T)$, and the scattering time $\tau_{\mathrm{tr}}^{1\mathrm{d}}$ is proportional to T^3. Therefore, the linewidth of EPR is also expected to be proportional to T^3.

On this basis of the experimental results, shown in Sect. 5.7.2, and assuming $n_0 = 0.3/\mathrm{Pt}$ ion and that the effective electron mass m^* equals the free electron mass, we obtain

$$\tau_{\mathrm{tr}}^{1\mathrm{d}} = \frac{3.2 \times 10^{-7} [\mathrm{s} \cdot \mathrm{K}^3]}{T^3} \quad . \tag{5.21}$$

Furthermore with the use of (5.16, 17, 21), the following equations were obtained [5.37]

$$\frac{1}{T_2^{\mathrm{s\text{-}p}}} = AT^3 \quad , \quad A = 3.1 \times 10^6 \, a \frac{\tau_{\parallel}}{\tau_{\perp}} \left(\frac{\lambda}{\Delta E} \right)^2 \quad . \tag{5.22}$$

b) Spin-Spin Relaxation

In KCP(Br) the distance between two 1d chains is much larger than the interatomic distance between Pt atoms within the chain. So the magnetic interaction between Pt^{3+} ions only has appreciable magnitude for ions within the same chain. The magnetic interaction is considered to be the magnetic dipolar interaction from neighbouring electron spins. The second moment of magnetic dipole interaction is given as follows for an anisotropic g factor [5.66]

$$\langle \Delta\omega^2 \rangle = \frac{1}{\hbar^2}\frac{s(s+1)}{3}\left(\frac{2g_{\parallel}^2+g_{\perp}^2}{2}\mu_{\mathrm{B}}^2\right)^2 \sum r_{ij}^{-6}\left(3\frac{g_{\parallel}^2}{g^2}\cos^2\theta-1\right)^2 \tag{5.23}$$

$$g^2 = g_{\parallel}^2\cos^2\theta + g_{\perp}^2\sin^2\theta \quad . \tag{5.24}$$

Here s is the electronic spin which equals $\frac{1}{2}$ in this case. So long as we are only concerned with the dipole interaction within the same chain, the vector $\boldsymbol{r}_{ij}$ between spins is parallel to the c axis. Therefore, the angle between $\boldsymbol{r}_{ij}$ and the magnetic field has a single value θ. Putting the observed g value into (5.23, 24), we can expect the minimum EPR linewidth at $\theta = 49.5°$ in the neighbourhood of the magic angle 54.7°. In Fig. 5.33, the observed angular dependence of EPR line width and the calculated value according

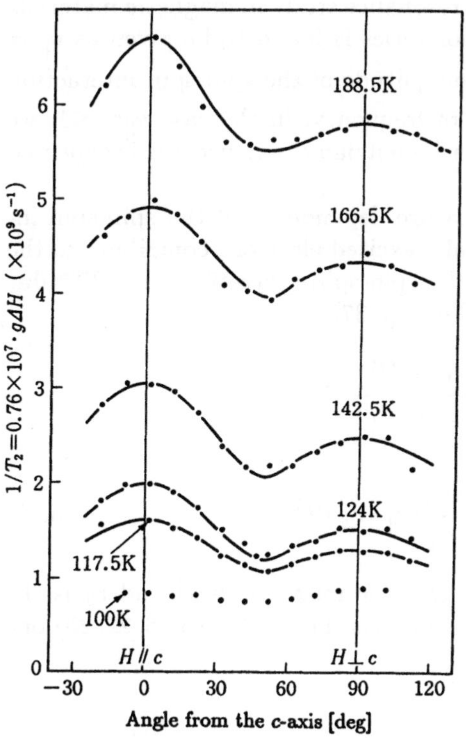

Fig. 5.33. Angular dependence of $1/T_2$ for the EPR line in KCP(Br), the solid line in the figure is calculated from (5.27) [5.37]

to (5.23) are shown by the dotted and solid lines respectively. According to the discussion of the NMR relaxation, the correlation time τ_c of the valence fluctuations in KCP(Br) is much shorter than the period of the magnetic dipole interaction between electron spins. So the extreme narrowing condition of the resonance line is fulfilled and the resonance linewidth is given as $\langle \Delta\omega^2 \rangle \tau_c$.

In a pure 1d system, the electron motion within a chain is limited to two directions – "up" and "down", and the probability that it returns to the original position is non-zero even after a long time and the so-called long-time tail effect is expected. In this case the extreme motional narrowing effect does not occur and the observed resonance line shape is of a non-Lorentzian type [5.67].

But in the case of KCP(Br), the observed resonance line shape is in fact Lorentzian in a wide magnetic field range, and so this cannot be taken as a pure 1d case. Therefore it becomes necessary to consider this system from the view point of the 3d model, in which electron hopping between chains is possible. In this case, the half linewidth η is given by

$$\eta = \langle \Delta\omega^2 \rangle (\tau_{\parallel} \cdot \tau_{\perp})^{1/2} \tag{5.25}$$

using the hopping frequency between chains, $1/\tau_{\perp}$ [5.68]. Here it is necessary to determine whether the present system is quasi 1d or 3d. This problem is discussed extensively in the magnetic resonance study in magnetic materials [5.68]. The parameter of the dimensionalities is found to be given as $\nu_{\parallel} = (1/2\pi)\sqrt{\langle \Delta\omega^2 \rangle}$, where $\nu_{\parallel}$ gives the magnitude of the spin-spin interaction within the same chain as a function of frequency. In the case $\nu_{\parallel}\tau_{\perp} \lesssim 1$, we can consider the system to be 3d and a Lorentzian resonance line is expected [5.68].

The next problem is the temperature dependence of the spin-spin interaction. Assuming that only thermally excited electrons contribute to the magnetic moment, we can estimate the spin-spin relaxation time $T_2^{\text{s-s}}$ due to the dipole-dipole interaction as follows [5.37]:

$$\frac{1}{T_2^{\text{s-s}}} = B\left[3\left(\frac{g_{\parallel}}{g}\right)^2 \cos^2\theta - 1\right]^2 \frac{\mathrm{e}^{-\Delta/T}}{1+\mathrm{e}^{-\Delta/T}}$$

$$B \equiv \frac{1}{\hbar^2}\frac{s(s+1)}{3}\left(\frac{2g_{\parallel}^2 + g_{\perp}^2}{2}\mu_{\mathrm{B}}^2\right)^2 \frac{0.6}{r_0^6}(\tau_{\parallel} \cdot \tau_{\perp})^{1/2} \tag{5.26}$$

where, θ is the angle between c axis and the magnetic field and r_0 is the nearest-neighbour distance c' between Pt ions. From Eqs. (5.16, 22, 26) one obtains

$$\frac{1}{T_2} = \frac{1}{T_2^{\text{s-p}}} + \frac{1}{T_2^{\text{s-s}}}$$

$$= AT^3 + B\left[3\left(\frac{g_{||}}{g}\right)^2 \cos^2\theta - 1\right]^2 \frac{e^{-\Delta/T}}{1+e^{-\Delta/T}} \quad . \tag{5.27}$$

Parameters A and B were determined experimentally as $A = 0.80 \times 10^3$ $[s^{-1}K^{-3}]$ and $B = 5.6 \times 10^9$ $[s^{-1}]$. The calculated values are shown as the solid lines in Fig. 5.33 and the characteristics of the experimental results are well reproduced by (5.27). On the basis of the observed g value and the two values, $\tau_{||}/\tau_{\perp}$ and $(\tau_{||} \cdot \tau_{\perp})^{1/2}$ are obtained as

$$\frac{\tau_{||}}{\tau_{\perp}} = 1.8 \times 10^{-3} \tag{5.28}$$

$$(\tau_{||} \cdot \tau_{\perp})^{1/2} = 7.2 \times 10^{-11}[\text{s}] \quad . \tag{5.29}$$

Here a is taken to be 30, as Elliot estimated for the case of alkali metals. The ratio $\tau_{||}/\tau_{\perp}$ is related to the dimensionality; small values correspond to the 1d case and values near to one to the 3d case. From the relation $\tau_{||}/\tau_{\perp} = (t_{\perp}/t_{||})^2$ deduced from (5.19). $(t_{\perp}/t_{||})_{\text{KCP}}$ is determined to be 0.043. This value accidentally coincides with that of 0.042 for TTF-TCNQ obtained by nuclear magnetic resonance [5.69]. Therefore, one can expect the degree of one-dimensionality of KCP(Br) is comparable with that of TTF-TCNQ. From (5.28, 29) $\tau_{||}$ and $\tau_{\perp}$ are determied as

$$\tau_{||} = 3.2 \times 10^{-12}\,\text{s} \quad , \quad \tau_{\perp} = 1.8 \times 10^{-9}\,\text{s} \quad . \tag{5.30}$$

Here we check the self-consistency conditions by using (5.30). The product $\nu_{||}\tau_{\perp}$ is found to have values 0.19 and 1.0 at 100 K and 200 K respectively, and it satisfies the three-dimensionality condition $\nu_{||}\tau_{\perp} \lesssim 1$. Equation (5.30) also satisfies the extreme narrowing condition of the resonance line $\sqrt{\langle \Delta\omega^2 \rangle \tau_{||} \cdot \tau_{\perp}} \ll 1$.

Next, on the basis of the inter-chain hopping time obtained by this analysis of the EPR linewidth, the electric conductivity perpendicular to Pt chain axis is discussed. In the case of the diffusive hopping conductivity, $\sigma_{\perp}$ is given by

$$\sigma_{\perp} = \frac{n a_0^2 e^2}{\tau_{\perp} k_B T} \quad , \tag{5.31}$$

where the interatomic distance $a = 9.87$ Å. The conductivity perpendicular to the 1d chain of KCP(Br) is calculated as $1.1 \times 10^{-4}\,\Omega^{-1}\,\text{cm}^{-1}$, and

$2.8 \times 10^{-3}\,\Omega^{-1}\,\text{cm}^{-1}$ at 100 K and 250 K respectively. These calculated values nearly coincide with the observed values [5.16] of $0.6 \times 10^{-4}\Omega^{-1}\text{cm}^{-1}$ and $2.7 \times 10^{-3}\,\Omega^{-1}\,\text{cm}^{-1}$ in spite of being only a rough estimate.

As shown above, the EPR line width consists of two terms, an isotropic term due to spin-lattice relaxation, and an anisotropic term due to spin-spin relaxation. The former corresponds to relaxation by the Elliot mechanism and has a strong correlation with the electric conductivity [5.37, 64].

5.7.5 Neutron Diffraction

In Sect. 5.5 the experimental neutron diffraction results have been discussed in terms of a 1d metallic band. Here we discuss whether or not the valence-fluctuation model of the mixed-valence states can explain the experimental results of the structural analysis.

At first, for simplicity, we consider this problem for $T = 0$, where there are no valence fluctuations. In this case, the XPS results suggest that 85 % of the Pt ions are divalent, and the remaining 15 % tetravalent. If we take into consideration the Coulomb interaction between electrons, tetravalent Pt^{4+} ions are expected to be distributed equidistantly. The average distance between tetravalent Pt^{4+} ions is expected to be $c'/0.15 = 6.67c' = 3.33c$. This periodicity in real space agrees with the wave number $2k_F = 2\pi/3.33c = 0.3c^*$ as discussed in Sect. 5.5 on the basis of the band picture. If the ionic radius differs in Pt^{2+} and Pt^{4+} ions, the arrangement of Pt^{2+} and Pt^{4+} ions is expected to be the origin of the modulation of the lattice periodicity with the wave number $2k_F$. The scattering peak of $Q = 0.3c^*$, $\omega = 0$ can be observed by the elastic neutron scattering or other diffraction methods.

In fact, the ionic radii of Sm^{2+} and Sm^{4+} ions differ by about 20 % in the mixed-valence state of the Sm compound. In this system, *Anderson* et al. [5.70] pointed out two possibilities; one is that the lattice distortion moves accompanying the valence fluctuations. The other is that the lattice distortion cannot catch up with the velocity of valence fluctuations, and then one can make use of the adiabatic approximation. In the other extreme slow valence fluctuations, the static approximation may be adopted. The metal-nonmetal transition in Sm compounds is discussed on the basis of the static approximation [5.70].

In the KCP(Br) system, the valence fluctuations are not so fast, and so the electron can move with the lattice distortion and a modulation of the lattice constant is expected corresponding to the periodic ordering of the valency. This phenomenon apparently coincides with the Peierls distortion at $T = 0$.

Next, we consider the problem at finite temperature. According to this model, the Pt^{3+} excited state is realized with a probability of

$$\mathrm{Pt}^{3+}(T) = \frac{e^{-\Delta/T}}{1+e^{-\Delta/T}} \quad . \tag{5.32}$$

The Pt^{3+} state changes its position with the hopping motion of electron, and thus Pt^{2+} or Pt^{4+} ions also move from one site to another by replacing Pt^{3+} ions. These electron hopping motions of the excited state within the Pt chain are the origin of the electric conductivity along the c axis. As yet however we have too little knowledge to elucidate this interesting conduction mechanism.

We list below some experimental results of neutron diffraction that can only be understood by the mixed-valence model:

i) The scattering peak at $Q = 0.3c^*$, $\omega = 0$ at room temperature

The excitation energy Δ for producing trivalent Pt ions is estimated to be about 600 K and thus the probability of an electron excited to Pt^{3+} at room temperature is about 0.1 accordig to (5.32). So the remaining 90 % the Pt ions are divalent or tetravalent at room temperature. ESR results have shown that the electron hopping time over one atomic distance, $\tau_{\|}$, is of the order of 10^{-12} s. Within this time longitudinal phonons can propagate along the c axis by only a few atomic distances, and therefore the valence fluctuation propagates accompanying the local lattice distortion in KCP(Br) as discussed previously for Sm compounds.

Neutrons see a nearly static Pt ionic distribution. At room temperature there remains ionic ordering of Pt^{2+} and Pt^{4+} ions, although they are slightly disturbed by the existence of Pt^{3+} ions. Therefore, one can expect a neutron sattering peak of wave number $Q = 0.3c^*$ and $\omega = 0$ at room temperature.

The detailed experimental results of the inelastic neutron scattering with constant-E scanning are shown in Sect. 5.5. A strong scattering peak was observed with wave vector $Q = 0.3c^*$, $\omega = 0$ at high temperature. The origin of this peak could not be understood by the band model as described previously, but according to the mixed-valence model one has a natural explanation for the experimental results.

ii) Strong elastic incoherent scattering

We consider the scattering of the incident neutron along the 1d c axis. The fluctuation of the lattice distortion through the motion of the trivalent Pt ions by electron hopping is statistically random. The disturbance of the periodicity broadens the scattering peaks with wave number $Q = 0.3c^*$. This random disturbance also contributes to increase the incoherent scattering. In fact, the raw experimental results [5.49, 52] of the Brookhaven group showed the existence of a large incoherent scattering component. In their analysis however, they neglected the large incoherent term and discussed only on the coherent neutron scattering as shown in Sect. 5.5.4.

iii) Inelastic scattering at $Q = 0.3c^*$ and the Kohn anomaly

According to the mixed-valence model, in which no conduction electrons exist, the softening of the $2k_F$ phonon, i.e. the Kohn anomaly, cannot be expected. But as shown in Sect. 5.5, the neutron scattering results (Fig. 5.31) clearly show a softening of the longitudinal phonon at $Q = 0.3c^*$ along the c axis, which corresponds to the Kohn anomaly. The wavelength of the $Q = 0.3c^*$ longitudinal phonon corresponds to the inter-ionic distance between tetravalent Pt ions. This periodic valence ordering of Pt ions implies the existence of a charge density wave. In terms of the electron-phonon interaction, the phonon with wave vector $Q = 0.3c^*$ couples with the static charge density wave of Pt^{2+} or Pt^{4+} ions and the phonon softening occurs at the wave vector $Q = 0.3c^*$. Therefore a phenomenon, similar to the Kohn anomaly is expected. But to what frequency the phonon will soften is not a simple problem. A more detailed knowledge of the valence fluctuations is essential for further progress. On the experimental side, the unpublished incoherent neutron scattering data at room temperature should be more extensively analyzed.

5.7.6 Raman Scattering

Assuming that a Pt ion can be either divalent or tetravalent at $T = 0$, then its neighbouring ions will experience a different interaction depending on its valence state. For example, the frequency of the CN stretching mode differs for the divalent or tetravalent Pt ions and thus the observed fine structure in the Raman scattering spectra may be understood. The fine structure of Raman spectra [5.30–32] were presented in Table 5.3. Two possibilities were suggested for the origin of this fine structure: one is the above-mentioned mixed-valence model and the other is the existence of two non equivalent CN bases depending on the distribution of Br^-, or H_2O molecules. But we have too little experimental evidence to solve this problem as yet – further studies are needed.

5.7.7 The Mixed-Valence State of Pt Ions

In this chapter, on the basis of the results of XPS experiments, we proposed the fluctuating mixed-valence model and tried to interpret the observed experimental results for KCP(Br) quantitatively. We have clearly found that various kinds of experimental results, which could not be understood by the simple 1d band model, can be successfully interpreted in terms of the valence-fluctuation model.

Here we return to the starting point of the mixed-valence model and discuss why divalent Pt^{2+} and tetravalent Pt^{4+} ions should be stable in KCP(Br). This problem of the stability of divalent and tetravalent Pt ions occurs not only in KCP(Br), but also throughout the Pt compounds in the field of inorganic chemistry. But it is not easy to understand the physical origin of these two stable valence states. The stability of the mixed-valence state is now being studied extensively and a decisive answer remains to be found.

On the mixed valence problem of KCP(Br), *Efetov* and *Larkin* [5.72] and *Sawatzky* and *Antonides* [5.60] proposed the existence of a large negative intra-atomic Coulomb interaction U in the Hubbard Hamiltonian. A large negative U means the existence in real space of strong attractive forces between the electrons. *Anderson* [5.73] showed that in the extreme limit of strong electron-phonon interactions, the attractive force between two electrons overcomes the on-site Coulomb repulsion, one can expect the spin singlet ground state of electron pairing in real space. The concept of on-site electron pairing, that is the negative U, has been applied to amorphous silicon by *Anderson* [5.73] and to $BaPb_{1-x}Bi_xO_3$ by *Rice* and *Sneddon* [5.74]. In KCP(Br), two electrons of Pt^{2+} occupy d_{z^2} orbitals and the strong long-range Coulomb repulsion between the rest of the Pt^{4+} ions makes Wigner crystallization [5.75] possible. So Pt^{4+} ions are arranged equidistantly at $6.67c'$ as discussed before. In Fig. 5.34, the arrangement of Pt^{4+} and Br^- ions is shown schematically. One can see that in the arrangement with wave number $2k_F$ for Pt^{4+}, $4k_F$ is expected for Br^-. If a charge density wave with large amplitude exists as discussed above, the higher harmonics of $2k_F$ wave vector are naturally expected. But these harmonics of the $2k_F$ mode could not be observed in KCP(Br), since the displacement along the c axis

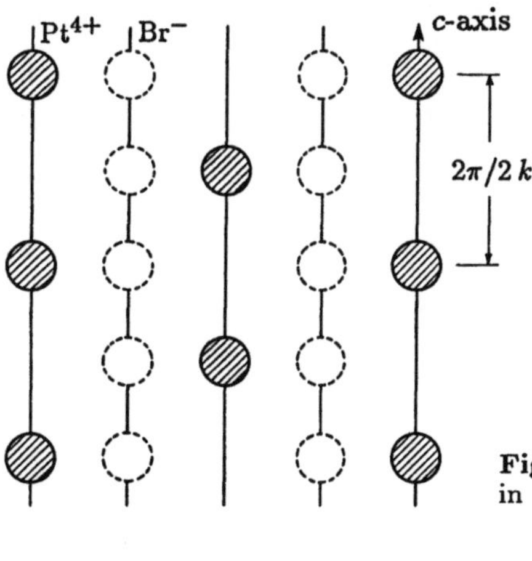

Fig. 5.34. Arrangement of Pt^{4+} and Br^- ions in KCP(Br)

is as small as 1 %. On the other hand, up to the third harmonics were observed in $K_{1.81}[Pt(C_2O_4)_2] \cdot 2H_2O$, $Rb_{1.67}[Pt(C_2O_4)_2] \cdot 1.5H_2O$ in which a large modulation of the interatomic distance of ~7 % along the c axis is found [5.11].

5.7.8 Comments on the Mixed-Valence Model

In Sects. 5.2–5, the experimental results were discussed on the basis of a 1d $5d_{z^2}$ band having 0.3 holes for each Pt ion. On the other hand in Sect. 5.7, we considered these experimental results again, in terms of the valence-fluctuation model in the mixed-valence state, where Pt ions are stable in both the divalent and tetravalent states and where the trivalent excited state plays an important role in the transport and magnetic properties of KCP(Br). These two models are in strikig contrast and correspond almost to the extreme limits of itinerancy and localization of electrons.

It is perhaps better to judge the models mainly accordig to their ability to explain the physical properties of KCP(Br), and not to judge which one is right or wrong. It would seem at present that both models are required for more quantitative and more detailed discussions.

Finally, the most important future problem for the "fluctuating mixed-valence model" is to elucidate the mechanism of electron transfer. The experimental results for electrical conductivity show the existence of a temperature dependent scattering proportional to T^3. This temperature dependence is not well understood. but it may feasibly be due to electron scatterig by lattice vibrations in quasi 1d KCP(Br), where the amplitude of lattice vibrations along the chain axis is much larger than in other directions.

The elucidation of the electron transfer mechanism (or the valence fluctuation) in systems with strong electron-phonon coupling is an important problem not only in KCP(Br) but also in other mixed-valence materials. Further detailed studies – both theoretical and experimental – are needed.

6. Properties of the Linear Chain Polymers $(CH)_x$ and $(SN)_x$

It is almost obvious that a linear-chain polymer should show an electronic one-dimensionality provided it is a good conductor. Most polymers, however, have a quite high resistivity; for example, polyethylene is used as a good electrical insulator. The reason for this is the lack of conduction electrons in these polymers. A carbon atom in polyethylene, $(CH_2)_x$, has four electrons in its outer shell; two of them are used to form bonds with hydrogen atoms and the remaining two form the σ-bonds between carbon atoms as shown in Fig. 6.1a. These electrons cannot move freely. One may expect that the polymer $(CH)_x$, in contrast to $(CH_2)_x$, has a free electron on each carbon atom. Moreover a polymer consisting of carbon atoms only, C_x, will have two excess electrons which are not used in the σ-bonding.[1] These excess electrons enter π-orbitals making a double bond between carbon atoms as shown in Fig. 6.1b and c. Another drawing, Fig. 6.1d, is also possible for the structure of $(CH)_x$. It is well known that the two states (Fig. 6.1b, d) form a resonance. In other words, the π-orbitals overlap along the chain giving rise to a π-band. In principle, any polymer can become metallic as long as it has π-electrons.

Among the many kinds of polymeric conductors, $(CH)_x$ and $(SN)_x$ have been investigated in the greatest detail. We shall make here a general survey of their properties.

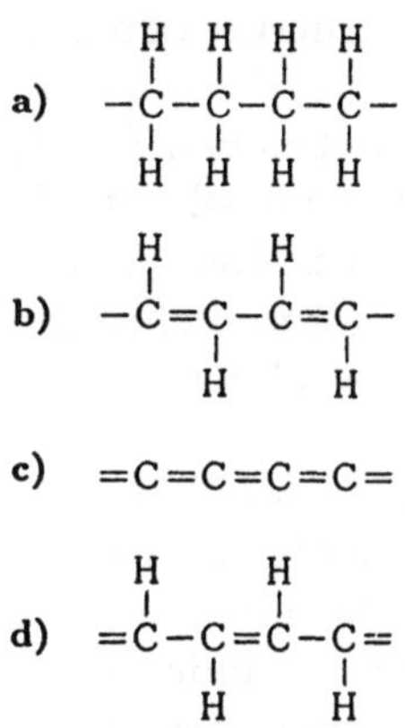

Fig. 6.1a–c. Atomic bonding in polymers of hydrocarbons. **(a)** Polyethylene $(CH_2)_x$. **(b)** Polyacetylene (correct name; polyvinylene) $(CH)_x$. **(c)** Carbyne C_x

[1] Properties of this material called carbyne are now being investigated.

6.1 Properties of $(CH)_x$

The linear-chain polymer $(CH)_x$ is usually called polyacetylene since it is obtained by polymerization of acetylene. Its correct name, however, is polyvinylene; the unit of $(CH)_x$ is $-CH=CH-$ and not CH. Sometimes it is called polyene, which is a general name for polymers consisting of hydrocarbon. The molecular weight of $(CH)_x$ is estimated to be 6000–8000; the value of x is 400–600 [6.1].

$(CH)_x$ is an intrinsic semiconductor. Doping with donors and acceptors which lie between the carbon chains can make $(CH)_x$ a p-type or an n-type semiconductor, respectively. The maximum content of dopants can reach 20 %. $(CH)_x$ shows metallic properties when highly doped. When the doping level is low, electrical conduction is believed to be well described in terms of a soliton model. $(CH)_x$ is expected to find application as a plastic semiconductor in engineering applications and many industries are trying to make electronic devices from $(CH)_x$. From a physical point of view, the soliton behaviour is very interesting.

6.1.1 Synthesis of $(CH)_x$ and Its Structure

When acetylene gas is introduced onto a dense solution of Ziegler-Natta's catalyst, which is frequently used in polymerization of acetylene, films of $(CH)_x$ grow on the surface of the solution. A powder form of $(CH)_x$ is obtained by the use of this catalyst [6.2]. Films of $(CH)_x$ were first made by *Shirakawa* and *Ikeda* [6.3]; they used a solution of the catalyst much denser than that used conventionally. Their success in obtaining films of $(CH)_x$ made it possible to investigate the physical and chemical properties of this material in detail. It is instructive to note that a success in material synthesis led directly to a new stage of research as was also the case with TTF-TCNQ. The photograph in the frontispiece shows the microscopic structure of a film of $(CH)_x$ made by *Shirakawa* et al.

Figure 6.2 shows the molecular structure of $(CH)_x$. It has two forms, the cis- and the trans-form. The trans-form is thermodynamically stable [6.4, 5]. The cis-form is obtained when the polymerization is done at low temperatures, for example −78° C. It turns into the trans-form upon heating up to about 180° C. When the polymerization takes place at about 150° C, the trans-form grows from the beginning.

The cis-form can be further classified into two; the cis-transoid and the trans-cisoid. Raman scattering suggests that only the former is obtained in the present procedure of polymerization [6.4]. Crystals of the cis-form have an orthorhombic symmetry and those of the trans-form are triclinic. It has also been argued that the trans-form has orthorhombic symmetry; exact analysis of the crystal structure is difficult because the bulk film is not a

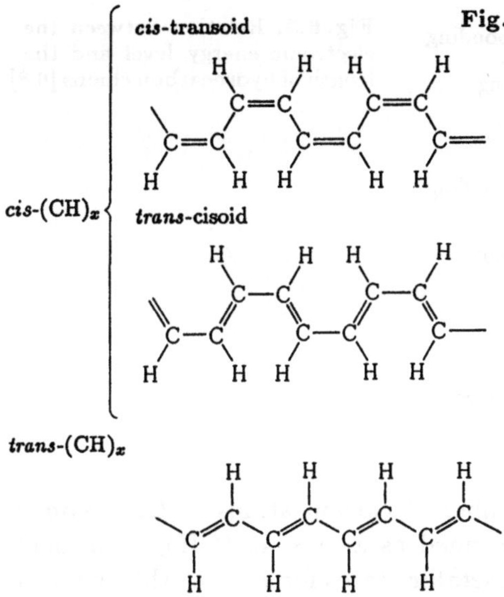

Fig. 6.2. Cis- and trans-form of $(CH)_x$ [6.4, 5]

single crystal [6.6]. Nevertheless it is sure that films of $(CH)_x$ are composed of CH chains aligned parallel to each other. The crystal forms a fibre with thickness of about 200 Å. A macroscopic film is composed of entangled fibres. About 50 % of the volume of the macroscopic films is occupied by $(CH)_x$, and the remainder is empty [6.3] (see photograph 2 in the frontispiece).

A film of the *cis*-$(CH)_x$ can be stretched by placing it under tension; the length reaches 2–2.5 times the original length. The fibres align nearly parallel in stretched films. On the other hand, the trans-form can hardly be stretched [6.7].

One can introduce acceptors such as halogen and AsF_5 or donors such as the alkali metals into $(CH)_x$. The material can be doped with nearly thirty kinds of acceptor, but donor-doping is rather difficult.

No solvent has yet been found for $(CH)_x$. Hence its molecular weight is estimated through measurements made on hydrogenated samples, which are soluble [6.1].

6.1.2 One-Electron States in the Band

We shall consider first electronic states in short-chain hydrocarbons before moving on to discuss those in $(CH)_x$ [6.8]. Figure 6.3a shows the structural formula of ethylene, C_2H_2, and the energy levels for the π-electrons in it. Two-electron wave functions form a bondong and an anti-bonding state; the former, since it has the lower energy, is occupied by two electrons. In case of butadiene, C_4H_6, two states lower in energy admit four π-electrons as shown in Fig. 6.3b. The separation in energy between neighbouring states

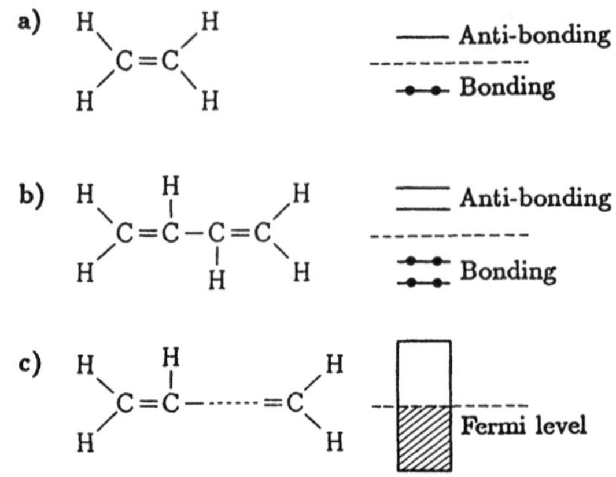

Fig. 6.3. Relation between the electronic energy level and the length of hydrocarbon chains [6.8]

becomes smaller with increasing number of carbon atoms in the chain; a band as shown in Fig. 6.3c will be formed as $x \to \infty$ in $(CH)_x$. Although one might expect the appearance of metallic states for $x \to \infty$, this is not in fact the case; the bond between carbon atoms has an intermediate nature between the double and the single bonds in the metallic state. If these bonds are formed alternately along the chain, a band gap appears at the Fermi level marked by the dashed line in Fig. 6.3c. The presence of the band gap causes a decrease in the electronic energy for a similar reason to that responsible for the Peierls transition. Thus, $(CH)_x$ for $x \to \infty$ becomes an intrinsic semiconductor having a band gap at the Fermi level. It has been pointed out that the mechanism for this is not the Peierls transition itself but is closely related to a Coulomb repulsion among electrons [6.9].

The dc electrical conductivity is of the order of $10^{-7}\,\Omega^{-1}\,m^{-1}$ for films of *cis*-$(CH)_x$ and about $10^{-3}\,\Omega^{-1}\,m^{-1}$ for those of the trans-form [6.10]. The width of the band gap is estimated from the temperature dependence of the conductivity to be 0.9 eV (1.4×10^{-19} J) for the cis- and 0.6 eV (0.96×10^{-19} J) for the trans-form.

A band calculation, however, gives an estimate of about 7 eV (1.1×10^{-18} J) for the band gap [6.11]. Moreover an analysis of the optical absorption edge gives the following; 1.78 eV (2.85×10^{-19} J) for the cis-form and 1.35 eV (2.16×10^{-19} J) for the trans-form [6.12]. The discrepancy among these results can be explained as follows: The intrinsic width of the band gap is several eV. The optical absorption edge is related however to the creation energy of excitons rather than to the band gap; this creation energy is given by subtracting the binding energy between an electron and a hole from the width of the band gap. The activation energy for the conductivity may have nothing to do with the band gap if the dc conduction is dominated by impurity conduction. Actually $(CH)_x$ doped with a small amount of ammonia has a conductivity lower than the original one by four

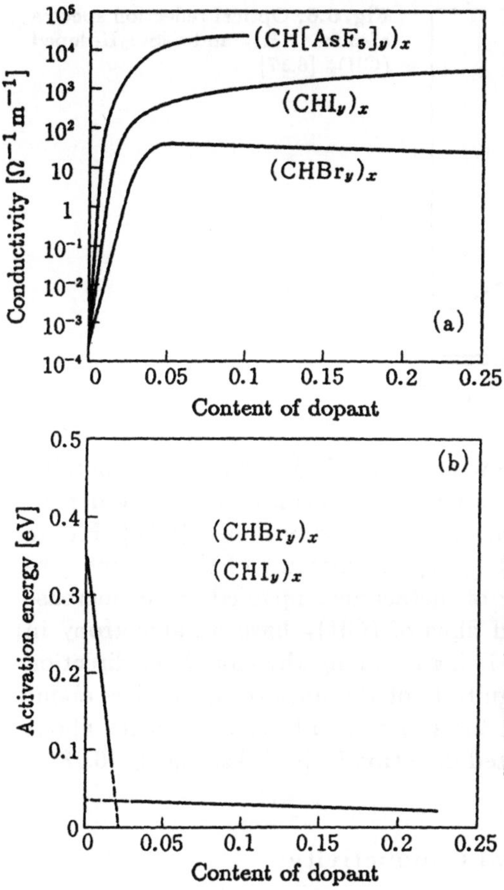

Fig. 6.4. Dc electrical conductivities and the activation energy of anion-doped $(CH)_x$ as functions of the content of dopants [6.14]

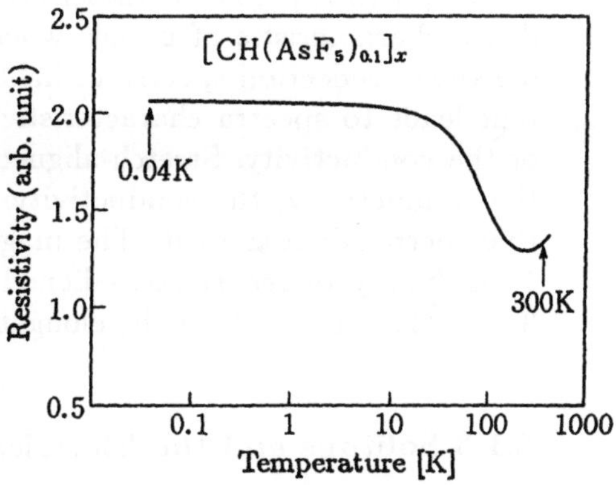

Fig. 6.5. Temperature dependence of the dc electrical conductivity of $(CH)_x$ doped with AsF_5 to 10 % [6.15]

orders of magnitude [6.13]. The ammonia will compensate acceptor impurities which are contained in as-grown $(CH)_x$ and which give rise to the impurity conduction.

Doping with donors and acceptors makes $(CH)_x$ an n-type or a p-type semiconductor, respectively. Figure 6.4 shows the conductivity and the activation energy as functions of the content of dopants [6.14]. The conductivity increases and the activation energy decreases rapidly with increasing the content. $(CH)_x$ is thought to have a metallic nature for dopant concentrations above about 3 %.

Figure 6.5 shows the temperature dependence of the resistivity measured in $[CH(AsF_5)_{0.1}]_x$. It suggests that carriers (holes) created by doping

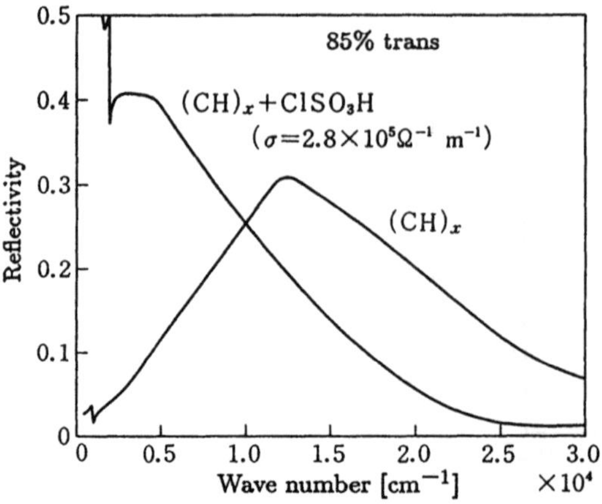

Fig. 6.6. Optical reflection spectra of pure $(CH)_x$ and $ClSO_3H$-doped $(CH)_x$ [6.17]

become a condensed Fermi gas below about 30 K giving rise to a metallic behaviour [6.15]. The thermoelectric power also has a temperature dependence characteristic of metals when $(CH)_x$ is highly doped [6.16]. Figure 6.6 shows reflection spectra of light [6.17]. An increase of the dopant content leads to spectra characteristic of metals accompanied by an increase of the conductivity. Stretch-aligned films of $(CH)_x$ have an anisotropy in the conductivity; the conductivity is higher along the elongated direction than perpendicular to it. The magnitude of the anisotropy reaches about 10 in highly doped stretch-aligned films. The conductivity reaches about $3\times10^5\,\Omega^{-1}\,m^{-1}$ along the elongated direction in $[CH(AsF_5)_{0.1}]_x$ [6.18].

6.1.3 Solitons and the Electrical Conductivity

Figure 6.7 shows the change in electrical and magnetic properties with varying content of the acceptor AsF_5 in $[CH(AsF_5)_y]_x$. The material has metallic properties in highly doped states.

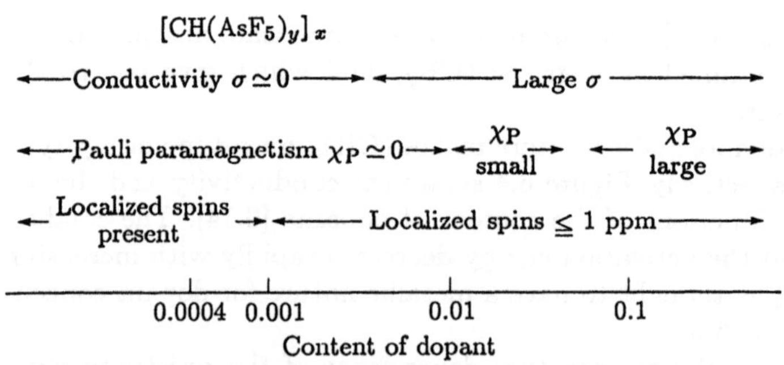

Fig. 6.7. Changes in the properties of $(CH)_x$ with increasing content of the dopant AsF_5

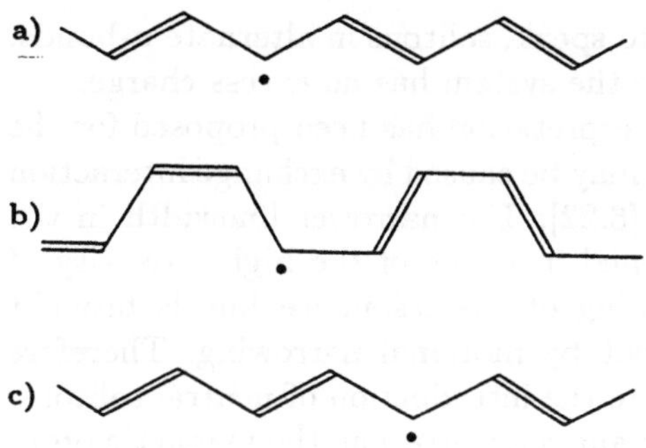

Fig. 6.8. Models of the neutral soliton in $(CH)_x$

In the regime of low doping, the conductivity is low and the magnetic susceptibility obeys Curie's law [6.19]; it is thought that localized spins are present. EPR measurements indicate that the resonance linewidth and the spin density are 6.6×10^{-4}T (6.6 G) and 1 spin/20 000 carbon atoms, respectively, in the cis-form while they are 1.5×10^{-4}T (1.5 G) and 1 spin/3000 carbon atoms in the trans-form [6.20]. A model of neutral solitons has been introduced in order to explain the difference between the cis- and the trans-forms. The trans-form can have isolated π-electrons in the carbon chain as shown in Fig. 6.8a; the regular arrangement of the double and single bonds changes its pattern in going from the left of the isolated electron to the right. This electron has a spin of $\frac{1}{2}$. On the other hand, when the arrangement of the bonds changes across the isolated electron in the *cis*-$(CH)_x$, as shown in Fig. 6.8b, the $(CH)_x$ to the left side and to the right must be cis-transoid and trans-cisoid or vice versa. Raman scattering, however, indicates the absence of the trans-cisoid form as already described. Hence isolated electrons will hardly appear in the cis-form because the trans-cisoid is thought to be unstable. This situation explains why the density of spins in the cis-form is lower than that in the trans-form.

The narrow width of the resonance line and, moreover, the even narrower one of the trans- compared to the the cis-form have been explained in terms of the so-called motional narrowing[2] [6.21]. The isolated spin is expected to move from the position shown in Fig. 6.8a to that in Fig. 6.8c. Observed spins in the cis-form are ascribed to small segments of chains where the trans-form is embedded in the apparent cis-form. The possible motion of spins in the apparent cis-form is confined to within the small segments of the trans-form giving rise to a broader resonance line than in the exclusively trans-form.

[2] The linewidth reflects the inhomogeneity of the local field acting on the spin. When the spin moves fast along the carbon chain, the inhomogeneous local field is averaged out, giving rise to a narrowing of the resonance line. This is the mechanism of motional narrowing.

The isolated π-electrons are, so to speak, solitons in alternate π-bonds. They are called neutral solitons since the system has no excess charge.

It is to be noted that another interpretation has been proposed for the narrow width of the resonance line; it may be caused by exchange interaction among spins of isolated π-electrons [6.22]. The narrower linewidth in the trans- than in the cis-form is explained in terms of the higher density of spins in the trans-form. The narrowing of the resonance line is brought about by exchange narrowing but not by motional narrowing. Therefore this interpretation does not necessitate the introduction of neutral solitons.

In the regime of intermediate dopant concentration, the magnetic properties indicate the absence of spins in spite of the rapid increase of the conductivity [6.23]; it is considered that the current carriers have no spin. This current carrier is called a charged soliton. It is equivalent to a neutral soliton whose electron is missing or which has an additional electron. When an acceptor is present near a carbon chain, the acceptor will take a π-electron away from a double bond as shown in Fig. 6.9a. There remains an isolated π-electron and a vacant π-orbital with a charge $+e$. The isolated π-electron propagates as a neutral soliton along the chain; it is caught by another acceptor leavig there a vacant positively charged orbital. We may understand this in another way by assuming that two neutral solitons produced by two acceptors encounter each other to form a stable π-bond.

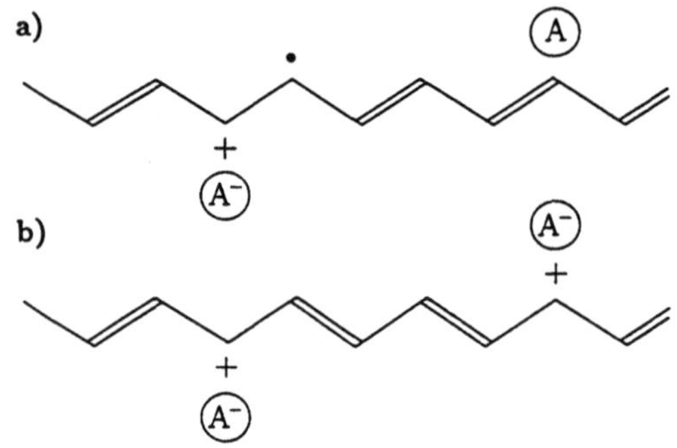

Fig. 6.9. A model of the charged soliton in $(CH)_x$ containing acceptors

In any case a vacant π-orbital charged positively with $+e$, is finally created. This is a kind of dislocation in the π-bonds; it is expected to move along the carbon chain. It is called a charged soliton. Donor doping results in a similar situation; it creates charged solitons with a negative charge $-e$ as shown in Fig. 6.10. Charged solitons, whilst they have electrical charges, have no spin.

The solitons shown in Figs. 6.9 and 6.10 appear as if they are localized on a particular carbon atom. A real soliton, however, will extend over several carbon atoms in the chain. A model calculation shows that the size of the soliton is about 20 times the interatomic distance along the chain

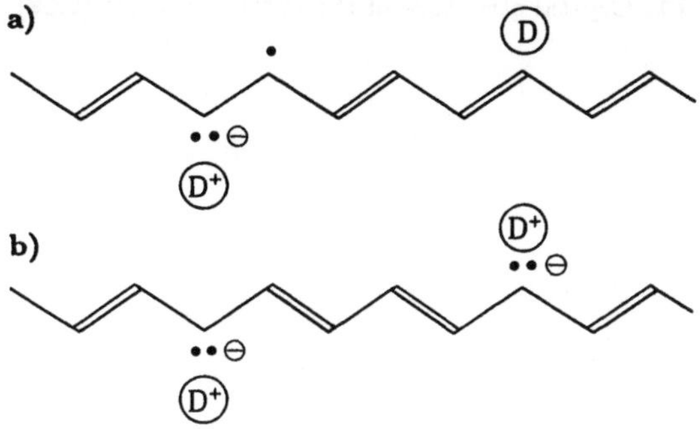

Fig. 6.10. A model of the charged soliton in $(CH)_x$ containing donors

direction [6.24]. Theoretical studies are actively investigating the structure and dynamics of solitons in $(CH)_x$ [6.24, 25].

It is to be noted that the necessity of the charged soliton model is also open to discussion as is the case for neutral solitons [6.22]. It could also be that intermediately doped $(CH)_x$ has metallic segments in the chain. This interpretation, however, is inadequate to explain the absence of the Pauli paramagnetism at intermediate doping [6.23].

6.2 Properties of $(SN)_x$

$(SN)_x$ is a chain-like polymer in which sulpher and nitrogen atoms alternate along the chain. Single crystals have a dc electrical conductivity of about $1.7 \times 10^5\,\Omega^{-1}\,m^{-1}$ along the 1d direction and the anisotropy is of the order of 10^3 [6.26]. As far as these values are concerned, $(SN)_x$ seems to belong to the class of 1d conductors like TTF-TCNQ and $K_2Pt(CN)_4Br_{0.3} \cdot 3H_2O$. Nevertheless a remarkable property is that it does not undergo a Peierls transition but turns instead into a superconductor below 0.3 K. The superconductivity of $(SN)_x$ was discovered in 1975; it is the first found among quasi 1d conductors and, moreover, $(SN)_x$ is the first superconductor that contains no metallic elements [6.27].

The band structure of $(SN)_x$ has been investigated more intensively than that of any other 1d conductor. Various of its properties, such as the absence of the Peierls transition, have been fairly well explained by the theoretical band-structure calculations.

6.2.1 The Synthesis and Structure of $(SN)_x$

We shall look first at an example of methods to make $(SN)_x$ [6.28]. A mixture of S_4N_4, S, NH_4Cl, etc. is obtained when ammonia gas is passed through a mixed solution of $CHCl_3$ and S_2Cl_2. Purified S_4N_4 is produced upon recrystallization of the mixture in a solution of chloroform. When it

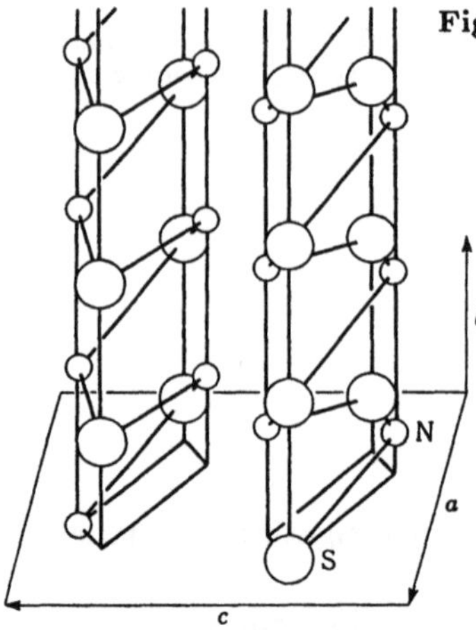

Fig. 6.11. Crystal structure of the $(SN)_x$ polymer [6.29]

is sublimated in vacuum, it changes into S_2N_2 through a reaction on the surface of a silver catalyst.

Crystals of $(SN)_x$, which have a luster like gold, are obtained by keeping S_2N_2 at room temperature for between several days and 10 weeks. A typical crystal has a size of about $1\,mm^3$. Films of $(SN)_x$ may also be obtained on a glass plate when the crystals are sublimated in vacuum. Although the crystals appear to be single ones, a precise inspection reveals that the bulk material is composed of fibrous crystals whose diameter is several μm. Figure 6.11 shows the crystal structure of $(SN)_x$ [6.29]. The fundamental unit is a spiral of sulpher and nitrogen atoms; two chains are contained in a unit cell of the crystal. The spiral axis is the electrically 1d direction. The unit cell has a monoclinic symmetry with the space group $P2_1/c$. Covalent-like bonding combines sulpher and nitrogen atoms in a chain while the distance between neighbouring chains is nearly the sum of the van der Waals's radii of two atoms. This is the origin of the one-dimensionality in $(SN)_x$.

When $(SN)_x$ is doped with bromine, it is thought that the Br atoms occupy sites between chains but not in the chains. In fact the doping causes an increase in the volume and density of crystals although the length along the chain direction does not change. Measurements such as Raman scattering show that Br is located between the chains in the form of Br_3^- or Br_5^-. Furthermore Br_2 can be present on the surface of fibrous crystals in bulk materials [6.30]. Br doping increases the electrical conductivity and also changes its temperature dependence. This is ascribed to the specific structure of the electronic band in $(SN)_x$ as will be described in Sect. 6.2.3.

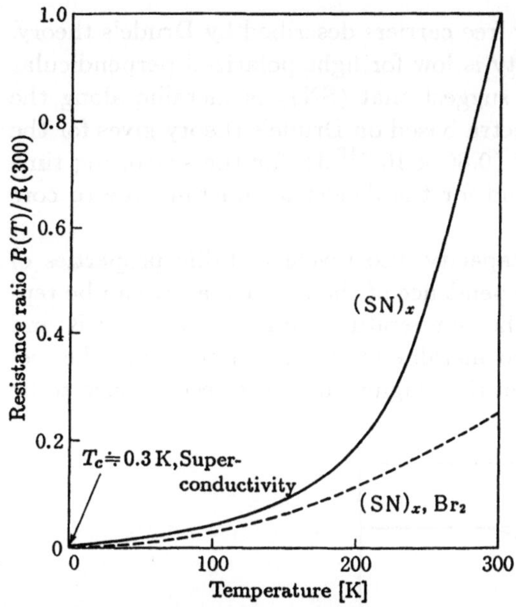

Fig. 6.12. Temperature dependence of the dc electrical resistivity of $(SN)_x$ and Br-doped $(SN)_x$ [6.26]

6.2.2 Electronic Properties

Figure 6.12 shows the temperature dependence of the dc electrical resistivity measured along the chain direction, i.e. along the 1d axis, of $(SN)_x$ [6.26]. It shows also the result obtained in Br-doped samples [6.31]. $(SN)_x$ is metallic in the sense that the resistivity decreases with decreasing temperature. When the temperature dependence is approximated as $\sigma \sim T^n$, the value of n is estimated to be 2 for pure $(SN)_x$ while n changes towards 1 with increasing Br content. As will be discussed in detail in the next section, such a temperature dependence may be explained in terms of two mechanisms, electron-hole and hole-phonon scattering in a semimetal with a quasi one-dimensionality.

Figure 6.13 shows reflection spectra of light [6.32]. The reflectivity is large for low-energy photons whose electric vector is polarized along the

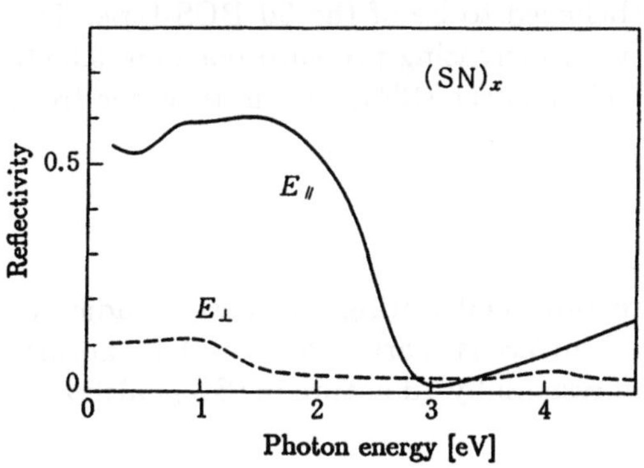

Fig. 6.13. Optical reflection spectra of $(SN)_x$ [6.32]

chain direction; this is typical for free carriers described by Drude's theory. On the other hand, the reflectivity is low for light polarized perpendicular to the chain axis. These results suggest that $(SN)_x$ is metallic along the chain axis. An analysis of the spectra based on Drude's theory gives for the plasma frequency $\hbar\omega_p = 5.99\,\mathrm{eV}$ (9.60×10^{-19} J), for the scattering time of carriers $\tau = 1.89 \times 10^{-15}$ s, and for the dielectric constant due to core electrons $\varepsilon_{core} = 5.55$ [6.33].

Measurements of the heat capacity also reveal metallic properties of $(SN)_x$ [6.34]. The temperature dependence of the heat capacity can be represented as $C = \gamma T + \beta T^3$ in the temperature range between 1.5 K and 10 K; the first term is ascribed to metallic electrons. In addition, the Debye temperature is estimated from the magnitude of the second term to be about 170 K.

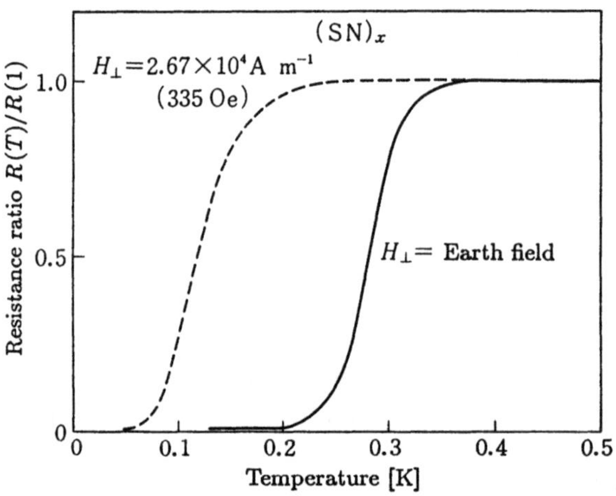

Fig. 6.14. Superconductivity transition and the magnetic field effect in $(SN)_x$ [6.27]

$(SN)_x$ undergoes a superconducting transition with $T_c = 0.3$ K at ambient pressure. Figure 6.14 shows the resistivity as a function of temperature [6.27]. $(SN)_x$ is a type II superconductor since T_c decreases with increasing magnetic field. As far as the data in Fig. 6.14 are concerned, the critical fluctuation in resistivity appears to be small; the essential mechanism of the superconductivity in $(SN)_x$ is believed to be of the 3d BCS-type. The critical temperature, T_c, increases with increasing pressure reaching 0.55 K at 10 kbar (1×10^9 Pa) [6.35]. In this respect $(SN)_x$ seems to differ from $(TMTSF)_2X$.

6.2.3 Band Structure

Two kinds of band structure were proposed during the early studies of $(SN)_x$. In the first proposal the Fermi surface is nearly planar with its normal parallel to the 1d axis; it suggests an essentially 1d nature of $(SN)_x$ although a weak three-dimensionality makes the Fermi surface warped [6.36]. The second suggestion is that pockets of electrons and holes at the boundaries of the Brillouin zone indicate a semimetallic character in three-dimensions [6.37].

The difference between these two ideas arises from a difference in the degree of mixing of the electronic wave functions in the band-structure calculations. The band is formed by $2s$ and $2p$ electrons of nitrogen and by $3s$ and $3p$ electrons of sulpher. A band of the first type described above is obtained when the mixing of the s and p states is small, but the second type is produced when the mixing is strong [6.33].

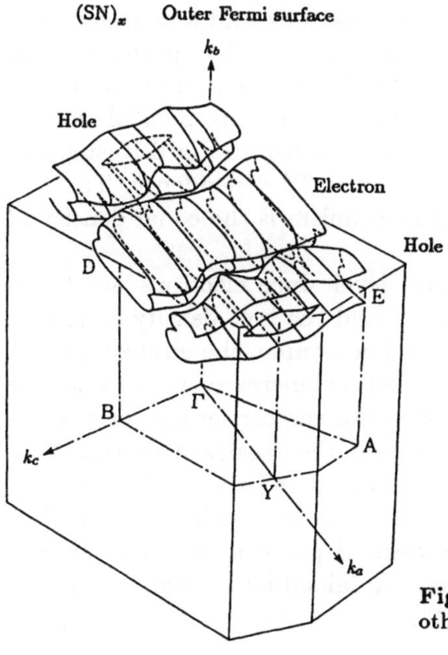

Fig. 6.15. A set of Fermi surfaces of $(SN)_x$. Another set is present inside these [6.38]

A more precise calculation performed recently provides a band structure which might be called intermediate between the two described above [6.38]. Figure 6.15 shows the shape of the Fermi surface. The Fermi surface is nearly planar along k_c, a direction perpendicular to the 1d axis, k_b. It is, however, closed along the other direction, k_a, indicating the presence of two-dimensionality. In the (k_a, k_b) plane the Fermi surface is squeezed along the k_b-direction; the electrical one-dimensionality can be explained by the smaller effective mass of the carriers along the k_b direction than along k_a. It is to be noted that Fig. 6.15 shows only the outer Fermi surfaces and that another set of them is present inside the outer one. The new band calculation described above allows a charge transfer from sulpher to nitrogen. The amount of the charge transfer is evaluated to be 0.4 electron/atom. This is in good agreement with results obtained by x-ray and ultraviolet photo-electron spectroscopies [6.39]; the results of this band-structure calculation are considered to be rather reliable.

Many kinds of experimental results are well explained by this band structure [6.38]: It explains better than the old ones the imaginary part of the complex dielectric constant measured by optical absorption in pure $(SN)_x$ and in Br doped samples. Moreover, the temperature dependence of the resistivity shown in Fig. 6.12, is systematically explained by this band structure.

We shall examine the temperature dependence of the resistivity in detail. It is dominated by the scattering time of carriers, τ. Two scattering mechanisms are important; a carrier-phonon scatterig and a carrier-carrier scattering. The T^2 dependence of the resistivity is valid in pure $(SN)_x$ in the temperature range 10–300 K although the Debye temperature is 170 K. Therefore the carrier-carrie scattering is considered to be responsible for the T^2 dependence. The shape of the Fermi surface shown in Fig. 6.15 suggests that an electron-hole scattering gives rise to the T^2 dependence. The reason for the dominance of this scattering mechanism is the equal density of electrons and holes in pure $(SN)_x$.

Bromine doping decreases the Fermi level since electrons are taken away by Br. The electron density decreases and the hole density increases. When the concentration of Br exceeds a certain value, the electron pocket in Fig. 6.15 will disappear and the hole pocket will increase in size. Thus the role of the carrier-phonon scattering, instead of the carrier-carrier scattering, must become dominant when the balance in the density of electrons and holes is broken. Scattering by phonons leads to a resistivity which is proportional to temperature when the temperature is not too low compared to the Debye temperature. Thus the temperature dependence of the resistivity shown in Fig. 6.12 is considered to be well explained by the band structure whose Fermi surface is shown in Fig. 6.15.

6.2.4 One-Dimensionality and the Kohn Anomaly

$(SN)_x$ does not undergo a Peierls transition. Materials undergoing no Peierls transition can nonetheless have CDWs as does $NbSe_3$ for example. In $(SN)_x$, however, no CDW is found although a very weak softening of the acoustic phonons has been observed by inelastic neutron scattering [6.40]. The onset of the Kohn anomaly and the CDW is dominated by the polarization function $\chi(Q)$ described in Chap. 1. $\chi(Q)$ shows no divergence when a Fermi surface has such a shape that the nesting is poor.

Among the band structures described in the preceding section, the first one with its quasi 1d Fermi surface has been examined in detail in relation to the functional form of $\chi(Q)$; the divergence of $\chi(Q)$ at $Q = 2k_F$

is suppressed by a warping of the Fermi surface due to interchain interactions [6.33]. The absence of CDWs in $(SN)_x$ is ascribed to the weak, but non-negligible, three-dimensionality due to interchain interactions. The interactions are brought about by an overlap of π-electron-like wave functions which are elongated perpendicular to the chain axis. In the second band model of a 3d semimental and in the quasi 1d model, CDWs are indeed expected to be absent due to the more pronounced three-dimensionality.

7. Properties of Linear-Chain Mercury Compounds

In this chapter we shall study the properties of the linear-chain mercury compound $Hg_{3-\delta}AsF_6$, where $\delta \simeq 0.18$. A unit cell of this crystal contains four chains of mercury atoms. The number of mercury atoms in the unit cell is not an integer and it changes a little when the temperature is varied. The mercury chains are responsible for the one-dimensional metallic properties although no CDW appears. This compound undergoes a superconducting transition at 4.1 K. However, there remain several unsolved problems relating to the superconductivity. $Hg_{3-\delta}AsF_6$ is a compound quite unusual in structure: Mass-density waves, instead of CDWs appear on the mercury chains and the chain behaves like a 1d liquid. In this chapter we focus on the properties of superconductivity and the mass-density wave.

7.1 Synthesis and Crystal Structure

The synthesis of $Hg_{3-\delta}AsF_6$ proceeds as follows: At 77 K AsF_5 is dissolved in SO_2 dehydrated by P_2O_5; mercury is then added to it. The amount of mercury added must be carefully controlled so that none remains in the liquid. Crystals are grown when the liquid is slowly heated up, whilst being stirred, to room temperature. The crystals are then washed in SO_2 and dried in vacuum. Special care must be taken not to expose the crystals to humid air. Another method can also be employed: $Hg_3(AsF_6)_2$, when it is available, is dissolved in SO_2 and then mercury is added. Single crystals with a typical size of 1 mm^3 are obtained in both methods [7.1, 2].

The crystal has a tetragonal symmetry with the space group $I4_1/amd$. The lattice parameters are $a = b = 7.538$ Å and $c = 12.339$ Å. A unit cell contains four formula units, $(Hg_{3-\delta}AsF_6)_4$. Figure 7.1 shows the crystal structure [7.1]. The coordinates of the As atoms, each of which lies at the centre of an octahedron formed by F atoms, are given by $(0, \frac{1}{4}, \frac{3}{8})$, $(0, \frac{3}{4}, \frac{5}{8})$, $(\frac{1}{2}, \frac{3}{4}, \frac{7}{8})$, and $(\frac{1}{2}, \frac{1}{4}, \frac{1}{8})$.

Four chains of mercury atoms lie perpendicular to the c-axis. Two of them are parallel to the a-axis and the remaining two to the b-axis; they do not intersect each other. Photographs of x-ray diffraction show the presence of diffuse sheets normal either to the a^*-axis or to the b^*-axis in addition to the usual Bragg spots. The diffuse sheets arise from the structural one-

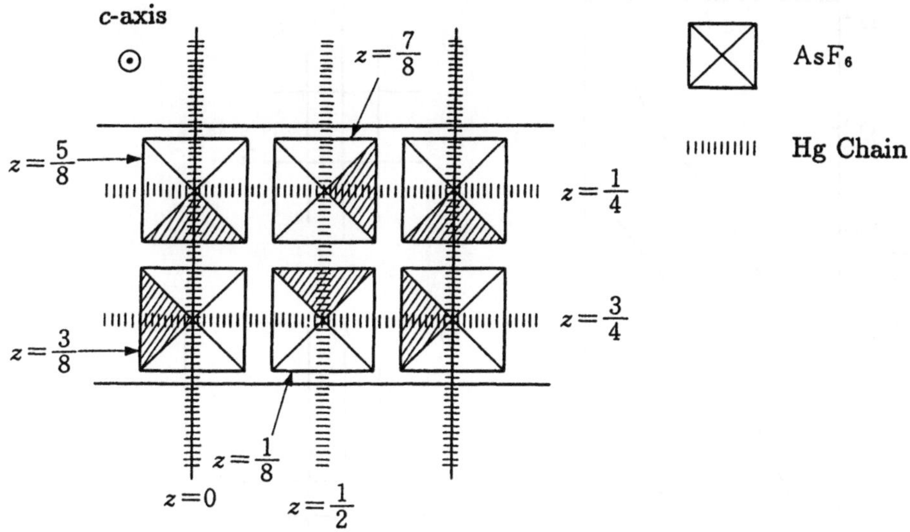

Fig. 7.1. Schematic view of the crystal structure of $Hg_{3-\delta}AsF_6$ [7.3]

dimensionality of the mercury chains. The average distance between neighbouring mercury atoms in a chain is estimated to be $a_{Hg} = 2.67\,\text{Å}$ from the location of the diffuse sheets on the photographs. This value is rather small compared to the interchain distance, 3.09 Å, and to the interatomic distance in mercury metal, 3.005 Å. The value of a_{Hg} gives an estimate of the content of mercury in the compound; $3 - \delta = 7.538/2.67 = 2.82$. This is in good agreement with the results of chemical analysis [7.1]. It is to be noted that the above result is obtained at room temperature. Its temperature dependence will be discussed in the next section.

7.2 Metallic Conductivity and Superconductivity

The resistivity along the a-axis is $\varrho_a \sim 10^{-6}\,\Omega\,\text{m}$ at room temperature [7.3]. Its temperature dependence suggests that this compound has a metallic nature as shown in Fig. 7.2. The metallic conduction will be brought about by conduction electrons on the mercury chains since AsF_6 is believed to be a negative ion, AsF_6^{-1}. The number of electrons belonging to the outer shell of the mercury atoms is given as $2 \times 2.82 \times 4 = 22.6$; i.e. 2 (6$s$-electrons/Hg), 2.82 (Hg atoms/formula unit) and 4 (formula units/unit cell). Four of these electrons localize on AsF_6 ions and 18.6 electrons remain. When each mercury chain is assumed to have 1d bands, two of them lower in energy are fully occupied by four electrons. Since the unit cell contains four mercury chains, 16 electrons are used to fill the lower bands. Thus the third band of

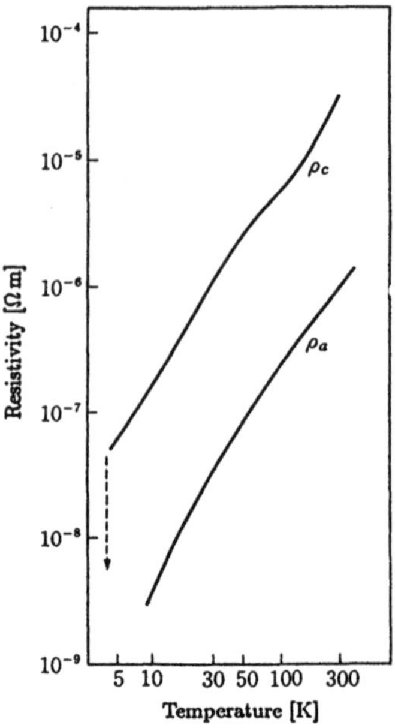

Fig. 7.2. Temperature dependence of the electrical resistivity of $Hg_{3-\delta}$ AsF_6 [7.3]

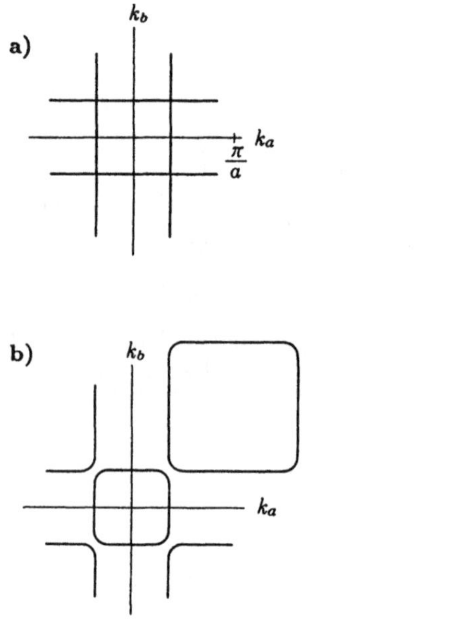

Fig. 7.3. A model of the Fermi surface of $Hg_{3-\delta}$ AsF_6 [7.4]

each chain accepts 2.6/4 (=(18.6−16)/4) electrons. The Fermi wave number is evaluated as $k_F = \pi(2.6/4a)/2$ using Eq. (2.8). When we assume the Fermi "point" of the 1d bands to be a plane as we did in Fig. 2.8, two kinds of Fermi surface are formed; one normal to the a^*-axis and the other normal to b^* as shown in Fig. 7.3a. Degeneracy on the crossing points in Fig. 7.3a is removed if some interaction exists between chains lying perpendicular to each other. Figure 7.3b shows the resulting Fermi surfaces; one of them has an electron character and the other a hole character. Their shape is square-column-like. They are Fermi surfaces of a 2d electron system possessing however some degree of one-dimensionality.

Electron systems with an ideal two-dimensionality have a cylindrical Fermi surface. Real 2d conductors usually have a Fermi surface with a modified cylindrical shape. On the other hand the columnar Fermi surface of the mercury compound is brought about by interchain interactions. Most parts of the Fermi surface are planar reflecting the intrinsic one-dimensionality.

Studies of the de Haas-van Alphen effect give six kinds of oscillations [7.4], as shown in Fig. 7.4, having an angular dependence $1/\cos\theta$ with respect to the angle between the c-axis and the direction of the applied magnetic field. Thus it is verified that the Fermi surface forms a columnar structure parallel to the c^*-axis because $1/\cos\theta$ is proportional to the area of

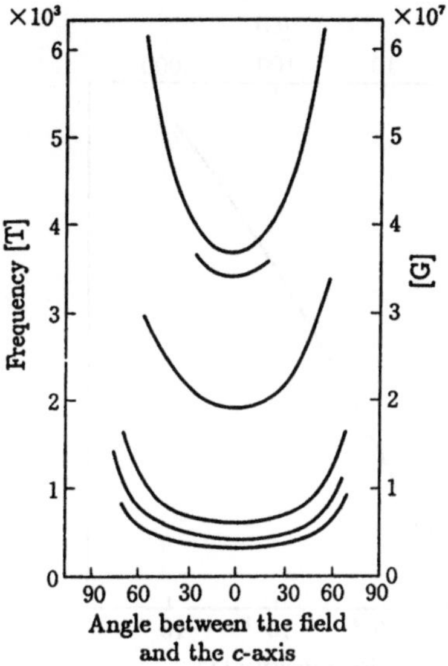

Fig. 7.4. De Haas-van Alphen frequency of $Hg_{3-\delta}$ AsF_6 [7.4]

the cross-section of a column cut by a plane whose normal makes an angle θ with the column axis.

Although Fig. 7.3b shows only two kinds of Fermi surface, the remaining four kinds of oscillations are explained when a distortion in the mercury chains, to be described in Sect. 7.3, is taken into account. The size of the Fermi surface estimated form the value δ at low temperatures agrees well with that derived from the minimal period of the oscillation.

The anisotropy in the resistivity, ϱ_c/ϱ_a, is about 100 reflecting the one-dimensinality of the conduction band. The anisotropy is also found in the optical properties [7.5]; crystals appear gold in a light polarized in the a-b plane, but look greenish blue when the illumination is polarized parallel to the c-axis since the reflectivity is low for this light in the long-wavelength region.

The temperature dependence of ϱ_a and ϱ_c shown in Fig. 7.2 is well represented as $T^{3/2}$. The proportionality predicted by the Bloch-Grüneisen law ($\varrho \propto T$) does not appear even at high temperatures of about four times the Debye temperature ($\theta_D = 78$ K, estimated from measurements of heat capacity at low temperatures [7.6]). The reason for this has not been found.

The superconductivity appears below 4.1 K but there is no CDW in $Hg_{3-\delta}AsF_6$ despite the presence of a 1d electron system [7.7]. It is to be noted that the resistivity vanishes only along the c-axis whereas ϱ_a decreases monotonically with decreasing temperature. Moreover ϱ_a appears to have no residual resistance in spite of the presence of AsF_6 vacancies with a concentration of several percent.

Fig. 7.5

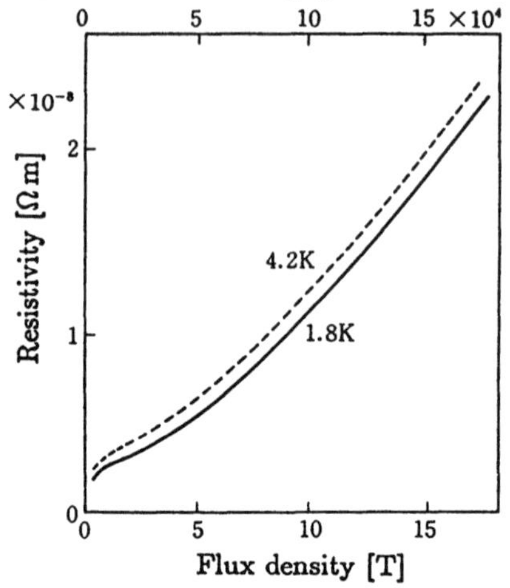

Fig. 7.6

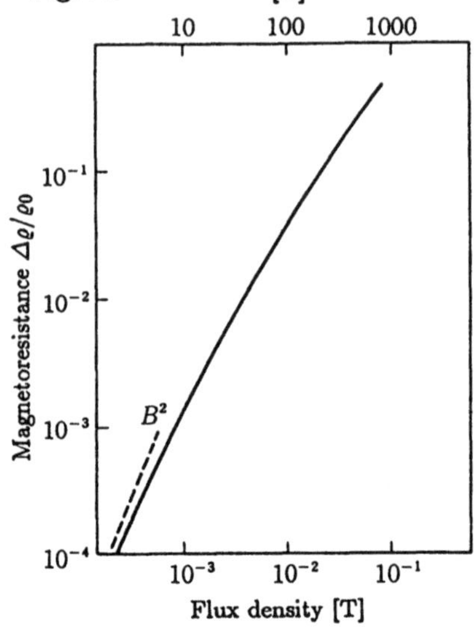

Fig. 7.5. Magnetoresistance of $Hg_{3-\delta}$ AsF_6 [7.8]

Fig. 7.6. Magnetoresistance of $Hg_{3-\delta}$ AsF_6 under low magnetic fields

Unusual behaviour is also found in the magnetoresistance of this compound. Figure 7.5 shows the resistivity ϱ_a as a function of applied magnetic field [7.8]. Ordinary magnetoresistance, $[\varrho(T,B) - \varrho(T,0)]/\varrho(T,0)$, is proportional to B^2 and increases with decreasing temperature. The B^2 dependence, however, is only observed in the limit of low magnetic field as shown in Fig. 7.6. Furthermore the effect of magnetic field on the resistivity is independent of temperature below 30 K. The sample appears to have a residual resistance whose magnitude depends on the magnetic field. In other words $[\varrho(T,B) - \varrho(4.2\,\mathrm{K},B)]$ is independent of B and agrees with ϱ_c shown in Fig. 7.2.

These unusual results were originally explained in terms of fluctuations of the superconductivity: The superconductivity of the mercury chains has no long-range order thus giving rise to a finite resistance ϱ_a, below 4.1 K. The residual resistance is expected to become zero for $T \to 0$ since the long-range order must be realized at absolute zero. The application of a magnetic field recovers the residual resistance because the superconductivity fluctuation will be suppressed by the field. Nevertheless this interpretation has difficulties in several respects: the field effet remaining up to 30 K, the field dependence of the residual resistance, and the difference between ϱ_a and ϱ_c.

Another interpretation has been proposed by *Chakraborty* et al. [7.8]. Figure 7.3b indicates that the Fermi surface consists of planar parts and curved corners which reflect a two-dimensionality. No contribution to the resistivity comes from those scattering processes which take an electron from one state on the planar part of the Fermi surface to another state on the same plane. It is the scattering of electrons on the curved corners that will give rise to the resistivity. When these two coexist, the latter plays a minor role. If however, a magnetic field is applied, electrons will go into the corners because of the Lorentz force; there they produce a strong scattering contributing to the resistivity. Thus the residual resistance is observed under the magnetic field; the two parts of the Fermi surface form, so to speak, a serial circuit. The field effect is expected to be small when the Lorentz force works along the column axis of the Fermi surface. In fact a field applied parallel to the c-axis gives little effect to the resistivity: $\Delta\varrho(4.2\,\mathrm{K}, 0.1\mathrm{T})/\varrho(4.2\,\mathrm{K}, 0) < 0.01$.

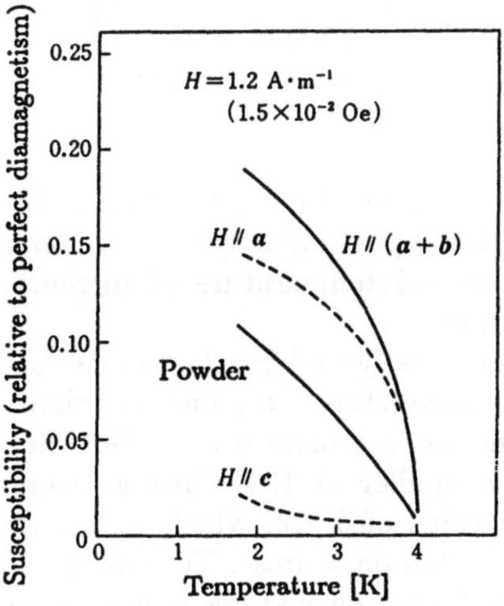

Fig. 7.7. Diamagentic susceptibility of $Hg_{3-\delta}AsF_6$ [7.9]

It is fairly evident that the superconductivity has nothing to do with the unusual behaviour of ϱ_a at low temperatures. At the same time the vanishing of ϱ_c below 4.1 K cannot be ascribed to a 3d superconductivity because ϱ_a does not vanish. Figure 7.7 shows the diamagentic susceptibility measured by ***Spal*** et al. [7.9]. The magnetization is measured as a function of temperature at a constant magnetic field. It has been also measured under an ac magnetic field in powder samples; the two results agree with each other. The important points concernig Fig. 7.7 are the following: (i) the anisotropy is large, (ii) no step-like increase is found on crossig the critical temperature from above under such a weak field, and (iii) a perfect diamagnetism is not observed for $T \to 0$. Nevertheless bulk samples are thought to show the superconductivity below 0.43 K. Magnetic susceptibilities measured at lower temperatures, Fig. 7.8, and the magnetization curve, Fig. 7.9, reveal the presence of another transition at 0.43 K, below which the system shows a perfect diamagnetism [7.10]. Therefore it is believed that these crystals have a bulk superconductivity below 0.43 K.

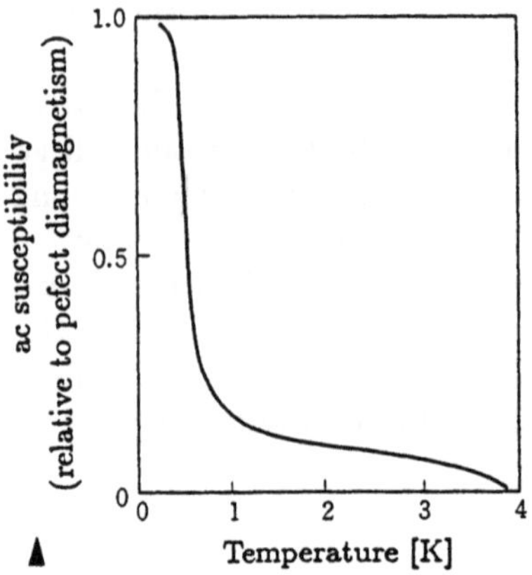

Fig. 7.9

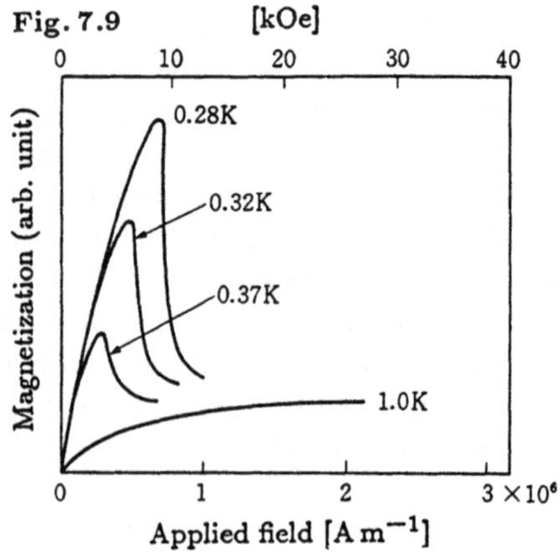

▲

Fig. 7.8. Ac diamagnetic susceptibility of powder samples of $Hg_{3-\delta}$ AsF_6 [7.10]
Fig. 7.9. Magnetization curve of $Hg_{3-\delta}$ AsF_6 $(H\|a)$ [7.10]

A question remains however about the superconductivity below 4.1 K. The possibility has been pointed out that this superconductivity is brought about by segregated mercury since the critical temperature of mercury metal, 4.15 K, is very close to the present one.

The temperature dependence of δ was investigated by *Hastings* et al. [7.11] and *Pouget* et al. [7.12]. The interatomic distance in a mercury chain is estimated to be 2.670 ± 0.005 Å independent of temperature. On the other hand the lattice parameters, a and c, are smaller at 10 K than at room temperature by about 1 %. Therefore the value of $3 - \delta$, which is 2.82 at room temperature, must become 2.79 at low temperatures[1]. The change of $3-\delta$ with temperature is reversible. The excess mercury at low temperature must thus remain somewhere in the sample. The first possibility is that it segregates to the surface of the crystal. In fact very small droplets of mercury are found on the surface at 77 K [7.7]. The amounts, however, are too small compared to the change of δ. Moreover various kinds of measurement have shown that the segregation to the surface does not occur when samples are kept free from humid air. A remaining possibility is the segregation of mercury within the crystals. Excess mercury atoms may enter between other atoms or occupy vacancies of the AsF_6 sites, or both.

From the above discussions we find two problems to be solved: (i) Why does the ground state of the mercury compound show superconductivity rather than the CDW state in spite of the presence of the 1d electron system? (ii) What is the origin of the imperfect superconductivity observed between 0.43 K and 4.1 K? Although in the electronic properties fundamental problems are left unsolved, the structural properties show very interestig features as described in the next section.

[1] This value, 2.79, is used in the analysis of results of the de Haas-van Alphen effect.

7.3 The One-Dimensional Lattice of Mercury Chains

7.3.1 One-Dimensional Liquid Mercury and Its Condensation

For a temperature variation between room temperature and 4 K, the interatomic distance in the mercury chain changes by less than 0.1 % whilst the lattice parameter changes by more than 1 % [7.12]. This fact suggests that the interaction between the lattice of AsF_6 and the chains of mercury is only weak. If this is the case, modes of lattice vibration will be separated into two classes: those of the AsF_6 lattice and those of the mercury chains. Moreover each mercury chain is expected to have independent vibrational states since the interchain distance is rather large. In other words one can discuss the lattice dynamics of the present compound on the basis of a naive model of linear atomic chains; this problem is dealt with in most text books on solid state physics. We will see later, however, that an elemental atomic chain cannot be a solid but behaves as a 1d liquid.

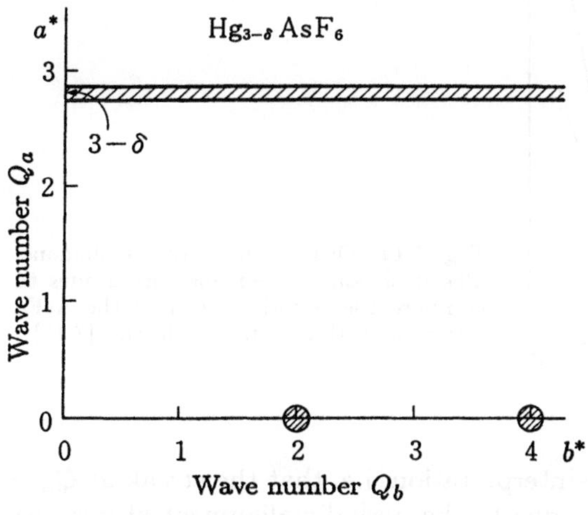

Fig. 7.10. Diffuse streaks and diffuse spots observed by x-ray scattering [7.12]

X-ray and neutron scatterig can probe the lattice dynamics of the mercury chains. Figure 7.10 shows the location of strong scattering observed in the a^*-b^* plane [7.12]. (Note that we introduce the notation b^* in order to discriminate one axis from the other equivalent one.) As we have learned in Chaps. 3–5, the scattering at $Q_a = (3-\delta)a^*$ indicates the presence of periodic structures specified by the wave number Q_a and the absence of any correlation between them along the b^*-axis. The wavelength of the periodic structure is given as $2\pi/Q_a = a/(3-\delta)$ and the quantitiy $a/(3-\delta)$ is nothing but the interatomic distance in the chain. Therefore the streak marked by hatching in Fig. 7.10 is something like a Bragg reflection coming from mercury chains whose axis is parallel to the a-axis; neighbouring chains have no correlation with each other.

In other words, when the atomic position is given as $x = x_i$, $x_i + a_{Hg}$, $x_i + 2a_{Hg}$ ... on the ith chain, x_i is perfectly random with respect to i. One chain in a bundle of chains must be able to slide along the chain direction without any friction. The scattering on the line $a^* = 0$ forms spots at $2b^*$, $4b^*$. The reason is that the scattering on this line has the scattering vector $\boldsymbol{K} = (0, k, 0)$ and, therefore, only the b-component of the correlation function, $G(r)$, is taken out. When we focus our attention on the b-direction only, the mercury chains are regularly present with the distance $b/2$ giving rise to usual Bragg reflections at $k = 2b^*, 4b^* \ldots$ thus we can ascribe the scattering marked by hatching to a set of chains parallel to the a-axis.

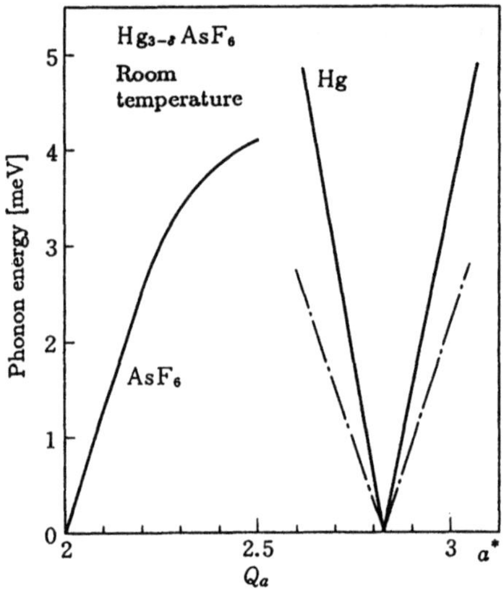

Fig. 7.11. Dispersion curves of phonons. The dash and dotted lines are guides to compare the sound velocity of the AsF_6 lattice with that of the Hg lattice [7.11]

When we accept the above interpretation, i.e. that the streak at $Q_a = (3 - \delta)a^*$ is the Bragg reflection due to the periodic alignment of mercury atoms along a chain, the position of $Q_a = (3 - \delta)a^*$ must be a reciprocal lattice point (i.e. the Γ-point of the Brillouin zone) of mercury chains. Inelastic neutron scattering provides a dispersion relation for the lattice vibrations around this point. Figure 7.11 shows the dispersion relation measured at room temperature [7.11]. The curve near $Q_a = 2a^*$ corresponds to phonons in the lattice of AsF_6. The other around $Q_a = 2.82a^*$ is ascribed to those of mercury chains. The dash-dotted curves are artificially drawn for the comparison of sound velocities; their slope indicates the velocity of phonons in the AsF_6 lattice. The sound velocity of the mercury chain is estimated as $v = 4 \times 10^3\,\mathrm{m\,s^{-1}}$ which is nearly twice as large as that of the AsF_6 lattice. The difference is a result of the short interatomic distance along the mercury chain.

A 3d ordering of the mercury chains will occur with decreasing temperature since interchain interactions or interactions between the chains and the AsF_6 lattice, or both will become large. Before returning to this point we shall show that the 1d chain is a liquid rather than a rigid lattice.

Emery and *Axe* have developed a theory to deal with the mercury chain [7.13]. The dynamics of a mercury chain is described by the following harmonic Hamiltonian,

$$H_0 = \frac{1}{2}\sum_i \left[\frac{\Pi_i^2}{m} + \frac{mv^2}{a_{\mathrm{Hg}}^2}(x_{i+1} - x_i - a_{\mathrm{Hg}})^2\right] \tag{7.1}$$

where x_i and Π_i denote the space and the momentum coordinates of the ith mercury atom, respectively; m is the mass and v the sound velocity. The standard deviation σ of the distance between two atoms on the chain is defined as

$$\langle (x_{i+l} - x_i - la_{\mathrm{Hg}})^2 \rangle = |l|\sigma^2 \quad . \tag{7.2}$$

In the present case, σ is given by

$$(\sigma/a_{\mathrm{Hg}})^2 = k_{\mathrm{B}}T/mv^2 \quad . \tag{7.3}$$

The right-hand side of (7.3) gives a small value, 6×10^{-4}, at room temperature since $mv^2 \gg k_{\mathrm{B}}T$; thus the harmonic approximation used in (7.1) may be valid. Nevertheless (7.2) tells us that $\langle (x_{i+l} - x_i - la_{\mathrm{Hg}})^2 \rangle$ diverges for $l \to \infty$. That is to say, the 1d lattice cannot be a crystal but a 1d liquid. This conclusion is essentially equivalent to that discussed in Sect. 2.3.1.

The point $Q_a = (3 - \delta)a^* = 2.82a^*$ cannot be called the reciprocal lattice point (i.e. the Γ-points); it is only the point where the phonon energy becomes zero[2]. For a liquid it is meaningless to define an exact Brillouin zone.

This conclusion is more easily understood when one investigates what happens at points nQ_a (n is an integer) in addition to Q_a. If the material is a crystal, similar Bragg reflections will be observed at $(1,0,0)$, $(2,0,0)$, ... On the other hand if it is a 1d liquid, the peak width κ_n of the nth Bragg reflection, $(n, 0, 0)$, is given by

$$\kappa_n a_{\mathrm{Hg}} = 2\pi^2 n^2 (\sigma/a_{\mathrm{Hg}})^2 \quad . \tag{7.4}$$

This can be rewritten using (7.3) as

$$\kappa_n \propto n^2 T \quad . \tag{7.5}$$

This theoretical prediction is verified by neutron scattering [7.14].

[2] The energy being zero means that a mercury chain can slide without any friction (cf. the zero energy of an acoustic phonon at the Γ-point). This situation is similar to that for the sliding motion of CDWs.

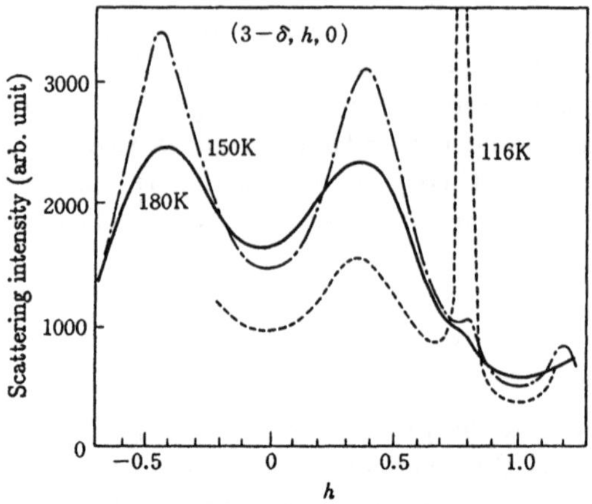

Fig. 7.12. X-ray critical scattering of $Hg_{3-\delta}$ AsF_6 [7.12]

The diffuse streak shown in Fig. 7.10 converges on the line $Q_a = (3 - \delta)a^*$ into broad spots as shown in Fig. 7.12 [7.12]. This scattering will be a critical scattering associated with a 3d ordering of chains. It is to be noted, however, that a sharp peak appears below 120 K at $\boldsymbol{Q} = (3-\delta,\ 1-\delta,\ 0)$ and not at the peak position of the critical scattering, $\boldsymbol{Q}_{cr} = (3-\delta,\ 0.4,\ 0)$. The 3d ordering at 120 K will be a liquid-solid phase transition brought about by a three-dimensionality due to interchain interactions. The difference of the peak positions between the critical scattering and the well-developed peak observed below the transition temperature suggests the presence of two types of interactions associated with the transition. According to the theory by Emery and Axe, the mechanism giving rise to the critical scattering is a 2d interaction among chains aligned parallel to each other, and the other, related to the real 3d ordering, is another interaction among chains perpendicular to each other [7.13]. The unusual result shown in Fig. 7.12 can be explained in terms of a difference of the two transition temperatures; the transition temperature due to the former mechanism is slightly lower than that due to the latter.

7.3.2 Mass-Density Waves

The 1d liquid of mercury condenses below 120 K giving rise to a 3d ordered structure. Nevertheless the ordered chains do not form a crystal but a mass-density wave akin to a CDW [7.15]. The indication for this is that the neutron diffraction Bragg peak is strong only at $\boldsymbol{Q} = (3-\delta,\ 1-\delta,\ 0)$; higher order diffraction has very weak intensity.

When the material is a crystal, the correlation function $G(r)$ given by (3.17) has a finite value for r =(interatomic distance) × (integer) only. Its Fourier transform, (3.18), gives the same value for $Q, 2Q, 3Q, \dots$. Therefore the strong intensity of diffraction only at Q leads to the conclusion that the

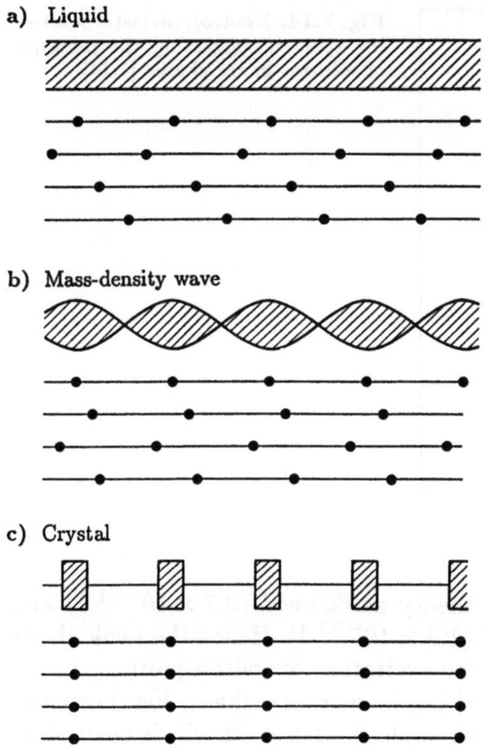

Fig. 7.13. Ordering of Hg chains in $Hg_{3-\delta}$ AsF_6 [7.15]

correlation function $G(r)$ has only one Fourier component; the alignment of atoms is described by a sinusoidal wave sin $(Q \cdot r)$.

A naive picture for this structure is shown in Fig. 7.13. Diffraction at $2Q$, $3Q$, ... becomes stronger with decreasig temperature suggesting a change of the state from the liquid to the crystal.

The phason energy at $q = 0$ becomes finite when the sliding motion of the CDW is forbidden by pinning or locking. Similar situations are expected to occur in the present system. The energy at $Q = (3 - \delta)a^*$ in Fig. 7.11 will become finite when the mercury chains form a three-dimensionally ordered structure. Interchain interactions may be assumed as the source of a pinning or locking mechanism. The interaction Hamiltonian has a similar form to that derived in Sect. 2.2.2 [7.5] for solitons. In fact a solution representing solitons is obtained here too [7.13]. The solution gives two possible states, a single-soliton state and a bound state of a soliton and an anti-soliton. The rest mass calculated for these quasi-particles gives the finite energy at $Q = (3 - \delta)a^*$ in Fig. 7.11. It agrees with that obtained by inelastic neutron scattering. Figure 7.14 shows the energy spectrum at $Q = (3-\delta)a^* = 2.79a^*$ measured at 70 K by neutron scattering [7.12]. The energy of the single-

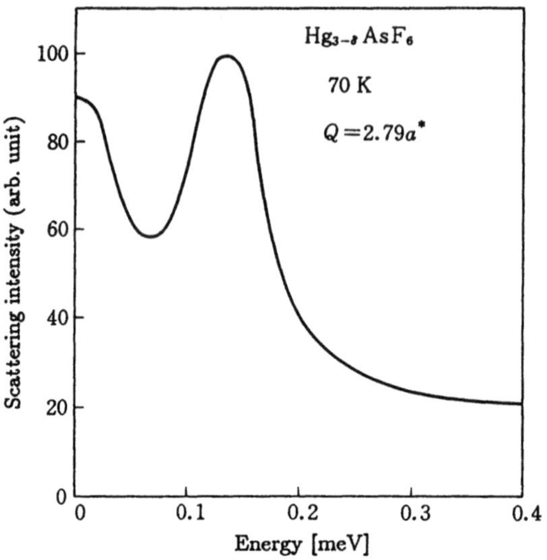

Fig. 7.14. Neutron inelastic scattering spectra suggesting the presence of solitons in Hg chains [7.14]

soliton state is evaluated from the theory as 60 meV (9.7×10^{-21} J) and that of the bound state as 0.08 meV (1.3×10^{-23} J). Hence the peak about 0.13 meV in Fig. 7.14 is ascribed to the excitation of soliton pairs.

Thus an unexpected state, a 1d liquid mercury chain, has been discovered in $Hg_{3-\delta}$ AsF_6. No other example is known in which such weak interactions are found between an element in a material and the host lattice. More advanced problems such as the behaviour of the order parameter are discussed in [7.13, 16].

Appendix

Some Formulae of Electromagnetism

$$D = \varepsilon_0 E + P = \varepsilon\varepsilon_0 E$$
$$B = \mu_0 H + \mu_0 M = \mu\mu_0 H$$

$\chi_m \equiv \dfrac{\mu_0 M}{H}$; magnetic susceptibility

$\overline{\chi}_m \equiv \dfrac{M}{H}$; relative magnetic susceptibility

$\mu_B \equiv \dfrac{e\hbar}{2m}$; Bohr magneton

Conversion of Units

Energy

$$\begin{aligned} 1\,\mathrm{eV} &= 8.0655 \times 10^3\,\mathrm{cm}^{-1} \\ &= 1.1604 \times 10^4\,\mathrm{K} \\ &= 1.6022 \times 10^{-12}\,\mathrm{erg} \\ &= 1.6022 \times 10^{-19}\,\mathrm{J} \end{aligned}$$

Pressure

$$\begin{aligned} 1\,\mathrm{bar} &= 1 \times 10^6\,\mathrm{dyn\,cm}^{-2} \\ &= 1 \times 10^5\,\mathrm{N\,m}^{-2} \\ &= 1 \times 10^5\,\mathrm{Pa} \\ &= 0.987\,\mathrm{atm} \end{aligned}$$

Magnetic induction field $\boldsymbol{B}$

$$\begin{aligned} 1\,\mathrm{T} &= 1\,\mathrm{Wb\,m}^{-2} \\ &= 10^4\,\mathrm{G} \end{aligned}$$

Magnetic field intensity $\boldsymbol{H}$

$$1\,\mathrm{A\,m}^{-1} = 4\pi \times 10^{-3}\,\mathrm{Oe}$$

Magnetic susceptibility $\overline{\chi}_m$

$$1\,\mathrm{mol}^{-1} = (4\pi)^{-1}\,\mathrm{emu\,mol}^{-1}$$

References

Chapter 2

2.1 As text books see for example, C. Kittel: *Introduction to Solid State Physics,* 5th ed. (Wiley, New York 1976) Chapters 6 and 7
J.M. Ziman: *Principles of the Theory of Solids,* 2nd ed. (Cambridge University Press, Cambridge 1972) Chapter 3
A.H. Wilson: *The Theory of Metals,* 2nd ed. (Cambridge University Press, Cambridge 1965) Chapters 1 and 2

2.2 J.M. Ziman: *Principles of the Theory of Solids,* 2nd ed. (Cambridge University Press, Cambridge 1972) Chapter 5

2.3 N.W. Ashcroft, N.D. Mermin: *Solid State Physics* (Holt, Rinehart and Winston, New York 1976) Chapter 17

2.4 M. Afanas'ev, Yu. Kagan: Sov. Phys. JETP **16**, 1030 (1963)

2.5 M.J. Rice: *Low Dimensional Cooperative Phenomena,* ed. H.J. Keller (Plenum, New York 1975) p.23

2.6 R.E. Peierls: *Quantum Theory of Solids* (Clarendon, Oxford 1955) Chapter 5

2.7 W. Kohn: Phys. Rev. Lett. **2**, 393 (1959)
E.J. Woll, Jr., W. Kohn: Phys. Rev. **126**, 1693 (1962)
M. Afanas'ev, Yu. Kagan: Sov. Phys. JETP **16**, 1030 (1963)

2.8 H. Fröhlich: Proc. R. Soc. London **A223**, 296 (1954)

2.9 See for example, G. Rickayzen: *Superconductivity,* Vol. 1, ed. R.D. Parks (Marcel Dekker, New York 1969) p. 73

2.10 J. Bardeen, L.N. Cooper, J.R. Schrieffer: Phys. Rev. **106**,162 (1957); ibid **108**, 1175 (1957)

2.11 P.A. Lee, T.M. Rice, P.W. Anderson: Solid State Commun. **14**, 703 (1974)

2.12 H. Fukuyama: J. Phys. Soc. Jpn. **41**, 513 (1976)

2.13 H. Fukuyama, P.A. Lee: Phys. Rev. B**17**, 535 (1978)

2.14 D. Allender, J.W. Bray, J. Bardeen: Phys. Rev. **B9**, 119 (1974)

2.15 See for example, [Ref. 2 p. 341]

2.16 D. Allender, J.W. Bray, J. Bardeen: Phys. Rev. **B9**, 119 (1974)
B.R. Patton, L.J. Sham: Phys. Rev. Lett. **33**, 638 (1974)
H. Fukuyama, T.M. Rice, C. Varma: Phys. Rev. Lett. **33**, 305 (1974)

2.17 M.J. Rice, A.R. Bishop, J.A. Krumhansl, S.E. Trullinger: Phys. Rev. Lett. **36**, 432 (1976)

2.18 See for example, L.D. Landau, E.M. Lifshitz: *Statistical Physics* (Pergamon, Oxford 1968) Sect. 152

2.19 P. Pincus: *Low Dimensional Cooperative Phenoman,* ed. H.J. Keller (Plenum Press, New York 1975) p. 1

2.20 P.A. Lee, T.M. Rice, P.W. Anderson: Phys. Rev. Lett. **31**, 462 (1973)

2.21 S. Nakajima, Y. Okabe: J. Phys. Soc. Jpn. **42**, 1115 (1977)

2.22 L.J. Sham, B.R. Patton: Phys. Rev. B**13**, 3151 (1976)
Y. Imry, S. Ma: Phys. Rev. Lett. **35**, 1399 (1975)
D.J. Bergman, T.M. Rice, P.A. Lee: Phys. Rev. B**15**, 1706 (1977)

2.23 J. Hubbard: Proc. R. Soc. (London) A**276**, 238 (1963); ibid A**281**, 401 (1964)

2.24 H. Sumi: Solid State Commun. **21**, 17 (1977)
J.B. Torrance: Phys. Rev. B**17**, 3099 (1978)

2.25 See for example, E.P. Wigner: Phys. Rev. **46**. 1002 (1934)
Proc. Intern. Conf. on Electronic Properties of Two-Dimensional Systems, ed. S. Kawaji, Surface Science **98** (1980)
2.26 J. Kondo, K. Yamaji: J. Phys. Soc. Jpn. **43**, 424 (1977)
2.27 J. Hubbard: Phys. Rev. B**17**, 494 (1978)
2.28 P.A. Lee, T.M. Rice, R.A. Klemm: Phys. Rev. B**15**, 2984 (1977)
V.J. Emery: Phys. Rev. Lett. **37**, 107 (1976)
2.29 I.S. Jacobs, J.W. Bray, H.R. Hart, Jr., L.V. Interrante, J.S. Kasper, G.D. Watkins: Phys. Rev. B**14**, 3036 (1976)
2.30 C. Kittel: *Quantum Theory of Solids* (Wiley, New York 1963) Chapter 4

Chapter 3

3.1 M.M. Labes, R. Sehr, M. Bose: J. Chem. Phys. **33**, 868 (1960)
H. Kuroda, M. Kobayashi, M. Kinoshita, S. Takemoto: J. Chem. Phys. **36**, 457 (1962)
M.S. Frant, R. Eiss: J. Electrochem. Soc. **110**, 769 (1963)
3.2 D.S. Acker, R.J. Harder, W.R. Hertler, W. Mahler, L.R. Melby, R.E. Benson, W.R. Mochel: J. Am. Chem. Soc. **82**, 6408 (1960)
D.S. Acker, W.R. Hertler: ibid **84**, 3370 (1962)
3.3 F. Wudl, G.M. Smith, E.J. Hufnagel: J. Chem. Soc. Chem. Commun. (1970) 1453
J.P. Ferraris, D.O. Cowan, V. Walatka, Jr., J.H. Perlstein: J. Am. Chem. Soc. **95**, 948 (1973)
3.4 L.B. Coleman, M.J. Cohen, D.J. Sandman, F.G. Yamagishi, A.F. Garito, A.J. Heeger: Solid State Commun. **12**, 1125 (1973)
3.5 M. Narita, C.V. Pittman: Synthesis 489 (1976)
3.6 H. Anzai, S. Saito: *Organic Metals,* Circulars of the Electrotechnical Laboratory No.194 (Electrotechnical Laboratory, Niihari-gun, Ibaraki-ken, Japan 1978)
G. Bahr, G. Schleitzer: Chem. Ber. **88**, 1771 (1955); ibid **90**, 438 (1957)
E. Klingsberg: J. Am. Chem. Soc. **86**, 5290 (1964)
F. Wudl, G.M. Smith, E.J. Hufnagel: Chem. Commun. 1453 (1970)
3.7 H. Anzai: J. Cryst. Growth **33**, 185 (1976); see also [3.6]
3.8 P. Chaudhari, B.A. Scott, R.B. Laibowitz, Y. Tomkiewicz, J.B. Torrance: Appl. Phys. Lett. **24**, 439 (1974)
3.9 T.E. Phillips, T.J. Kistenmacher, J.P. Ferraris, D.O. Cowan: J. Chem. Soc. Chem. Commun. 471 (1973)
T.J. Kistenmacher, T.E. Phillips, D.O. Cowan: Acta Crystallogr. B**30**, 763 (1974)
3.10 P. Coppens: Phys. Rev. Lett. **35**, 98 (1975)
3.11 W.D.Grobman, B.D. Silverman: Solid State Commun. **19**, 319 (1976)
3.12 M.J. Cohen, L.B. Coleman, A.F. Garito, A.J. Heeger: Phys. Rev. B**10**, 1298 (1974)
S. Etemad: Phys. Rev. B**13**, 2254 (1976)
T. Ishiguro, S. Kagoshima, H. Anzai: J. Phys.Soc. Jpn. **41**, 351 (1976)
M.J. Cohen, A.J. Heeger: Phys. Rev. B**16**, 688 (1977)
3.13 S. Etemad: Phys. Rev. B**13**, 2254 (1976)
3.14 A.A. Bright, A.F. Garito, A.J. Heeger: Solid State Commun. **13**, 943 (1973)
D.B. Tanner, C.S. Jacobsen, A.F. Garito, A.J. Heeger: Phys. Rev. B**13**, 3381 (1976)
C.S. Jacobsen: *Quasi One-Dimensional Conductors I,* ed. by S. Barisic, A. Bjelis, J.R. Cooper, B. Leontic, Lecture Notes in Phys. Vol. 95 (Springer, Berlin, Heidelberg 1979) p. 223
3.15 See for example, J.M. Ziman: *Principles of the Theory of Soilds,* 2nd ed. (Cambridge University Press, Cambridge 1972) Chapter 8
3.16 J.F. Kwak, P.M. Chaikin, A.A. Russel, A.F. Garito, A.J. Heeger: Solid State Commun. **16**, 729 (1975)
P.M. Chaikin, R.L. Greene, S. Etamed, E.M. Engler: Phys. Rev. B**13**, 1627 (1976)
3.17 T. Ishiguro, H. Sumi, S. Kagoshima, K. Kajimura, H. Anzai: J. Phys. Soc. Jpn. **48**, 456 (1980)

G. Soda, D. Jerome, M. Weger, J. Alizon, J. Gallice, J.M. Fabre, L. Giral: J. de Phys. **38**, 931 (1977)
3.18 D.B. Tanner, C.S. Jacobsen, A.F. Garito, A.J. Heeger: Phys. Rev. Lett. **32**, 1301 (1974)
A.A. Bright, A.F. Garito, A.J. Heeger: Phys. Rev. **B10**, 1328 (1974)
C.S. Jacobsen: In *Quasi One-Dimensional Conductors I*, ed. by S. Barisic, A. Bjelis, J.R. Cooper, B. Leontic, Lecture Notes in Physics, Vol. 95 (Springer, Berlin, Heidelberg 1979) p. 223
3.19 J.E. Eldrige: Solid State Commun. **19**, 607 (1976)
3.20 M.J. Cohen, P.R. Newman, A.J. Heeger: Phys. Rev. Lett. **37**, 1500 (1976)
M.J. Cohen, A.J. Heeger: Phys. Rev. **B16**, 688 (1977)
3.21 Y. Tomkiewicz, B.A. Scott, L.J. Tao, R.S. Title: Phys. Rev. Lett. **32**, 1363 (1974)
3.22 J.C. Scott, A.F. Garito, A.J. Heeger: Phys. Rev. **B10**, 3131 (1974)
3.23 See for example, C. Kittel: *Introduction to Solid State Physics*, 5th ed. (Wiley, New York 1976) Chapter 14
3.24 J.B. Torrance, Y. Tomkiewicz, B.D. Silverman: Phys. Rev. **B15**, 4738 (1977)
3.25 P.A. Lee, T.M. Rice, R.A. Klemm: Phys. Rev. **B15**, 2984 (1977)
3.26 Y. Tomkiewicz, E.M. Engler, T.D. Schultz: Phys. Rev. Lett. **35**, 456 (1975)
Y. Yafet: *Solid State Physics*, ed. by H. Ehrenfreund, F. Seitz, D. Turnbull (Academic, New York 1963) Vol. 14
3.27 Y. Tomkiewicz, A.R. Taranko, J.B. Torrance: Phys. Rev. Lett. **36**, 751 (1976)
3.28 Y. Tomkiewicz, A.R. Taranko, J.B. Torrance: Phys. Rev. **B15**, 1017 (1977)
3.29 See for example, A. Abragam: *Nuclear Magnetism* (Oxford University Press, Oxford 1961) Chapter VI
3.30 E.F. Rybaczewski, A.F. Garito, A.J. Heeger, E. Ehrenfreund: Phys. Rev. Lett. **34**, 524 (1975)
3.31 E.F. Rybaczewski, L.S. Smith, A.F. Garito, A.J. Heeger, B.G. Silbernagel: Phys. Rev. **B14**, 2746 (1976)
3.32 G. Soda, D. Jérome, M. Weger, J. Alizon, J. Gallice, H. Robert, J.M. Fabre, L. Giral: J. de Phys. **38**, 931 (1977)
3.33 See for example, R.W. James: *The Optical Principles of the Diffraction of X-Rays* (B. Bell and Sons, London, new ed. 1950) Chapter V
A.A. Maradudin, E.W. Montrol, G.H. Weiss, I.P. Ipatova: *Solid State Physics*, suppl. 3, ed. H. Ehrenfreund et al. (Academic Press, New York 1971) Chapter VII
3.34 S. Kagoshima, H. Anzai, K. Kajimura, T. Ishiguro: J. Phys. Soc. Jpn. **39**, 1143 (1975)
S. Kagoshima, T. Ishiguro, H. Anzai: J. Phys. Soc. Jpn. **41**, 2061 (1976)
3.35 F. Denoyer, R. Comès, A.F. Garito, A.J. Heeger: Phys. Rev. Lett. **35**, 445 (1975)
J.P. Pouget, S.K. Khanna, F. Denoyer, R. Comès, A.F. Garito, A.J. Heeger: Phys. Rev. Lett. **37**, 437 (1976)
S.K. Khanna, J.P. Pouget, R. Comès, A.F. Garito, A.J. Heeger: Phys. Rev. **B16**, 1468 (1976)
R. Comès, G. Shirane: *Highly Conducting One-Dimensional Solids*, ed. by J.T. Devreese, R.P. Evrard, V.E. van Doren (Plenum, New York 1979) p. 17
3.36 G. Shirane, S.M. Shapiro, R. Comès, A.F. Garito, A.J. Heeger: Phys. Rev. **B14**, 2325 (1976)
J.P. Pouget, S.M. Shapiro, G. Shirane, A.F. Garito, A.J. Heeger: Phys. Rev. **B19**, 1792 (1979)
3.37 R. Comès, S.M. Shapiro, G. Shirane, A.F. Garito, A.J. Heeger: Phys. Rev. Lett. **35**, 1518 (1975)
W.D. Ellenson, R. Comès, S.M. Shapiro, G. Shirane, A.F. Garito, A.J. Heeger: Solid State Commun. **20**, 53 (1976)
R. Comès, G. Shirane, S.M. Shapiro, A.F. Garito, A.J. Heeger: Phys. Rev. **B14**, 2376 (1976)
W.D. Ellenson, S.M. Shapiro, G. Shirane, A.F. Garito: Phys. Rev. **B16**, 3244 (1977); see also [3.34, 5]
3.38 P. Bak, V.J. Emery: Phys. Rev. Lett. **36**, 978 (1976)

3.39 T.D. Schultz, S. Etemad: Phys. Rev. B**13**, (1976) 4928
T.D. Schultz: Solid State Commun. **22**, 289 (1977)
3.40 A. Bjelis, S. Barisic: Phys. Rev. Lett. **37**, 1517 (1976)
M. Weger, J. Friedel: J. de Phys. **38**, 241 (1977)
P. Bak: Phys. Rev. Lett. **37**, 1071 (1976)
B. Horovitz, D. Mukamel: Solid State Commun. **23**, 285 (1977)
3.41 J. Yamauchi, K. Sato, S. Iwabuchi, Y. Nagaoka: J. Phys. Soc. Jpn. **44**, 460 (1978)
K. Sato, S. Iwabuchi, J. Yamauchi, Y. Nagaoka: J. Phys. Soc. Jpn. **45**, 515 (1978)
3.42 C. Weyl, E.M. Engler, K. Bechgaard, G. Jehanno, S. Etemad: Solid State Commun. **19**, 925 (1976)
3.43 S. Kagoshima: J. Phys. Soc. Jpn. **49**, 857 (1980) suppl. A
3.44 A.J. Berlinsky, J.F. Carolan, L. Weiler: Solid State Commun. **15**, 795 (1975)
T.D. Schultz, R.A. Craven: *Highly Conducting One-Dimensional Solids,* ed. J.T. Devreese (Plenum Press, New York, 1978) p. 147
3.45 A. Andrieux, H.J. Schulz, D. Jérome, K. Bechgaard: J. de Phys. Lett. **40**, L-385 (1979)
A. Andrieux, H.J. Schulz, D. Jérome, K. Bechgaard: Phys. Rev. Lett. **43**, 227 (1979)
3.46 S. Megtert, R. Comès, C. Vettier, R. Pynn, A.F. Garito: Solid State Commun. **31**, 977 (1979)
3.47 R.H. Friend, M. Miljak, D. Jérome: Phys. Rev. Lett. **40**, 1048 (1978)
3.48 Y. Suzumura, H. Fukuyama: J. Phys. Soc. Jpn. **49**, 915 (1980); ibid **49**, 2081 (1981)
3.49 J.F. Thomas, D. Jérome: Solid State Commun. **36**, 813 (1980)
3.50 H. Fukuyama, P.A. Lee: Phys. Rev. B**17**, 535 (1978)
3.51 T.D. Schultz, R.A. Craven: *Highly Conducting One-Dimensional Solids,* ed. J.T. Devreese (Plenum Press, New York 1978) p. 147
3.52 P.M. Chaikin, R.L. Greene, E.M. Engler: Solid State Commun. **25**, 1009 (1978)
3.53 S. Kagoshima, T. Ishiguro, T.D. Schultz, Y. Tomkiewicz: Solid State Commun. **28**, 485 (1978)
3.54 S. Megtert, A.F. Garito, J.P. Pouget, R. Comès: *Quasi One-Dimensional Conductors I,* ed. S. Barisic, A. Bjelis, J.R. Cooper, B. Leontic, Lecture Notes in Physics, Vol. 95 (Springer, Berlin, Heidelberg 1979) p. 57
3.55 R.L. Greene, J.J. Mayerle, R. Schumaker, G. Castro, P.M. Chaikin, S. Etemad, S.J. Laplaca: Solid State Commun. **20**, 943 (1976)
J.R. Cooper, M. Weger, D. Jérome, D. Lefur, K. Bechgaard, A.N. Bloch, D.O. Cowan: Solid State Commun. **19**, 749 (1976)
3.56 T.E. Phillips, T.J. Kistenmacher, A.N. Bloch, D.O. Cowan: J. Chem. Soc. Chem. Commun. 334 (1976)
3.57 J.P. Pouget, S. Megtert, R. Comès: *Quasi One-Dimensional Conductors I,* ed. S. Barisic, A. Bjelis, J.R. Cooper, B. Leontic, Lecture Notes in Physics (Springer, Berlin, Heidelberg 1979) p. 14
3.58 A. Bjelis, S. Barisic: J. de Phys. Lett. **39**, L437 (1978)
3.59 C.J. Fritchie: Acta Cryst. **20**, 892 (1966)
B. Morosin: Phys. Lett. **53A**, 455 (1975)
3.60 A.J. Epstein, S. Etemad, A.F. Garito, A.J. Heeger: Phys. Rev. B**5**, 952 (1972)
3.61 A.J. Epstein, E.M. Conwell, D.J. Sandman, J.S. Miller: Solid State Commun. **23**, 355 (1977)
3.62 J.P. Pouget, S. Megtert, R. Comès, A.J. Epstein: Phys. Rev. B**21**, 486 (1980)
3.63 J.P. Pouget, R. Comès, K. Bechgaard: *Low-Dimensional Solids,* ed. L. Alcacer (Reidel, Dordrecht, Holland 1980) p. 113
3.64 J.R. Andersen, K. Bechgaard, C.S. Jacobsen, G. Rindorf, H. Soling, N. Thorup: Acta Cryst. B**34**, 1901 (1978)
3.65 A. Andrieux, P.M. Chaikin, C. Duroure, D. Jérome, C. Weyl, K. Bechgaard, J.R. Andersen: J. de Phys. **40**, 1199 (1979)
3.66 K. Bechgaard, C.S. Jacobsen, K. Mortensen, H.J. Pedersen, N. Thorup: Solid State Commun. **33**, 1119 (1980)

3.67 D. Jérome, A. Mazaud, M. Ribault, K. Bechgaard: J. de Phys. Lett. **41**, L95 (1980)
M. Ribault, G. Benedek, D. Jérome, K. Bechgaard: J. de Phys. Lett. **41**, L397 (1980)
3.68 M. Ribault, J.P. Pouget, D. Jérome, K. Bechgaard: J. de Phys. Lett. **41**, L607 (1980)
3.69 S.S.P. Parkin, M. Ribault, D. Jérome, K. Bechgaard: to be published
3.70 N. Thorup, T. Rindorf, H. Soling, K. Bechgaard: Acta Crystallogr. **B37**, 1236 (1981)
3.71 C.S. Jacobsen, D.B. Tanner, K. Bechgaard: Phys. Rev. Lett. **46**, 1142 (1981)
3.72 H.J. Schulz, D. Jérome, M. Ribault, A. Mazaud, K. Bechgaard: J. de Phys. Lett. **42**, L51 (1981)
R.L. Greene, E.M. Engler: Phys. Rev. Lett. **45**, 1587 (1980)
3.73 K. Andres, F. Wudl, D.B. McWhan, G.A. Thomas, D.Nalewajek, A.L. Stevens: Phys. Rev. Lett. **45**, 1449 (1980)
3.74 T. Ishiguro, K. Kajimura, H. Bando, K. Murata, H. Anzai: Mol. Cryst. Liq. Cryst. **119**, 19 (1985)
3.75 J.P. Pouget: *Proceedings of the International Conference on Low-Dimensional Synthetic Metal*, Helsingor, Denmark, Chem. Sci. **17**, 85 (1980)
3.76 J.C. Scott, H.J. Pedersen, K. Bechgaard: Phys. Rev. Lett. **45**, 2125 (1980)
3.77 H.J. Pedersen, J.C. Scott, K. Bechgaard: Solid State Commun. **35**, 207 (1980)
W.M. Walsch, F. Wudl, G.A. Thomas, D. Nalewajek, J.J. Hauser, P.A. Lee, T. Poehler: Phys. Rev. Lett. **45**, 829 (1980)
3.78 A. Andrieux, D. Jérome, K. Bechgaard: J. de Phys. Lett. **42**, L87 (1981)
3.79 R. Moret, J.P. Pouget, R. Comès, K. Bechgaard: Phys. Rev. **49**, 1008 (1982)
3.80 J.P. Pouget, G. Shirane, K. Bechgaard, J.M. Fabre: Phys. Rev. **B27**, 5203 (1983)
3.81 R. Moret, J.P. Pouget, R. Comès, K. Bechgaard: J. de Phys. **44**, C3-967 (1983)
3.82 T. Takahashi, D. Jérome, K. Bechgaard: J. de Phys. Lett. **43**, L566 (1982)
3.83 S. Kagoshima, T. Yasunaga, T. Ishiguro, H. Anzai, G. Saito: Solid State Commun. **46**, 867 (1983)
3.84 V.J. Emery: J. de Phys. Colloq. **44**, C3-977 (1983)
H. Gutfreund, B. Horovitz, M. Weger: ibid C3-983
3.85 K. Yamaji: J. Phys. Soc. Jpn. **51**, 2787 (1982)
3.86 S.S.P. Parkin, E.M. Engler, R.R. Schumaker, R. Lagier, V.Y. Lee, J. Voiron, K. Carneiro, J.C. Scott, R.L. Greene: J. de Phys. Colloq. **44**, C3-791 (1983)
3.87 R.P. Shibaeva, V.F. Kaminskii, E.B. Yagubskii: Mol. Cryst. Liq. Cryst. **119**, 361 (1985)
P.C.W.Leung, T.J. Emge, A.J. Schultz, M.A. Beno, K.D. Carlson, H.H. Wang, M.A. Firestone, J.M. Williams: Solid State Commun., **57**, 93 (1986)
3.88 E.B. Yagubskii, I.F. Shchegolev, V.N. Laukhin, P.A. Kononovich, M.V. Kartsovnic, A.V. Zvarykina, L.I. Buravov: JETP Lett. **39**, 12 (1984)
K. Murata, M. Tokumoto, H. Anzai, H. Bando, G. Saito, K. Kajimura, T. Isiguro: J. Phys. Soc. Jpn. **54**, 1236 (1985); ibid 2084
3.89 M. Tokumoto, K. Murata, H. Bando, H. Anzai, G. Saito, K. Kajimura, T. Ishiguro: Solid State Commun. **54**, 1031 (1985)

Chapter 4

4.1 J.A. Wilson, A.D. Yoffe: Adv. Phys. **18**, 193 (1969)
4.2 J.A. Wilson, F.J. DiSalvo, S. Mahajan: Adv. Phys. **24**, 117 (1975)
4.3 H. Schäfer: *Chemical Transport Reactions* (Academic, New York 1964)
4.4 F. Hulliger: *Structural Chemistry of Layer-Type Phases* (D. Reidel Publishing Co., 1976) p.247
4.5 J.L. Hodeau, M. Marezio, C. Roucau, R. Ayroles, A. Meerschaut, J. Rouxel, P. Monceau: J. Phys. **C11**, 4117 (1978)
4.6 J.A. Wilson: Phys. Rev. **B19**, 6456 (1979)
4.7 E. Bjerkelund, J.H. Fermor, A. Kjekshus: Acta Chem. Scand. **20**, 1836 (1966)

4.8 A. Meershaut, L. Guemas, J. Rouxel: J. Solid State Chem. **36**, 118 (1981)
4.9 E. Bjerkelund, A. Kjekshus: Acta Chem. Scand. **19**, 701 (1965)
4.10 S. Kikkawa, N. Ogawa, M. Koizumi, Y. Ohnuki: J. Solid State Commun. **41**, 315 (1982)
4.11 R.M. Fleming, J.A. Polo, Jr., R.V. Coleman: Phys. Rev. B**17**, 1634 (1978)
4.12 P. Monceau: Solid State Commun. **24**, 331 (1977)
4.13 D.W. Bullett: J. Phys. C**15**, 3069 (1982)
R. Hoffmann, S. Shaik, J.C. Scott, M. Whangbo, M.J. Foshee: J. Solid State Chem. **34**, 263 (1980)
N. Shima: J. Phys. Soc. Jpn. **51**, 11 (1982)
4.14 see texts on solid state theory, for example; N.F. Mott, H. Jones: *The Theory of the Properties of Metals and Alloys* (Dover 1958) Chapter 7
4.15 For example; A.A. Abrikosov: *Introduction to the Theory of Metals* (Academic Press, 1972) Chapter 11
4.16 K. Tsutsumi, T. Takagaki, M. Yamamoto, Y. Shiozaki, M. Ido, T. Sambongi, K. Yamaya, Y. Abe: Phys. Rev. Lett. **34**, 1675 (1977)
4.17 R.M. Fleming, D.E. Moncton, D.B. McWhan: Phys. Rev. B**18**, 5560 (1978)
4.18 W.L. McMillan: Phys. Rev. B**14**, 1496 (1976)
4.19 K. Nakanishi, H. Shiba: J. Phys. Soc. Jpn. **45**, 1147 (1978)
4.20 R.M. Fleming: J. de Phys. C**3**, 1651 (1983)
4.21 H.C. Montgomery: J. Appl. Phys. **42**, 2971 (1971)
4.22 N.P. Ong, J.W. Brill: Phys. Rev. B**15**, 5265 (1978)
4.23 M. Ido: Mol. Cryst. Liq. Cryst. **77**, 809 (1981)
4.24 N.P. Ong, P. Monceau: Solid State Commun. **26**, 487 (1978)
4.25 T. Takagaki, M. Ido, T. Sambongi: J. Phys. Soc. Jpn. **45**, 2039 (1978)
R.H. Dee, P.M. Chaikin, N.P. Ong: Phys. Rev. Lett. **42**, 1234 (1979)
4.26 S. Tomic, K. Biljakovic, D. Djurek, J.R. Cooper, P. Monceau, A. Meerschaut: Solid State Commun. **38**, 109 (1981)
4.27 J.D. Kulick, J.C. Scott: Solid State Commun. **32**, 217 (1979)
4.28 R.M. Fleming, C.C. Grimes: Phys. Rev. Lett. **42**, 1423 (1979)
4.29 K. Kawabata, M. Ido, T. Sambongi: J. Phys. Soc. Jpn. **50**. 739 (1981)
4.30 J. Richard, P. Monceau: Solid State Commun. **33**, 635 (1980)
4.31 N.P. Ong, P. Monceau: Phys. Rev. B**16**, 3433 (1977)
4.32 R.M. Fleming: Phys. Rev. B**22**, 5606 (1980)
4.33 G.X. Tessema, N.P. Ong: Phys. Rev. B**23**, 5607 (1981)
4.34 P.A. Lee, T.M. Rice: Phys. Rev. B**19**, 3970 (1979)
4.35 J.W. Bril, N.P. Ong, J.C. Eckert, J.W. Savage, S.K. Khanna, R.B. Somoano: Phys. Rev. B**23**, 1517 (1981)
4.36 W.W. Fuller, G. Grunner, P.M. Chaikin, N.P. Ong: Phys. Rev. B**23**, 6259 (1981)
4.37 P. Monceau, J. Richard, M. Renard: Phys. Rev. B**25**, 931 (1982)
4.38 G. Gruner, L.C. Tippie, J. Sanny, W.G. Clark, N.P. Ong: Phys. Rev. Lett. **45**, 935 (1980)
4.39 G. Gruner, A. Zawadowski, P.M. Chaikin: Phys. Rev. Lett. **46**, 511 (1981)
4.40 P. Monceau, J. Richard, M. Renard: Phys. Rev. Lett. **45**, 43 (1980)
4.41 J. Bardden: Mol. Cryst. Liq. Cryst. **81**, 1 (1981)
4.42 B.A. Hubermann, J.P. Crutchfield: Phys. Rev. Lett. **43**, 1743 (1979)
4.43 N.P. Ong, G.Verma: Phys. Rev. B**27**, 4495 (1982)
4.44 J.C. Gill: Solid State Commun. **39**, 1203 (1981)
4.45 A. Briggs, P. Monceau, M. Nunez-Regueiro, J. Peyrard, M. Ribault, J. Richard: J. Phys. C**13**, 2117 (1980)
4.46 C.M. Bastuscheck, C.M. Buhrman, R.A. Kulick, J.C. Scott: Solid State Commun. **36**, 983 (1980)
4.47 K. Kawabata, M. Ido: Solid State Commun. **44**, 1539 (1982)
4.48 K. Kawabata, M. Ido, T. Sambongi: J. de Phys. C**3**, 1647 (1983)
4.49 W.W. Fuller, P.M. Chaikin, N.P. Ong: Phys. Rev. B**24**, 1333 (1981)
4.50 T. Sambongi, K. Tsutsumi, Y. Shiozaki, M. Yamamoto, K. Yamaya, Y. Abe: Solid State Commun. **22**, 729 (1977)

4.51 K. Tsutusimi, T. Sambongi, S. Kagoshima, T. Ishiguro: J. Phys. Soc. Jpn. **44**, 1735 (1978)
4.52 C. Roucau: J. de Phys. C3, 1725 (1983)
4.53 M. Ido, K. Tsutsumi, T. Sambongi, N. Mori: Solid State Commun. **29**, 399 (1979)
4.54 T. Takoshima, M. Ido, K. Tsutsumi, T. Sambongi, S. Homma, K. Yamaya, Y. Abe: Solid State Commun. **35**, 911 (1980)
4.55 J.C. Gill: Solid State Commun. **47**, 737 (1983)
4.56 P.A. Lee, T.M. Rice, P.W. Anderson: Solid State Commun. **14**, 703 (1974)
4.57 P. Monceau, H. Salva, Z.Z. Wang: J. de Phys. C3, 1639 (1983)
4.58 G. Gruner, A. Zettl: Phys. Rep. **119**, 117 (1985)
4.59 G. Hutiray, J. Solyom (ed.): *Charge density waves in Solids*, Lect. Notes Phys. (Springer, Berlin, Heidelberg 1985)
4.60 P. Monceau (ed.): *Electronic Properties of Inorganic quasi One-Dimensional Metals* (D. Reidel, 1985)
4.61 K. Yamaya, T.H. Geballe, J.F. Kwak, R.L. Greene: Solid State Commun. **31**, 627 (1979)
4.62 W.L. McMillan: Phys. Rev. **167**, 331 (1968)
4.63 P.B. Allen, B. Mitrovic: in *Solid State Physics* Vol. 37, ed. H. Ehrenreich, F. Seitz, and D. Turnbull (Academic Press) 1982
4.64 K. Yamaya, G. Ohmi: J. Phys. Soc. Jpn. **51**, 3512 (1982); **52**, 1886 (1983)
4.65 Y. Yamamoto: J. Phys. Soc. Jpn. **45**, 431 (1978)
4.66 M. Tinkham: *Introduction to Superconductivity* (McGraw Hil, 1975) 113
4.67 C.M. Bastuscheck, R.A. Buhrman, J.D. Kulick, J.C. Scott: Mol. Cryst. Liq. Cryst. **81**, 155 (1982)
4.68 V.M. Pan, I.E. Bulakh, A.L. Kasatkin, A.D. Shevchenko: JETP Lett. **27**, 595 (1978)
4.69 R.M. Fleming, F.J. DiSalvo, R.J. Cava, J.V. Waszczak: Phys. Rev. B**24**, 2850 (1981)

Chapter 5

5.1 W. Knop: Justus Liebig's Ann. Chem. **43**, 111 (1842)
5.2 K. Krogmann, H.D. Hausen: Z. Anorg. Allg. Chem. **358**, 67 (1968)
5.3 L.A. Levy: J. Chem. Soc. (1912) 1081
5.4 K. Krogmann, G. Ringwald: Z. Naturforsch. **23B**, 1112 (1968)
5.5 K. Krogmann, H.D. Hausen: Z. Naturforsch. **23B**, 1111 (1968)
5.6 K. Krogmann: Angew. Chem. Int. Edn. **8**, 35 (1969)
5.7 K. Krogmann: Z. anorg. Chem. **358**, 97 (1968)
5.8 J.M. Williams, R.K. Brown: Inorg. Chem. **17**, 2607 (1978)
5.9 J.M. Williams, D.D. Gerrity, A.J. Schultz: J. Amer. Chem. Soc. **99:5**, 1668 (1977)
5.10 J.M. Williams, A.J. Schultz, K.B. Cornett, R.E. Besinger: J. Amer. Chem. Soc. **100:17**, 5572 (1978)
5.11 A. Kobayashi, Y. Sasaki, H. Kobayashi: Bull. Chem. Soc. Japan **52**, 3682 (1979)
5.12 A.J. Schultz, G.D. Stucky, J.M. Williams, T.R. Koch, R.L. Maffly: Solid State Commun. **21**, 197 (1977)
5.13 H.J. Pedersen, A.E. Underhill: Solid State Commun. **33**, 289 (1980)
5.14 A.E. Underhill, D.M. Watkins, C.S. Jacobsen: Solid State Commun. **36**, 477 (1980)
5.15 R.K. Brown, J.M. Williams: Inorg. Chem. **18**, 801 (1979)
5.16 J. Bernasconi, P. Brüesch, H.R. Zeller: J. Phys. Chem. Solid **35**, 145 (1974)
H.R. Zeller, A. Beck: J. Phys. Chem. Solid **35**, 77 (1974)
5.17 M.J. Minot, J.H. Perlstein, T.J. Kistenmacher: Solid State Commun. **13**, 1319 (1973)
H.J. Guggenheim, D. Bahnck: J. Cryst. Growth **26**, 29 (1974)
5.18 R.B. Saillant, R.C. Jaklevic, C.D. Bedford: Mat. Res. Bull. **9**, 289 (1974)
5.19 H.R. Zeller: Adv. in Solid State Phys. (Festkörperproblem) **13**, 31 (1973)
5.20 P. Brüesch: Solid State Commun. **13**, 13 (1973)

5.21 C. Peter, C.E. Eagan: Phys. Rev. Lett. **34**, 1132 (1975)
5.22 G. Heger, H.J. Deiseroth, H. Schulz: Acta Cryst. B**34**, 725 (1978)
5.23 D. Brinkmann: Solid State Commun. **19**, 1099 (1976)
5.24 J.M. Williams, J.L. Petersen, H.M. Gerdes, S.W. Peterson: Phys. Rev. Lett. **38**, 1077 (1974)
J.M. Williams, M. Iwata, S.W. Peterson, K. Leslie, H.J. Guggenheim: Phys. Rev. Lett. **34**, 1653 (1975)
5.25 L.V. Interrante, R.P. Messmer: Chem. Phys. Lett. **26**, 225 (1974)
5.26 A.S. Berenblyum, L.I. Baravov, M.D. Khidekel, I.F. Shchegolev, E.B. Yakimov: JETP Lett. **13**, 440 (1971)
I.F.Shchegolev: Phys. Status Solidi A**12**, 9 (1972)
5.27 A.N. Bloch, R.B. Weissmann, C.M. Verma: Phys. Rev. Lett. **28**, 753 (1972)
5.28 H.P. Geserich, H.D. Hausen, K. Krogmann, P. Stampel: Phys. Status Solidi A**9**, 187 (1972)
5.29 D. Kuse, H.R. Zeller: Phys. Rev. Lett. **27**, 1060 (1971)
H. Wagner, H.P. Geserich, R.V. Baltz, K. Krogmann: Solid State Commun. **13**, 659 (1973)
5.30 D.L. Rousseau, M.A. Butler, H.J. Guggenheim, R.B. Weissmann, A.N. Bloch: Phys. Rev. B**10**, 2281 (1974)
5.31 J.R. Ferraro, L.J. Basile, J.M. Williams: J. Chem. Phys. **64**, 732 (1976)
5.32 E.F. Steigmeier, R. Loudon, G. Harbeke, H. Auderset: Solid State Commun. **17**, 1447 (1975)
5.33 A. Menth, M.J. Rice: Solid State Commun. **11**, 1025 (1972)
5.34 P.I. Kuindersma, G.A. Swatzky: Solid State Commun. **13**, 39 (1973)
5.35 T. Takahashi, H. Akagawa, H. Doi, H. Nagasawa: Solid State Commun. **23**, 809 (1977)
5.36 F. Mehran, B.A. Scott: Phys. Rev. Lett. **31**, 1347 (1973)
5.37 T. Takahashi, H. Doi, H. Nagasawa: J. Phys. Soc. Jpn. **48**, 423 (1980)
5.38 H.H. Rupp: Z. Naturforsch. A**26**, 1937 (1971)
H.J. Keller, H.H. Rupp: Z. Naturforsch. A**26**, 785, 2066 (1971)
5.39 C. Slichter: *Principles of Magnetic Resonances* (Harper & Row, New York 1968)
5.40 A. Lee, T.M. Rice, P.W. Anderson: Phys. Rev. Lett. **31**, 462 (1973)
5.41 H. Nagasawa: J. Phys. Soc. Jpn. **45**, 701 (1978)
5.42 H. Niedoba, H. Launois, D. Brinkmann, R. Brugger, H.R. Zeller: Phys. Status Solidi B**58**, 309 (1973)
5.43 H. Niedoba, H. Launois, D. Brinkmann, H.V. Keller: J. de Phys. Lett. **35**, 251 (1974)
5.44 H. Fukuyama: J. Phys. Soc. Jpn. **45**, 1266 (1978)
5.45 E. Ehrenfreund, E.F. Rybaszewski, A.F. Garito, A.J. Heeger: Phys. Rev. Lett. **28**, 873 (1972)
5.46 R.E. Peierls: *Quantum Theory of Solids.* (Oxford University Press, Oxford 1955) p. 108
5.47 R. Comès, M. Lambert, H.R. Zeller: Phys. Status Solidi B**58**, 587 (1973)
R. Comès, M. Lambert, H. Launois, H.R. Zeller: Phys. Rev. B**8**, 571 (1973)
5.48 H.J. Deiseroth, H. Schulz: Phys. Rev. Lett. **33**, 963 (1974)
J.M. Williams, M. Iwata, F.K. Ross, J.L. Petersen, S.W. Peterson: Mat. Res. Bull. **10**, 411 (1975)
5.49 J.W. Lynn, M. Iizumi, G. Shirane, S.A. Werner, R.B. Saillant: Phys. Rev. B**12**, 1154 (1975)
5.50 B. Renker, L. Pintchovius, W. Glaser, H. Rietschel, R. Comès, L. Liebert, W. Drexel: Phys. Rev. Lett. **32**, 836 (1974)
5.51 J.C. Phillips: Phys. Status Solidi B**77**, 259 (1976)
J.C. Phillips: Phys. Status Solidi B**79** 111 (1977)
5.52 K. Carneiro, G. Shirane, S.A. Werner, S. Kaiser: Phys. Rev. B**13**, 4258 (1976)
5.53 P. Brüesch, H.R. Zeller: Solid State Commun. **14**, 1037 (1974)
5.54 R. Comès, G. Shirane: *Highly Conducting One-Dimensional Solids.* ed. J.T. Devreese, R.P. Evrard, V.E. van Doren (Plenum, New York 1979) p. 17

5.55 H. Niki, H. Doi, H. Nagasawa: J. Phys. Soc. Jpn. **51**, 2470 (1982)
5.56 S. Drosdziok, M. Engbrodt: Solid State Commun. **17**, 1339 (1975)
S. Drosdziok, M. Engbrodt: Phys. Status Solidi B**72**, 739 (1975)
G. Czycholl, N. Grewe, H.J. Leder, S. Drosdziok: Solid State Commun. **22**, 67 (1977)
5.57 H.Doi, H. Nagasawa, T.Ishiguro, S. Kagoshima: Solid State Commun. **24**, 729 (1977)
5.58 S. Kurihara, H. Fukuyama, S. Nakajima: J. Phys. Soc. Jpn. **47**, 1403 (1979)
5.59 G.E. Pake: J. Chem. Phys. **41**, 327 (1948)
5.60 G.A. Sawatzky, E. Antonides: J. de Phys. C**4**, 117 (1976)
P.I. Kuindersma: PhD Thesis (University of Groningen, 1975)
5.61 W.Ruegg, D. Kuse, H.R. Zeller: Phys. Rev. B**8**, 952 (1973)
5.62 Y. Yafet, V. Jaccarino: Phys. Rev. **133A**, 1630 (1964)
5.63 R. Kubo, K. Tomita: J. Phys. Soc. Jpn. **9**, 888 (1954)
5.64 H. Nagasawa: Phys. Status Solidi B**109**, 749 (1982)
5.65 R.J. Elliot: Phys. Rev. **96**, 266 (1954)
5.66 A. Abragham, B. Bleaney: *Elecltron Paramagnetic Resonance of Transition Ions* (Clarendon, Oxford 1970)
5.67 R.E. Dietz, F.R. Merrit, R. Dingle, D. Hone, B.G. Silbernagel, P.M. Richards: Phys. Rev. Lett. **26**, 1186 (1971)
5.68 M.J. Hennessy, C.D. McElwee, P.M. Richards: Phys. Rev. B**7**, 930 (1973)
5.69 G. Soda, D. Jérome, M. Weger, J.A. Alizon, J. Gallice, R. Robert, J.M. Fabre, L. Giral: J. de Phys. **38**, 931 (1977)
5.70 P.W. Anderson, S.T. Chui: Phys. Rev. B**9**, 3229 (1974)
5.71 Ed. R.D. Parks: *Valence Instabilities and Related Narrow Band Phenomena* (Plenum, New York 1977)
5.72 K.B. Efetov, A.I. Larkin: Sov. Phys. JETP **42**, 390 (1976)
5.73 P.W. Anderson: Phys. Rev. Lett. **34**, 953 (1975)
5.74 T.M. Rice, L. Sneddon: Phys. Rev. Lett. **47**, 689 (1981)
5.75 J. Kondo, K. Yamaji: J. Phys. Soc. Jpn. **43**, 424 (1977)

Chapter 6

6.1 H. Shirakawa, M. Sato, A. Hamamo, S. Kawakami, K. Soga, S. Ikeda: Macromolecules **13**, 457 (1980)
6.2 G. Natta, G. Mazzanti, P. Corradini: Rend. Accad. Naz. Lincei [8] **25**, 3 (1958)
6.3 T. Ito, H. Shirakawa, S. Ikeda: J. Polym. Sci., Polym. Chem. Ed. **12**, 11 (1974)
6.4 H. Shirakawa, S. Ikeda: Polym. J. **2**, 231 (1971)
H. Shirakawa, T. Ito, S. Ikeda: ibid **4**, 460 (1973)
6.5 T. Ito, H. Shirakawa, S. Ikeda: J. Polym. Sci., Polym. Chem. Ed. **13**, 1943 (1975)
6.6 R.H. Baughman, S.L. Hsu, G.P. Pez, A.J. Signorelli: J. Chem. Phys. **68**, 5405 (1978)
R.H. Baughman, S.L. Hsu, L.R. Anderson, G.P. Pez, A.J. Signorelli: *Molecular Metals,* ed. W.E. Hatfield (Plenum, New York 1979) p. 187
T. Akaishi, K. Miyasaka, K. Ishikawa, H. Shirakawa, S. Ikeda: Pep. Porg. Polym. Phys. Jpn. **22**, 125 (1979)
6.7 H. Shirakawa, S. Ikeda: Synthetic Metals **1**, 175 (1979/80)
6.8 see for example, L. Salem: *Molecular Orbital Theory of Conjugated Systems* (Benjamin, New York 1966)
6.9 J. Kondo: Physica **98B**, 176 (1980)
6.10 H. Shirakawa, T. Ito, S. Ikeda: Makromol. Chem. **179**, 1565 (1978)
6.11 A. Karpfen, J. Petkov: Theort. Chim. Acta (Berl.) **53**, 65 (1979)
T.Yamabe, K. Tanaka, H. Terama-e, K. Fukui, A. Imamura, H. Shirakawa, S. Ikeda: J. Phys. C**12**, L257 (1979)
6.12 H. Shirakawa, T. Ito, S. Ikeda: Polym. J. **4**, 460 (1973)

6.13 C.K. Chiang, C.R. Fincher, Jr., Y.W. Park, A.J. Heeger, H. Shirakawa, E.J. Louis, S.C. Gau, A.G. MacDiarmid: Phys. Rev. Lett. **39**, 1098 (1977)
6.14 C.K. Chiang, S.C. Gau, C.R. Fincher, Jr., Y.W. Park, A.G. MacDiarmid, A.J. Heeger: Appl. Phys. Lett. **33**, 18 (1978)
C.K. Chiang, Y.W. Park, A.J. Heeger, H. Shirakawa, E.J. Louis, A.G. MacDiarmid: J. Chem. Phys. **69**, 5098 (1978)
6.15 J.F. Kwak, T.C. Clarke, R.L. Greene, G.B. Street: Solid State Commun. **31**, 355 (1979)
6.16 Y.W. Park, A. Denenstein, C.K. Chiang, A.J. Heeger, A.G. MacDiarmid: Solid State Commun. **29**, 747 (1979)
6.17 M. Tanaka, A. Watanabe, J. Tanaka: Bull. Chem. Soc. Jpn. **53**, 3430 (1980)
6.18 Y.W. Park, M.A. Druy, C.K. Chiang, A.G. MacDiarmid, A.J. Heeger, H. Shirakawa, S. Ikeda: J. Polym. Sci., Polym. Lett. Ed. **17**, 195 (1979)
6.19 B.R. Weinberger, J. Käufer, A. Pron, A.J. Heeger, A.G. MacDiarmid: Phys. Rev. B**20**, 223 (1979)
6.20 H. Shirakawa, T. Ito, S. Ikeda: Makromol. Chem. **179**, 1965 (1978)
6.21 I.B. Goldberg, H.R. Crowe, P.R. Newman, A.J. Heeger, A.G. MacDiarmid: J. Chem. Phys. **70**, 1132 (1979)
B.R. Weinberger, E. Ehrenfreund, A. Pron, A.J. Heeger, A.G. MacDiarmid: J. Chem. Phys. **70**, 4749 (1980)
M. Nechtenschein, F. Devreux, R.L. Greene, T.C. Clarke, G.B. Street: Phys. Rev. Lett. **44**, 356 (1980)
6.22 Y. Tomkiewicz, T.D. Schultz, H.B. Broom, T.C. Clarge, G.B. Street: Phys. Rev. **43**, 1532 (1979)
6.23 S. Ikeda, J. Käufer, T. Woerner, A. Pron, M.A. Druy, A. Sivak, A.J. Heeger, A.G. MacDiarmid: Phys. Rev. Lett. **45**, 1123 (1980)
6.24 W.P. Su, J.R. Schrieffer, A.J. Heeger: Phys. Rev. Lett. **42**, 1698 (1979)
6.25 M.J. Rice: Phys. Lett. **71A**, 152 (1979)
H. Takayama, Y.R. Lin-Liu, K. Maki: Phys. Rev. B**21**, 2388 (1980)
S.A. Brazovskii: Sov. Phys. JETP Lett. **28**, 606 (1978)
B. Horovitz, J.A. Krumhansl: Solid State Commun. **26**, 1698 (1979)
6.26 V.V. Walatka, M.M. Labes, J.H. Perlstein: Phys. Rev. Lett. **31**, 1139 (1973)
C.K. Chiang, M.J. Cohen, A.F. Garito, A.J. Heeger, C.M. Mikulski, A.G. MacDiarmid: Solid State Commun. **18**, 1451 (1976)
R.H. Friend, D. Jérome, S. Rehmatullah, A.D. Yoffe: J. Phys. C**10**, 1001 (1976)
K. Kaneto, M. Yamamoto, K. Yoshino, Y. Inuishi: J. Phys. Soc. Jpn. **47**, 167 (1979)
6.27 R.L. Greene, G.B. Street, L.J. Suter: Phys. Rev. Lett. **34**, 577 (1975)
6.28 M.H.M. Arnold, J.A.C. Hugill, J.M. Hutson: J. Chem. Soc. 1645 (1936)
C.M. Mikulski, P.J. Russo, M.S. Saran, A.G. MacDiarmid, A.F. Garito, A.J. Heeger: J. Amer. Chem. Soc. **97**, 6358 (1975)
6.29 M. Boudelle, P. Michel: Acta Cryst. A**28**, S199 (1972)
M.J. Cohen, A.F. Garito, A.J. Heeger, A.G. MacDiarmid, C.M.Mikulski, M.S. Saran, J. Kleppinger: J. Amer. Chem. Soc. **98**, 3844 (1976)
6.30 Z. Igbal, J. Sharma, R.H. Baughman, M. Akhtar, A.G. MacDiarmid: In *Quasi One-Dimensional Conductors II*, ed. S. Barisic, A. Bjelis, J.R. Cooper, B. Leontic, Lecture Notes in Phys., Vol. 96 (Springer, Berlin, Heidelberg 1979) p. 432
H. Morawitz, W.D. Gill, P.M. Grant, G.B. Street, D. Sayers: ibid p. 390
6.31 K. Kaneto, K. Yoshino, Y. Inuishi: *Proc. Symposium on Design of Inorganic and Organic Materials of Technological Importance*, Kyoto, 1979
6.32 P.M. Grant, R.L. Greene, G.B. Street: Phys. Rev. Lett. **35**, 1743 (1975)
H. Kamimura, A.J. Grant, F. Levy, A.D. Yoffe, G.D. Pitt: Solid State Commun. **17**, 49 (1975)
6.33 H. Kamimura, A.J. Grant, F. Levy, A.D. Yoffe, G.D. Pitt: Solid State Commun. **17**, 49 (1975)
6.34 R.L. Greene, P.M. Grant, G.B. Street: Phys. Rev. Lett. **34**, 89 (1975)
6.35 W.D. Gill, R.L. Greene, G.B. Street, W.A. Little: Phys. Rev. Lett. **35**, 1732 (1975)

6.36 H. Kamimura, A.M. Glazer, A.J. Grant, Y. Natsume, M. Schreiber, A.D. Yoffe: J. Phys. C**9**, 291 (1976)
6.37 W.I. Friesen, A.J. Berlinsky, B. Bergesson, L. Weiler, T.M. Rice: J. Phys. C**8**, 3549 (1975)
W. Rudge, P.M. Grant: Phys. Rev. Lett. **35**, 1799 (1975)
6.38 A. Oshiyama, H. Kamimura: J. Phys. C**14**, 5091 (1981)
A. Oshiyama: J. Phys. C**14**, 5109 (1981)
6.39 W.R. Salaneck, J.W.-P Lin, A.J. Epstein: Phys. Rev. B**13**, 5574 (1976)
P. Mengel, I.B. Ortenburger, W.E. Rudge, P.M. Grant: *Proc. Int. Conf. Organic Conductors and Semiconductors*, Siofok, 1976 (Springer, Berlin, Heidelberg 1977) p. 591
6.40 L. Pintschovius, R. Pynn: In *Quasi One-Dimensional Conductors II*, ed. by S. Barisic, A. Bjelis, J.R. Cooper, B. Leontic, Lecture Notes in Phys., Vol. 96 (Springer, Berlin, Heidelberg 1979) p. 421

Chapter 7

7.1 I.D. Brown, B.D. Cutforth, C.G. Davies, R.J. Gillespie, P.R. Ireland, J.E. Vekris: Canad. J. Chem. **52**, 791 (1974)
7.2 B.D. Cutforth, W.R. Daters, A. van Schyndel, R.J. Gillespie: Solid State Commun. **21**, 377 (1977)
7.3 C.K. Chiang, R. Spal, A. Denenstein, A.J. Heeger, N.D. Miro, A.G. MacDiarmid: Solid State Commun. **22**, 293 (1977)
7.4 F.S. Razavi, W.R. Daters, D. Chartier, R.J. Gillespie: Phys. Rev. Lett. **42**, 1182 (1979)
7.5 D.L. Peebles, C.K. Chiang, M.J. Cohen, A.J. Heeger, N.D. Miro, A.G. MacDiarmid: Phys. Rev. B**15**, 4607 (1977)
7.6 T. Wei, A.F. Garito, C.K. Chiang, N.D. Miro: Phys. Rev. B**16**, 3373 (1977)
7.7 N.D. Miro, A.G. MacDiarmid, A.J. Heeger, A.F. Garito, C.K. Chiang, A.J. Schultz, J.M. Williams: J. Inorg. Nucl. Chem. **40**, 1351 (1978)
7.8 D.P. Chakraborty, R. Spal, A.M. Denenstein, K.B. Lee, A.J. Heeger, M.Ya. Azbel: Phys. Rev. Lett. **43**, 1832 (1979)
7.9 R. Spal, C.K. Chiang, A. Denenstein, A.J. Heeger, N.D. Miro, A.G. MacDiarmid: Phys. Rev. Lett. **39**, 650 (1977)
7.10 R. Spal, D.P. Chakraborty, C.K. Chiang, A. Denenstein, A.J. Heeger, A.G. MacDiarmid: *Quasi One-Dimensional Conductors II*, ed. S. Barisic, A. Bjelis, J.R. Cooper, B. Leontic, Lecture Notes in Phys., Vol. 96 (Springer, Berlin, Heidelbeg 1979) p. 145
7.11 J.M. Hastings, J.P. Pouget, G. Shirane, A.J. Heeger, N.D. Miro, A.G. MacDiarmid: Phys. Rev. Lett. **39**, 1484 (1977)
7.12 J.P. Pouget, G. Shirane, J.M. Hastings, A.J. Heeger, N.D. Miro, A.G. MacDiarmid: Phys. Rev. B**18**, 3645 (1978)
7.13 V.J. Emery, J.D. Axe: Phys. Rev. Lett. **40**, 1507 (1978)
7.14 I.U. Heilman, J.D. Axe, J.M. Hastings, G. Shirane, A.J. Heeger, A.G. MacDiarmid: Phys. Rev. B**20**, 751 (1979)
7.15 G. Shirane: Butsuri (Bulletin of the Phys. Soc. Japan) **34**, 406 (1979)
7.16 V.J. Emery, G. Shirane: *Molecular Metals*, ed. by W.E. Hatfield (Plenum, New York 1979) p. 455

Subject Index

MIX
Papier aus verantwortungsvollen Quellen
Paper from responsible sources
FSC® C105338

If you have any concerns about our products,
you can contact us on
ProductSafety@springernature.com

In case Publisher is established outside the EU,
the EU authorized representative is:
Springer Nature Customer Service Center GmbH
Europaplatz 3, 69115 Heidelberg, Germany

Printed by Libri Plureos GmbH
in Hamburg, Germany